Analyses of the Effects of Global Change on Human Health and Welfare and Human Systems

FEDERAL EXECUTIVE TEAM

Director, Climate Change Science Program..William J. Brennan

Director, Climate Change Science Program Office ....Peter A. Schultz

Lead Agency Principal Representative to Climate Change Science Program,
National Program Director for the Global Change Research Program,
U.S. Environmental Protection Agency..Joel D. Scheraga

Product Lead, Global Ecosystem Research and Assessment Coordinator,
Global Change Research Program, U.S. Environmental Protection Agency..........Janet L. Gamble

Chair, Synthesis and Assessment Product Advisory Group
Associate Director, National Center for Environmental
Assessment, U.S. Environmental Protection Agency .... Michael W. Slimak

Synthesis and Assessment Product Coordinator,
Climate Change Science Program Office...Fabien J.G. Laurier

Special Advisor, National Oceanic and
Atmospheric Administration ....Chad McNutt

EDITORIAL AND PRODUCTION TEAM

Editor, U.S. Environmental Protection Agency...Janet L. Gamble

Technical Advisor, Climate Change Science Program Office ....David J. Dokken

Technical Editor, ICF International...Melinda Harris

Technical Editor, ICF International ....Toby Krasney

Reference Coordinator, ICF International ....Paul Stewart

Reference Coordinator, ICF International ....Dylan Harrison-Atlas

Reference Coordinator, ICF International ....Sarah Shapiro

Logistical and Technical Support, ICF International ....Lauren Smith

Analyses of the Effects of Global Change on Human Health and Welfare and Human Systems

Final Report, Synthesis and Assessment Product 4.6
Report by the U.S. Climate Change Science Program
and the Subcommittee on Global Change Research

Convening Lead Author
Janet L. Gamble, Ph.D., U.S. Environmental Protection Agency
Lead Authors[1]
Kristie L. Ebi, Ph.D., ESS LLC
Anne E. Grambsch, U.S. Environmental Protection Agency
Frances G. Sussman, Ph.D., Environmental Economics Consulting
Thomas J. Wilbanks, Ph.D., Oak Ridge National Laboratory

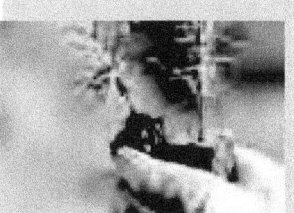

1 Contributing authors are acknowledged in individual chapters.

July 2008

Members of Congress:

On behalf of the National Science and Technology Council, the U.S. Climate Change Science Program (CCSP) is pleased to transmit to the President and the Congress this Synthesis and Assessment Product (SAP), *Analyses of the Effects of Global Change on Human Health and Welfare and Human Systems.* This is part of a series of 21 SAPs produced by the CCSP aimed at providing current assessments of climate change science to inform public debate, policy, and operational decisions. These SAPs are also intended to help the CCSP develop future program research priorities. This SAP is issued pursuant to Section 106 of the Global Change Research Act of 1990 (Public Law 101-606).

The CCSP's guiding vision is to provide the Nation and the global community with the science-based knowledge needed to manage the risks and capture the opportunities associated with climate and related environmental changes. The SAPs are important steps toward achieving that vision and help to translate the CCSP's extensive observational and research database into informational tools that directly address key questions being asked of the research community.

This SAP focuses on the effects of global change on human health and welfare and human systems. It was developed with broad scientific input and in accordance with the Guidelines for Producing CCSP SAPs, the Federal Advisory Committee Act, the Information Quality Act, Section 515 of the Treasury and General Government Appropriations Act for fiscal year 2001 (Public Law 106-554), and the guidelines issued by the Environmental Protection Agency pursuant to Section 515.

We commend the report's authors for both the thorough nature of their work and their adherence to an inclusive review process.

Sincerely,

Carlos M. Gutierrez
Secretary of Commerce

Chair, Committee on
Climate Change Science
and Technology Integration

Samuel W. Bodman
Secretary of Energy

Vice Chair, Committee on
Climate Change Science
and Technology Integration

John H. Marburger III
Director, Office of Science
and Technology Policy

Executive Director, Committee
on Climate Change Science and
Technology Integration

ACKNOWLEDGMENTS

This report has been peer reviewed in draft form by individuals identified for their diverse perspectives and technical expertise. The expert review and selection of reviewers followed the Office of Management and Budget's Information Quality Bulletin for Peer Review. The purpose of this independent review is to provide candid and critical comments that will assist the Climate Change Science Program in making this published report as sound as possible and to ensure that the report meets institutional standards. The peer review comments, draft manuscript, and response to the peer review comments are publicly available at: www.climatescience.gov/Library/sap/sap4-6/default.php.

Environmental Protection Agency Internal Reviewers

We wish to thank the following individuals from the U.S. Environmental Protection Agency for their review of the first and, in some cases, later drafts of the report. Reviewers from within EPA included:

Lisa Conner	Adam Daigneault	Benjamin De Angelo
Barbara Glenn	Doug Grano	Matthew Heberling
Ju-Chin Huang	Stephen Newbold	Jason Samenow
Sara Terry		

National Center for Environmental Assessment, Global Change Research Program

We extend our thanks to our colleagues in the Global Change Research Program who contributed thoughtful insights, reviewed numerous drafts, and helped with the production of the report.

John Thomas	Christopher Weaver

Federal Agency Reviewers

Likewise, we thank the reviewers from within the federal "family." Reviewers from across the federal agencies provided their comments during the public comment period.

Brigid DeCoursey	Department of Transportation
Mary Gant	National Institutes of Environmental Health Sciences/NIH
Indur M. Goklany	Department of Interior
Charlotte Skidmore	Office of Management and Budget
Samuel P. Williamson	Office of the Federal Coordinator for Meteorology

Public Reviewers

We also extend our thanks to the reviewers who provided their comments during the public comment period, included individuals from the public.

William Fang	Edison Electric Institute
Katherine Farrell	AACDH
Hans Martin Fuessel	PICIR
Eric Holdsworth	Edison Electric Institute
John Kinsman	Edison Electric Institute
Kim Knowlton	Natural Resources Defense Council
Sabrina McCormick	Michigan State University
J. Alan Roberson	American Water Works Association
Gina Solomon	Natural Resources Defense Council

Human Impacts of Climate Change Advisory Committee (HICCAC)

Finally, we are indebted to the thoughtful review provided by a Federal Advisory Committee convened by the U.S. Environmental Protection Agency to provide an independent expert review of the SAP 4.6. The HICCAC panel met in October 2007 to discuss their findings and recommendations for the report. Following extensive revisions to the report, the HICCAC reconvened by teleconference in January 2008 to review the authors' response to comments. The panel's review of the report has contributed to a markedly improved document.

Chair	Tom Dietz	Michigan State University
Co-chair	Barbara Entwisle	University of North Carolina
Members	Howard Frumkin	Centers for Disease Control and Prevention
	Peter Gleick	Pacific Institute
	Jonathan Patz	University of Wisconsin
	Roger Pulwarty	NOAA
	Eugene Rosa	Washington State University
	Susan Stonich	University of California at Santa Barbara

Thanks are also in order to Joanna Foellmer, the Designated Federal Official from within the National Center for Environmental Assessment who organized and managed the HICCAC.

ICF International

We thank our colleagues at ICF International for their support—logistical and technical—in preparing the report. We wish to extend special thanks to Melinda Harris and Randy Freed.

Summary

It has been an honor and a pleasure to work with all of the people named above as well as the many colleagues we have encountered in the process of preparing this report. We hope that this document will be a positive step forward in our efforts to assess the impacts of climate change on human systems and to evaluate opportunities for adaptation.

AUTHOR TEAM FOR THIS REPORT

Executive Summary   **Convening Lead Author:** Janet L. Gamble, U.S. EPA
Lead Authors: Kristie L. Ebi, ESS LLC; Frances G. Sussman, Environmental
Economics Consulting; Thomas J. Wilbanks, Oak Ridge National Laboratory

Contributing Authors: Colleen Reid, ASPH Fellow; John V. Thomas, U.S. EPA;
Christopher P. Weaver, U.S. EPA; Melinda Harris, ICF International; Randy Freed, ICF
International

Chapter 1   **Convening Lead Author:** Janet L. Gamble, U.S. EPA

Lead Authors: Kristie L. Ebi, ESS LLC; Anne Grambsch, U.S. EPA; Frances G.
Sussman, Environmental Economics Consulting; Thomas J. Wilbanks, Oak Ridge
National Laboratory

Contributing Authors: Colleen E. Reid, ASPH Fellow; Katharine Hayhoe, Texas Tech
University; John V. Thomas, U.S. EPA; Christopher P. Weaver, U.S. EPA

Chapter 2   **Lead Author:** Kristie L. Ebi, ESS LLC

Contributing Authors: John Balbus, Environmental Defense; Patrick L. Kinney,
Columbia University; Erin Lipp, University of Georgia; David Mills, Stratus Consulting;
Marie S. O'Neill, University of Michigan; Mark Wilson, University of Michigan

Chapter 3   **Lead Author:** Thomas J. Wilbanks, Oak Ridge National Laboratory

Contributing Authors: Paul Kirshen, Tufts University; Dale Quattrochi, NASA/
Marshall Space Flight Center; Patricia Romero-Lankao, NCAR; Cynthia Rosenzweig,
NASA/Goddard; Matthias Ruth, University of Maryland; William Solecki, Hunter
College; Joel Tarr, Carnegie Mellon University

Contributors: Peter Larsen, University of Alaska-Anchorage; Brian Stone, Georgia Tech

Chapter 4   **Lead Author:** Frances G. Sussman, Environmental Economics Consulting

Contributing Authors: Maureen L. Cropper, University of Maryland at College Park;
Hector Galbraith, Galbraith Environmental Sciences LLC.; David Godschalk, University
of North Carolina at Chapel Hill; John Loomis, Colorado State University; George Luber,
Centers for Disease Control and Prevention; Michael McGeehin, Centers for Disease
Control and Prevention; James E. Neumann, Industrial Economics, Inc.; W. Douglass
Shaw, Texas A&M University; Arnold Vedlitz, Texas A&M University; Sammy Zahran,
Colorado State University

Chapter 5   **Convening Lead Author:** Janet L. Gamble, U.S. EPA

Lead Authors: Kristie L. Ebi, ESS LLC; Frances G. Sussman, Environmental
Economics Consulting; Thomas J. Wilbanks, Oak Ridge National Laboratory

Contributing Authors: Colleen E. Reid, ASPH Fellow; John V. Thomas, U.S. EPA;
Christopher P. Weaver, U.S. EPA

RECOMMENDED CITATIONS

For the Report as a Whole:

CCSP, 2008: Analyses of the effects of global change on human health and welfare and human systems. A Report by the U.S. Climate Change Science Program and the Subcommittee on Global Change Research. [Gamble, J.L. (ed.), K.L. Ebi, F.G. Sussman, T.J. Wilbanks, (Authors)]. U.S. Environmental Protection Agency, Washington, DC, USA.

For Executive Summary:

Gamble, J.L., K.L. Ebi, F.G. Sussman, T.J. Wilbanks, C. Reid, J.V. Thomas, C.P. Weaver, M. Harris, and R. Freed, 2008: Executive Summary. In: Analyses of the effects of global change on human health and welfare and human systems. A Report by the U.S. Climate Change Science Program and the Subcommittee on Global Change Research. [Gamble, J.L. (ed.), K.L. Ebi, F.G. Sussman, T.J. Wilbanks, (Authors)]. U.S. Environmental Protection Agency, Washington, DC, USA, p. 1–11.

For Chapter 1:

Gamble, J.L., K.L. Ebi, A. Grambsch, F.G. Sussman, T.J. Wilbanks, C.E. Reid, K. Hayhoe, J.V. Thomas, and C.P. Weaver, 2008: Introduction. In: Analyses of the effects of global change on human health and welfare and human systems. A Report by the U.S. Climate Change Science Program and the Subcommittee on Global Change Research. [Gamble, J.L. (ed.), K.L. Ebi, F.G. Sussman, T.J. Wilbanks, (Authors)]. U.S. Environmental Protection Agency, Washington, DC, USA, p. 13–37.

For Chapter 2:

Ebi, K.L., J. Balbus, P.L. Kinney, E. Lipp, D. Mills, M.S. O'Neill, and M. Wilson, 2008: Effects of Global Change on Human Health. In: Analyses of the effects of global change on human health and welfare and human systems. A Report by the U.S. Climate Change Science Program and the Subcommittee on Global Change Research. [Gamble, J.L. (ed.), K.L. Ebi, F.G. Sussman, T.J. Wilbanks, (Authors)]. U.S. Environmental Protection Agency, Washington, DC, USA, p. 39–87.

For Chapter 3:

Wilbanks, T.J., P. Kirshen, D. Quattrochi, P. Romero-Lankao, C. Rosenzweig, M. Ruth, W. Solecki, and J. Tarr, 2008: Effects of Global Change on Human Settlements. In: Analyses of the effects of global change on human health and welfare and human systems. A Report by the U.S. Climate Change Science Program and the Subcommittee on Global Change Research. [Gamble, J.L. (ed.), K.L. Ebi, F.G. Sussman, T.J. Wilbanks, (Authors)]. U.S. Environmental Protection Agency, Washington, DC, USA, p. 89–109.

For Chapter 4:

Sussman, F.G., M.L. Cropper, H. Galbraith, D. Godschalk, J. Loomis, G. Luber, M. McGeehin, J.E. Neumann, W.D. Shaw, A. Vedlitz, and S. Zahran, 2008: Effects of Global Change on Human Welfare. In: Analyses of the effects of global change on human health and welfare and human systems. A Report by the U.S. Climate Change Science Program and the Subcommittee on Global Change Research. [Gamble, J.L. (ed.), K.L. Ebi, F.G. Sussman, T.J. Wilbanks, (Authors)]. U.S. Environmental Protection Agency, Washington, DC, USA, p. 111–168.

For Chapter 5:

Gamble, J.L., K.L. Ebi, F.G. Sussman, T.J. Wilbanks, C. Reid, J.V. Thomas, and C.P. Weaver, 2008: Common Themes and Research Recommendations. In: Analyses of the effects of global change on human health and welfare and human systems. A Report by the U.S. Climate Change Science Program and the Subcommittee on Global Change Research. [Gamble, J.L. (ed.), K.L. Ebi, F.G. Sussman, T.J. Wilbanks, (Authors)]. U.S. Environmental Protection Agency, Washington, DC, USA, p. 169–176.

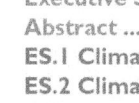

TABLE OF CONTENTS

Executive Summary
Abstract I
ES.1 Climate Change and Vulnerability I
ES.2 Climate Change and Human Health 3
ES.3 Climate Change and Human Settlements...................................8
ES.4 Climate Change and Human Welfare 10

CHAPTER

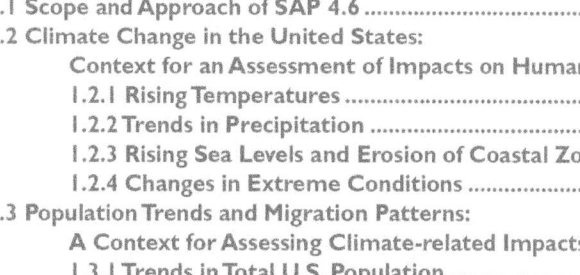

1 Introduction..13
1.1 Scope and Approach of SAP 4.6 13
1.2 Climate Change in the United States:
    Context for an Assessment of Impacts on Human Systems 16
    1.2.1 Rising Temperatures 17
    1.2.2 Trends in Precipitation 18
    1.2.3 Rising Sea Levels and Erosion of Coastal Zones....................19
    1.2.4 Changes in Extreme Conditions 20
1.3 Population Trends and Migration Patterns:
    A Context for Assessing Climate-related Impacts 22
    1.3.1 Trends in Total U.S. Population 22
    1.3.2 Migration Patterns..24
1.4 Complex Linkages: The Role of Non-climate Factors 26
    1.4.1 Economic Status 26
    1.4.2 Technology 27
    1.4.3 Infrastructure 27
    1.4.4 Human and Social Capital and Behaviors 28
    1.4.5 Institutions..29
    1.4.6 Interacting Effects...29
1.5 Reporting Uncertainty in SAP 4.6...30
1.6 References..32

2 Effects of Global Change on Human Health 39
2.1 Introduction...39
2.2 Observed Climate-sensitive Health Outcomes in the United States 40
    2.2.1 Thermal Extremes: Heat Waves...40
    2.2.2 Thermal Extremes: Cold Waves 42
    2.2.3 Extreme Events: Hurricanes, Floods, and Wildfires...................42
    2.2.4 Indirect Health Impacts of Climate Change............................44
2.3 Projected Health Impacts of Climate Change in the United States 50
    2.3.1 Heat-related Mortality 50
    2.3.2 Hurricanes, Floods, Wildfires and Health Impacts 52
    2.3.3 Vector-borne and Zoonotic Diseases 53
    2.3.4 Water- and Food-borne Diseases 53
    2.3.5 Air Quality Morbidity and Mortality 53
2.4 Vulnerable Regions and Subpopulations 61
    2.4.1 Vulnerable Regions..61
    2.4.2 Specific Subpopulations at Risk..62
2.5 Adaptation 65
    2.5.1 Actors and Their Roles and Responsibilities for Adaptation......66
    2.5.2 Adaptation Measures to Manage
    Climate Change-Related Health Risks 68
2.6 Conclusions..68
2.7 Expanding the Knowledge Base...73
2.8 References..75

TABLE OF CONTENTS

3 Effects of Global Change on Human Settlements...89
3.1 Introduction...89
    3.1.1 Purpose...89
    3.1.2 Background 89
    3.1.3 Current State of Knowledge...90
3.2 Climate Change Impacts and the
    Vulnerabilities of Human Settlements 90
    3.2.1 Determinants of Vulnerability 90
    3.2.2 Impacts of Climate Change on Human Settlements.................92
    3.2.3 The Interaction of Climate Impacts
    with Non-climate Factors 96
    3.2.4 Realizing Opportunities from
    Climate Change in the United States...98
    3.2.5 Examples of Impacts on
    Metropolitan Areas in the United States 98
3.3 Opportunities for Adaptation of
    Human Settlements to Climate Change...100
    3.3.1 Perspectives on Adaptation by Settlements 101
    3.3.2 Major Categories of Adaptation Strategies 102
    3.3.3 Examples of Current Adaptation Strategies...........................103
    3.3.4 Strategies to Enhance Adaptive Capacity 104
3.4 Conclusions...104
3.5 Expanding the Knowledge Base...105
3.6 References..106

4 Effects of Global Change on Human Welfare...111
4.1 Introduction..111
4.2 Human Welfare, Well-being, and Quality of Life 112
    4.2.1 Individual Measures of Well-being..114
    4.2.2 The Social Indicators Approach...115
    4.2.3 A Closer Look at Communities...120
    4.2.4 Vulnerable Populations, Communities, and Adaptation...........123
4.3 An Economic Approach to Human Welfare 124
    4.3.1 Economic Valuation 126
    4.3.2 Impacts Assessment and Monetary Valuation 127
    4.3.3 Human Health..128
    4.3.4 Ecosystems 133
    4.3.5 Recreational Activities and Opportunities...............................140
    4.3.6 Amenity Value of Climate 147
4.4 Conclusions...151
4.5 Expanding the Knowledge Base...153
4.6 References..154
4.7 Appendix 163

TABLE OF CONTENTS

5 Common Themes and Research Recommendations 169
5.1 Synthesis and Assessment Product 4.6: Advances in the Science 169
  5.1.1 Complex Linkages and a Cascading
  Chain of Impacts Across Global Changes...169
  5.1.2 Changes in Climate Extremes and Climate Averages.............170
  5.1.3 Vulnerable Populations and Vulnerable Locations...................171
  5.1.4 The Cost of and Capacity for Adaptation...............................172
  5.1.5 An Integrative Framework...172
5.2 Expanding the Knowledge Base...173
  5.2.1 Human Health Research Gaps..174
  5.2.2 Human Settlements Research Gaps...175
  5.2.3 Human Welfare Research Gaps...176

6 Glossary and Acronyms...177

Executive Summary

Convening Lead Author: Janet L. Gamble, U.S. Environmental Protection Agency

Lead Authors: Kristie L. Ebi, ESS, LLC; Frances G. Sussman, Environmental Economics Consulting; Thomas J. Wilbanks, Oak Ridge National Laboratory

Contributing Authors: Colleen Reid, ASPH Fellow; John V. Thomas, U.S. Environmental Protection Agency; Christopher P. Weaver, U.S. Environmental Protection Agency; Melinda Harris, ICF International; Randy Freed, ICF International

Climate change, interacting with changes in land use and demographics, will affect important human dimensions in the United States, especially those related to human health, settlements, and welfare. The challenges presented by population growth, an aging population, migration patterns, and urban and coastal development will be affected by changes in temperature, precipitation, and extreme climate-related events. In the future, with continued global warming, heat waves and heavy downpours are very likely to further increase in frequency and intensity. Cold days and cold nights are very likely to become much less frequent over North America. Substantial areas of North America are likely to have more frequent droughts of greater severity. Hurricane wind speeds, rainfall intensity, and storm surge levels are likely to increase. Other changes include measurable sea level rise and increases in the occurrence of coastal and riverine flooding. The United States is certainly capable of adapting to the collective impacts of climate change. However, there will still be certain individuals and locations where the adaptive capacity is less and these individuals and their communities will be disproportionally impacted by climate change.

This report— Synthesis and Assessment Product 4.6 (SAP 4.6)—focuses on impacts of global climate change, especially impacts on three broad dimensions of the human condition: human health, human settlements, and human welfare. SAP 4.6 has been prepared by a team of experts from academia, government, and the private sector in response to the mandate of the U.S. Climate Change Science Program's Strategic Plan (2003). The assessment examines potential impacts of climate change on human society, opportunities for adaptation, and associated recommendations for addressing data gaps and near- and long-term research goals.

ES.1 CLIMATE CHANGE AND VULNERABILITY

Climate variability and change challenge even the world's most advanced societies. At a very basic level, climate affects the costs of providing comfort in our homes and work places. A favorable climate can provide inputs for a good life: adequate fresh water supplies; products from the ranch, the farm, the forests, the rivers, and the coasts; pleasure derived from tourist destinations and from nature, biodiversity, and outdoor recreation. Climate not only supports the provision of many goods and services, but also affects the spread of some diseases and the prevalence of other health problems. It

is also associated with threats from extreme events and natural disasters such as tropical storms, riverine and coastal flooding, wildfires, droughts, wind, hail, ice, heat, and cold.

This report examines the impacts on human society of global change, especially those associated with climate change. The impact assessments in this report do not rely on specific emissions or climate change scenarios but, instead, rely on the existing scientific literature with respect to our understanding of climate change and its impacts on human health, settlements, and well-being in the United States. Because climate change forecasts are generally not specific enough for the scale of

local decision-making, this report adopts a vulnerability perspective in assessing impacts on human society.

A vulnerability approach focuses on estimating risks or opportunities associated with possible impacts of climate change, rather than on estimating (quantitatively) the impacts themselves, which would require far more detailed information about future conditions. Vulnerabilities are shaped not only by existing exposures, sensitivities, and adaptive capacities but also by responses to risks. For example, Boston is generally more vulnerable to heat waves than Dallas because there are fewer air-conditioned homes in Boston than in Dallas. At the same time, human responses (e.g., the elderly not using air-conditioning) also are an important determinant of impacts. This leads to our conclusion that climate change will result in regional differences in impacts in the United States not only due to a regional pattern of changes in climate but the regional nature of our communities in adapting to these changes.

In the United States, we are observing the evidence of long-term changes in temperature and precipitation consistent with global warming. Changes in average conditions are being realized through rising temperatures, changes in annual and seasonal precipitation, and rising sea levels. Observations also indicate there are changes in extreme conditions, such as an increased frequency of heavy rainfall (with some increase in flooding), more heat waves, fewer very cold days, and an increase in areas affected by drought. There have been large fluctuations in the number of hurricanes from year to year, which make it difficult to discern trends. Evidence suggests that the intensity of Atlantic hurricanes and tropical storms has increased over the past few decades. However, changes in frequency are currently too uncertain for confident projection.

Changes in the size of the population, including especially sensitive sub-populations, and their geographic distribution across the landscape need to be accounted for when assessing climate variability and change impacts. According to the Census Bureau's middle series projection, by 2100 the U.S. population will increase to some 570 million people. Moreover, the elderly population is increasing rapidly and many health assessments identify them as more vulnerable than younger age groups to a range of health impacts associated with climate change. Although numbers produced by population projections are important, nearly all trends point to more Americans living in areas that may be especially vulnerable to the effects of climate change. Many rapidly growing cities and towns in the Mountain West may also experience decreased snow pack during winter and earlier spring melting, leading to lower stream flows, particularly during the high-demand period of summer. Similarly, coastal areas are projected to continue to increase in population, with associated increases in population at risk over the next several decades.

Climate is only one of a number of global changes that affect human well-being. Non-climate processes and stresses interact with climate change, determining the overall severity of climate impacts. Socioeconomic factors that can influence exposures, vulnerability, and impacts include population, economic status, technology, infrastructure, human capital and social context and behaviors, and institutions. Trends in these factors alter anticipated impacts from climate because they fundamentally shape the nature and scope of human vulnerability. Understanding the impacts of climate change and variability on the quality of life in the United States implies knowledge of how these factors vary by location, time, and socioeconomic group.

Climate change will seldom be the sole or primary factor determining a population's or a location's well-being. Ongoing adaptation also can significantly influence climate impacts. For example, emergency warning systems have generally reduced deaths and death rates from extreme events, while greater access to insurance and broader, government-funded safety nets for people struck by natural disasters have ameliorated the hardships they face. While this assessment focuses on how climate change could affect future health, well-being, and settlements in the United States, the extent of any impacts will depend on an array of non-climate factors and adaptive measures. Finally, the effects of climate change very often spread from directly impacted areas and sectors to other areas and sectors through extensive and complex linkages. In summary, the importance of climate change depends on the directness of the climate impact coupled with demographic, social, economic, institutional, and political factors, including, the degree of preparedness.

Consistent with all of the Synthesis and Assessment Products being prepared by the CCSP, this report includes statements regarding uncertainty. Each author team assigned likelihood judgments that reflect their assessment of the current consensus of the science and the quality and amount of evidence. The likelihood terminology and the corresponding values that are used in this report are consistent with the latest IPCC Fourth Assessment and are further explained in Chapter 1 of this report. As the focus of this report is on impacts, it is important to note that these likelihood statements refer to the statement of the impact, not statements related to underlying climatic changes.

Table ES.1 provides examples of climate change impacts that are identified in the chapters for human health, settlements, and human welfare and includes potential adaptation strategies.

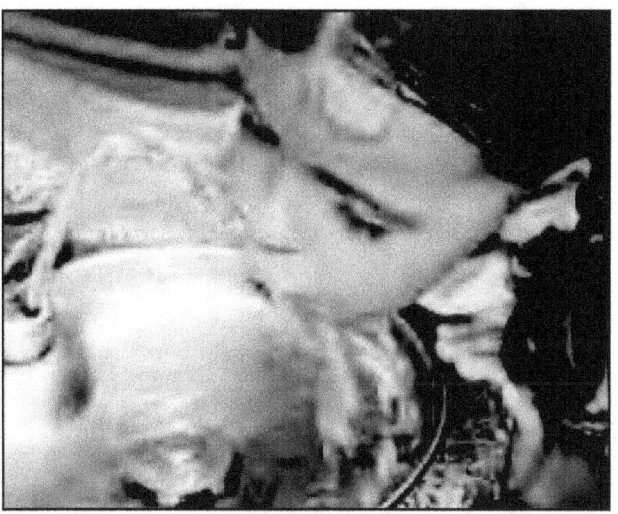

The list of impacts is not comprehensive, but rather includes those that the available evidence suggests may occur. It is important to note that not all effects have been equally well-studied. The effects identified for welfare, in particular, should be taken as examples of effects about which we have some knowledge, rather than a complete listing of all welfare effects.

ES.2 CLIMATE CHANGE AND HUMAN HEALTH

The United States is a highly developed country with a wide range of climates. While there may be fewer cases of illness and death associated with climate change in the United States than in the developing world, we nevertheless anticipate increased costs to human health and well being. Greater wealth and a more developed public health system and infrastructure (e.g., water treatment plants, sewers, and drinking water systems; roads, rails, and bridges; and flood control structures) will continue to enhance our capacity to respond to climate change. Similarly, governments' capacities for disaster planning and emergency response are key assets that should allow the United States to adapt to many of the health effects associated with climate change.

It is very likely that heat-related morbidity and mortality will increase over the coming decades. According to the U.S. Census, the U.S. population is aging; the percent of the population over age 65 is projected to be 13 percent by 2010 and 20 percent by 2030 (more than 50 million

Table ES.1 Examples of Possible Impacts (present to 2050) of Climate Variability and Change on Human Health, Settlements, and Welfare in the United States and Potential Adaptation Strategies

Focus Area	Climate Event	Examples of Possible Impacts	Likelihood of Impact Given Climate Event Occurs[a]	Potential Adaptation Strategies
HUMAN HEALTH				
	Extreme temperatures More heat waves and higher maximum temperatures Fewer cold waves and higher minimum temperatures	Heat stress/stroke or hyperthermia Uncertain impacts on mortality[b]	Very likely in Midwest and Northeast urban centers	Early watch and warning systems and installation of cooling systems in residential and commercial buildings
	Changes in precipitation, especially extreme precipitation	Contaminated water and food supplies with associated gastrointestinal illnesses, including salmonella and giardia	Likely in areas with outdated or over-subscribed water treatment plants	Improve infrastructure to guard against combined sewer overflow; public health response to include "boil water" advisories
	Hurricane and storm surge	Injuries from flying debris and drowning/exposure to contaminated flood waters and to mold and mildew/ exposure to carbon monoxide poisoning from portable generators	Likely in coastal zones of the southeast Atlantic and the Gulf Coast	Increase knowledge and awareness of vulnerability to climate change (e.g., maps showing areas vulnerable to storm surges); public health advisories in immediate aftermath of storms; coordinate storm relief efforts to insure that people receive necessary information for safeguarding their health
	Temperature-related effects on ozone[c]	Ozone concentrations more likely to increase than decrease; possible contribution to cardiovascular and pulmonary illnesses, including exacerbation of asthma and chronic obstructive pulmonary disorder (COPD) if current regulatory standards are not attained	Likely in urban centers in the mid-Atlantic and the Northeast	Public warning via air quality action days; encourage public transit, walking, and bicycling to decrease emissions
	Wildfires	Degraded air quality, contributing to asthma and COPD aggravated	Likely in California, the intermountain West, the Southwest and the Southeast	Public health air quality advisories

HUMAN SETTLEMENTS

Focus Area	Climate Event	Examples of Possible Impacts	Likelihood of Impact Given Climate Event Occurs[a]	Potential Adaptation Strategies
	Extreme temperatures More heat waves and higher maximum temperatures Fewer cold waves and higher minimum temperatures	Increased net energy demand for peak cooling Reduced cold-related stresses and costs	Very likely	Expand capacity for cooling through public utilities; invest in alternative energy sources
	Drought	Strain on municipal and agricultural water supplies	Very likely in intermountain West, desert Southwest, and Southeast	Reallocate water among current users; develop water markets to encourage more efficient allocation; identify new sources; encourage conservation of water for personal and public use; develop drought resistant crops
	Hurricane and storm surge	Disruption of infrastructure, including levee systems, river channels, bridges, and highway systems; disruption of residential neighborhoods	Very likely in southern Atlantic Coast and Gulf Coast	Increase knowledge and awareness of climate impacts (e.g., maps showing areas vulnerable to storm surges); harden coastal zones or retreat or relocate; insure against catastrophic loss due to flooding and high winds
	Wildfires	Disruption of communities and property destruction	Very likely in intermountain West, desert Southwest, and Southeast	Clear vegetation away from buildings; issue emergency evacuation orders, prescribed burns, thinning of combustible matter
	Late snow fall and early snow melt	Disruption of water supplies for municipal and agricultural use	Very likely in intermountain West	Build reservoirs; conserve water supplies; divert supply from agricultural to municipal use; modify operation of existing infrastructure to account for changes in hydrology; develop drought resistant crops, water prices at replacement cost, enable trading by working with states to develop property rights

Focus Area	Climate Event	Examples of Possible Impacts	Likelihood of Impact Given Climate Event Occurs[a]	Potential Adaptation Strategies
HUMAN WELFARE	Extreme temperatures More heat waves and higher maximum temperatures Fewer cold waves and higher minimum temperatures	Discomfort; limit some outdoor activities/recreation Limit some snow- and cold-related recreational opportunities; substantial economic disruption to recreation industry	Very likely in more northern latitudes of the United States and in Alaska Very likely in intermountain West, northern New England and the Upper Great Lakes	Public health watch/warning advisories Engage in alternative recreation activities
	Late autumn snow fall and early spring snow melt	Limit some snow-related recreational opportunities; substantial economic disruption to recreation industry	Very likely in intermountain West, northern New England and the Upper Great Lakes	Engage in alternative recreation activities
	Extreme precipitation events	Local flooding and contamination of water supplies	Very likely nationwide	Issue flood advisories/warnings
	Hurricane and coastal storms	At-risk properties experience flood and wind damage; individuals experience disruption to daily life	Very likely in coastal zone of the Gulf Coast and the southern Atlantic	Relocate dwellings and business, and reinforce structures and infrastructure to reduce disruptions

a Based on impacts identified in the published, peer-reviewed literature and expert opinion. Does not include an evaluation of likelihood of the climate event. May include some adaptation (e.g., in the baseline estimate) but generally does not account for additional changes or developments in adaptive capacity.

b Many factors contribute to winter mortality, making highly uncertain how climate change could affect mortality. No projections have been published for the United States that incorporates critical factors, such as the influence of influenza outbreaks.

c If areas remain in compliance with National Ambient Air Quality Standards, people will not be exposed to unhealthy air (i.e., cardiovascular and pulmonary illnesses will not occur). More stringent emissions controls may be required to remain in compliance although this is uncertain and additional study is needed.

people). Older adults, very young children, and persons with compromised immune systems are vulnerable to temperature extremes. This suggests that temperature-related morbidity and mortality are likely to increase. Similarly, heat-related mortality affects poor and minority populations disproportionately, in part due to lack of air conditioning. The concentration of poverty in inner city neighborhoods leads to disproportionate adverse effects associated with urban heat islands.

There is considerable speculation concerning the balance of climate change-related decreases in winter mortality compared with increases in summer mortality. Net changes in mortality are difficult to estimate because, in part, much depends on complexities in the relationship between mortality and the changes associated with global change. Few studies have attempted to link the epidemiological findings to climate scenarios for the United States, and studies that have done so have focused on the effects of changes in average temperature, with results dependent on climate scenarios and assumptions of future adaptation. Moreover, many factors contribute to winter mortality, making highly uncertain how climate change could affect mortality. No projections have been published for the United States that incorporate critical factors, such as the influence of influenza outbreaks.

The impacts of higher temperatures in urban areas and likely associated increases in tropospheric ozone concentrations can contribute to or exacerbate cardiovascular and pulmonary illness if current regulatory standards are not attained. In addition, stagnant air masses related to climate change are likely to degrade air quality in some densely populated areas. It is important to recognize that the United States has a well-developed and successful national regulatory program for ozone, PM2.5, and other criteria pollutants. That is, the influence of climate change on air quality will play out against a backdrop of ongoing regulatory control that will shift the baseline concentrations of air pollutants. Studies to date have typically held air pollutant emissions constant over future decades (i.e., have examined the sensitivity of ozone concentrations to climate change rather than projecting actual future ozone concentrations). Physical features

of communities, including housing quality and green space, social programs that affect access to health care, aspects of population composition (level of education, racial/ethnic composition), and social and cultural factors are all likely to affect vulnerability to air quality.

Hurricanes, extreme precipitation resulting in floods, and wildfires all have the potential to affect public health through direct and indirect health risks. SAP 3.3 indicates that there is evidence for increased sea surface temperatures in the tropical Atlantic and there is a strong correlation to Atlantic tropical storm frequency, duration, and intensity. However, a valid assessment will require further studies. The health risks associated with such extreme events are thus likely to increase with the size of the population and the degree to which it is physically, mentally, or financially constrained in its ability to prepare for and respond to extreme weather events. For example, coastal evacuations prompted by imminent hurricane landfall are only moderately successful. Many of those who are advised to flee to higher ground stay behind in inadequate shelter. Surveys find that the public is either not aware of the appropriate preventive actions or incorrectly assesses the extent of their personal risk.

There will likely be an increase in the spread of several food and water-borne pathogens among susceptible populations depending on the pathogens' survival, persistence, habitat range, and transmission under changing climate and environmental conditions. While the United States has successful programs to protect water quality under the Safe Drinking Water Act and the Clean Water Act, some contamination pathways and routes of exposure do not fall under regulatory programs (e.g., dermal absorption from floodwaters, swimming in lakes and ponds with elevated pathogen levels, etc.). The primary climate-related factors that affect these pathogens include temperature, precipitation, extreme weather events, and shifts in their ecological regimes. Consistent with our understanding of climate change on human health, the impact of climate on food and water-borne pathogens will seldom be the only factor determining the burden of human injuries, illness, and death.

Health burdens related to climate change will vary by region. For the continental United States, the northern latitudes are likely to experience the largest increases in average temperatures; they will also bear the brunt of increases in ground-level ozone and other air-borne pollutants. Because Midwestern and Northeastern cities are generally not as well adapted to the heat as Southern cities, their populations are likely to be disproportionately affected by heat related illnesses as heat waves increase in frequency, severity, and duration. The range of many vectors is likely to extend northward and to higher elevations. For some vectors, such as rodents associated with Hantavirus, ranges are likely to expand, as the precipitation patterns under a warmer climate enhance the vegetation that controls the rodent population. Forest fires, with their associated decrements to air quality and pulmonary effects, are likely to increase in frequency, severity, distribution, and duration in the Southeast, the Intermountain West and the West. Table ES.2 summarizes regional vulnerabilities to a range of climate impacts.

Finally, climate change is very likely to accentuate the disparities already evident in the American health care system. Many of the expected health effects are likely to fall disproportionately on the poor, the elderly, the disabled, and the uninsured. The most important adaptation to ameliorate health effects from climate change is to support and maintain the United States' public health infrastructure.

ES.3 CLIMATE CHANGE AND HUMAN SETTLEMENTS

Effects of climate change on human settlements are likely to vary considerably according to location-specific vulnerabilities, with the most vulnerable areas likely to include Alaska with increased permafrost melt, flood-risk in coastal zones and river basins, and arid areas with associated water scarcity. The main climate impacts have to do with changes in the intensity, frequency, and location of extreme weather events and, in some cases, water availability rather than temperature change.

Changes in precipitation patterns will affect water supplies nationwide, with precipitation varying across regions and over time. Likely reductions in snow melt, river flows, and groundwater levels, along with increases in saline intrusion into coastal rivers and groundwater will reduce fresh water supplies. All things held constant, population growth will increase the demand for drinking water even as changes in precipitation will change the availability of water supplies. Moreover, storms, floods, and other severe weather events are likely to affect infrastructure such as sanitation, transportation, supply lines for food and energy, and communication. Some of the nation's most expensive infrastructure, such as exposed structures like bridges and utility networks, are especially vulnerable. In many cases, water supply networks and stressed reservoir capacity interact with growing populations (especially in coastal cities and in the Mountain and Pacific West). The complex interactions of land use, population growth, and dynamics of settlement patterns further challenge supplies of water for municipal, industrial, and agricultural uses. In the Pacific Northwest the electricity base dominated by hydropower is directly dependent upon water flows from snow melt. Reduced hydropower would mean the need for supplemental electricity sources, resulting in a wide variety of negative ripple effects to the economy and to human welfare. Similarly, along the West Coast, communities are likely to experience greater demands on water supplies even as regional precipitation declines and average snow packs decrease.

Communities in risk-prone regions, such as coastal zones, have reason to be concerned about potential increases in severe weather events. The combined effects of severe storms and sea level rise in coastal areas or increased risks of fire in more arid areas are examples of how climate change may increase the magnitude of challenges already facing risk-prone regions. Vulnerabilities may be especially pronounced for rapidly growing and/or larger metropolitan areas, where the potential magnitude of both impacts and coping requirements are likely to be very large. On the other hand, such regions have greater opportunity to adapt infrastructure and to make decisions that limit vulnerability.

Table ES.2 Summary of Regional Vulnerabilities to Climate-Related Impacts[a]

United States Census Regions	Early Snow Melt	Degraded Air Quality	Urban Heat Island	Wildfires	Heat Waves	Drought	Tropical Storms	Extreme Rainfall with Flooding	Sea Level Rise
New England ME VT NH MA RI CT	•	•	•		•	•		•	•
Middle Atlantic NY PA NJ DE	•	•	•		•	•	•	•	•
East North Central WI MI IL IN OH	•	•	•		•	•		•	
West North Central ND MN SD IA NE KS MO	•		•		•	•		•	
South Atlantic WV VA MD NC SC GA FL DC		•	•	•	•	•	•	•	•
East South Central KY TN MS AL					•	•	•		•
West South Central TX OK AR LA		•	•	•	•	•			•
Mountain MT ID WY NV UT CO AZ NM	•	•	•	•	•	•			
Pacific AK CA WA OR HI	•	•	•	•	•	•	•	•	•

Climate-Related Impacts

[a] Based on impacts identified in the published, peer-reviewed literature and expert opinion.

Warming is virtually certain to increase energy demand in U.S. cities for cooling in buildings while it reduces demands for heating in buildings (see SAP 4.5 **Effects of Climate Change on Energy Production and Use in the United States***).* Demands for cooling during warm periods could jeopardize the reliability of service in some regions by exceeding the supply, especially during periods of unusually high temperatures. Higher temperatures also affect costs of living and business operation by increasing costs of climate control in buildings.

Climate change has the potential not only to affect communities directly but also indirectly through impacts on other areas linked to their economies. Regional economies that depend on sectors highly sensitive to climate such as agriculture, forestry, water resources, or recreation and tourism could be affected either positively or negatively by climate change. Climate change can add to stress on social and political structures by increasing management and budget requirements for public services such as public health care, disaster risk reduction, and even public safety. As sources of stress grow and combine, the resilience of social and political structures is expected to be challenged, especially in locales with relatively limited social and political capital.

Finally, population growth and economic development are occurring in those areas that are likely to be vulnerable to the effects of climate change. Approximately half of the U.S. population, 160 million people, live in one of 673 coastal counties. Coastal areas—particularly those on gently sloping coasts and zones with gradual land subsidence—will be at risk from sea level rise, impacts especially those related to severe storms and storm surges.

ES.4 CLIMATE CHANGE AND HUMAN WELFARE

The terms human welfare, quality of life, and well-being are often used interchangeably, and by a number of disciplines as diverse as psychology, economics, health science, geography, urban planning, and sociology. There is a shared understanding that all three terms refer to aspects of individual and group life that involve living conditions and chances of injury, stress, and loss.

Human well-being is typically defined and measured as a multi-dimensional concept. Taxonomies of place-specific well-being or quality of life typically converge on six dimensions: 1) economic conditions, 2) natural resources and amenities, 3) human health, 4) public and private infrastructure, including transportation systems, 5) government and public safety and 6) social and cultural resources. Climate change will likely have impacts across all of these dimensions—both positive and negative. In addition, the positive and negative effects of climate change will affect broader communities, as networks of households, businesses, physical structures, and institutions are located together across space and time.

Quantifying impacts of climate change on human well-being requires linking effects in quality of life to the projected[1] physical effects of climate change and the consequent effects on human and natural systems. Economic analyses provide a means of quantifying and, in some cases, placing dollar values on welfare effects. However, even in cases where welfare effects have been quantified, it is difficult to compare and aggregate a range of effects across a number of sectors.

This report examines four types of effects on economic welfare: those on ecosystems, human health, recreation, and amenities associated with climate. Many of the goods and services affected by climate are not traded in markets; as a result, they can be difficult to value. For example, ecologists have already identified a number of ecological impacts of climate change, including the shifting, break up, and loss of certain ecological communities; plant and animal extinctions and a loss in biodiversity; shifting ranges of plant and animal populations; and changes in ecosystem processes, such as

---

1 A climate projection is the calculated response of the climate system to emissions or concentration scenarios of greenhouse gases and aerosols, or radiative forcing scenarios, often based on simulations by climate models. Climate projections are distinguished from climate predictions, in that the former critically depend on the emissions/concentration/radiative forcing scenario used, and therefore on highly uncertain assumptions of future socioeconomic and technological development.

nutrient cycling and decomposition. While ecosystems provide a variety of services to humans, including food and fiber, regulating air and water quality, support services such as photosynthesis, and cultural services such as recreation and aesthetic or spiritual values, these typically are not traded in markets.

Little research has been done linking these ecological changes to changes in services, and still less has been done to quantify, or place dollar values on, these changes. Ecosystem impacts also extend beyond the obvious direct effects within the natural environment to indirect effects on human systems. For instance, nearly 90 percent of Americans take part in outdoor recreation. The length of season of some of these activities, such as hiking, boating, or golfing, may be favorably affected by slightly increased temperatures. However, snow and ice sport seasons are likely to be shortened, resulting in lost recreation opportunities. The net effect is unclear as decrements associated with snow-based recreation may be more than outweighed by increases in other outdoor activities.

An agenda for understanding the impacts of climate change on human welfare may require taking steps both to develop a framework for addressing welfare, and to address the data and methodological gaps inherent in the estimation and quantification of effects. To that end, the study of climate change on human welfare is still developing, and, to our knowledge, no study has made a systematic survey of the full range of welfare impacts associated with climate change, much less attempted to quantify them.

CHAPTER 1

Introduction

Convening Lead Author: Janet L. Gamble, U.S. Environmental Protection Agency

Lead Authors: Kristie L. Ebi, ESS, LLC; Anne Grambsch, U.S. Environmental Protection Agency; Frances G. Sussman, Environmental Economics Consulting; Thomas J. Wilbanks, Oak Ridge National Laboratory

Contributing Authors: Colleen E. Reid, ASPH Fellow; Katharine Hayhoe, Texas Tech University; John V. Thomas, U.S. Environmental Protection Agency; Christopher P. Weaver, U.S. Environmental Protection Agency

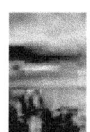

1.1 SCOPE AND APPROACH OF SAP 4.6

The Global Change Research Act of 1990 (Public Law 101-606) calls for the periodic assessment of the impacts of global environmental change for the United States. In 2001, a series of sector and regional assessments were conducted by the U.S. Global Change Research Program as part of the First National Assessment of the Potential Consequences of Climate Variability and Change on the United States. Subsequently, the U.S. Climate Change Science Program developed a *Strategic Plan* (CCSP, 2003) calling for the preparation of 21 synthesis and assessment products (SAPs) to inform policy making and adaptive management across a range of climate-sensitive issues. Synthesis and Assessment Product 4.6 examines the effects of global change on human systems. This product addresses Goal 4 of the five strategic goals set forth in the CCSP *Strategic Plan* to "understand the sensitivity and adaptability of different natural and managed ecosystems and human systems to climate and related global changes" (CCSP, 2003). The "global changes" assessed in this report include: climate variability and change, evolving patterns of land use within the United States, and changes in the nation's population.

While the mandate for the preparation of this report calls for evaluating the impacts of *global* change, the emphasis is on those impacts associated with climate change.

Collectively, global changes are human problems, not simply problems for the natural or the physical world. Hence, this SAP examines the vulnerability of human health and socioeconomic systems to climate change across three foci, including: human health, human settlements, and human welfare. The three topics are fundamentally linked but unique dimensions of global change.

Human health is one of the most basic and direct measures of human welfare. Following past assessments of climate change impacts on human health, SAP 4.6 focuses on human morbidity and mortality associated with extreme weather, vector-, water- and food-borne diseases, and changes in air quality in the United States. However, it should be noted that climate change in other parts of the world could impact human health in the United States. (*e.g.*, by affecting migration into the U.S., the safety of food imported into the U.S., etc.). Adaptation is a key component to evaluating human health vulnerabilities, including consideration of public health interventions (such as prevention, response, and treatment strategies) that could be revised, supplemented, or implemented to protect human health and determine how much adaptation could be achieved.

Settlements are where people live. Humans live in a wide variety of settlements in the United States, ranging from small villages and towns with a handful of people to metropolitan regions with millions of inhabitants. In particular,

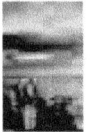

SAP 4.6 focuses on urban and highly developed population centers in the United States. Because of their high population density, urban areas multiply human health risks, and this is compounded by their relatively high proportions of the very old, the very young, and the poor. In addition, the components of infrastructure that support settlements, such as energy, water supply, transportation, and waste disposal, have varying degrees of vulnerability to climate change.

Welfare is an economic term used to describe the state of well-being of humans on an individual or collective basis. Human welfare is an elusive concept, and there is no single, commonly accepted definition or approach to thinking about welfare. There is, however, a shared understanding that increases in human welfare are associated with improvements in individual and communal conditions in areas such as political power, individual freedom, economic power, social contacts, health and opportunities for leisure and recreation, along with reductions in injury, stress, and loss. The physical environment, with climate as one aspect, is among many factors that can affect human welfare via economic, physical, psychological, and social pathways that influence individual perceptions of quality of life. Some core aspects of quality of life are expressed directly in markets (*e.g.*, income, consumption, personal wealth, etc.). The focus in SAP 4.6 is on non-market effects, although, these aspects of human welfare are often difficult to measure and value (Mendelsohn *et al.*, 1999; EPA, 2000).

The other Synthesis and Assessment Products related to CCSP's Goal 4 include reports on climate impacts on sea level rise (SAP 4.1), ecosystem changes (SAP 4.2), agricultural production (SAP 4.3), adaptive options for climate sensitive ecosystems (SAP 4.4), energy use (SAP 4.5), and transportation system impacts along the Gulf Coast (SAP 4.7). Collectively, these reports provide an overview of climate change impacts and adaptations related to a range of human conditions in the United States.

The audience for this report includes research scientists, public health practitioners, resource managers, urban planners, transportation planners, elected officials and other policy makers, and concerned citizens. A recent National Research Council analysis of global change assessments argues that the best assessments have an audience asking for them and a broad range of stakeholders (U.S. National Research Council, 2007). This report clearly identifies the pertinent audience and what decisions it will inform.

Chapters 2–4 describe the impacts of climate change on human systems and outline opportunities for adaptation. SAP 4.6 addresses the questions of how and where climate change may impact U.S. socio-economic systems. The challenge for this project is to derive an assessment of risks associated with health, welfare, and settlements and to develop timely adaptive strategies to address a range of vulnerabilities. Risk assessments evaluate impacts of climate change across an array of characteristics, including: the magnitude of risk (both baseline and incremental risks); the distribution of risks across populations (including minimally impacted individuals as compared to maximally exposed individuals); and the availability, difficulty, irreversibility, and cost of adaptation strategies. While the state of science limits the ability to conduct formal, quantitative risk assessments, it is possible to develop information that is useful for formulating adaptation strategies. Primary goals for adaptation to climate variability and change include the following:

- Avoid maladaptive responses;
- Establish protocols to detect and measure risks and to manage risks proactively when possible;

- Leverage technical and institutional capacity;

- Reduce current vulnerabilities to climate change;

- Develop adaptive capacity to address new climate risks that exceed conventional adaptive responses; and,

- Recognize and respond to impacts that play out across time. (Scheraga and Grambsch, 1998; WHO, 2003; IPCC, 2007b).

The issue of co-benefits is central in the consideration of adaptation to climate change. Many potential adaptive strategies have co-benefits. Along with helping human populations cope with climate change, adaptive strategies produce additional benefits. For example:

- Creating and implementing early warning systems and emergency response plans for heat waves can also improve those services for other emergency responses while improving all-hazards preparedness; (Glantz, 2004)

- Improving the infrastructure and capacity of combined sewer systems to avoid overflows due to changes in precipitation patterns also has the added benefit of decreasing contaminant flows that cause beach closings and impact the local ecology; (Rose *et al.*, 2001)

- A key adaptation technique for settlements in coastal zones is to promote maintenance or reconstruction of coastal wetlands ecosystems, which has the added benefit of creation or protection of coastal habitats (Rose *et al.*, 2001); and,

- Promotion of green building practices has added health and welfare benefits as improving natural light in office space and schools has been shown to increase productivity and mental health (Edwards and Torcellini, 2002).

Chapter 2 assesses the potential impacts of climate change on human health in the United States. Timely knowledge of human health impacts may support our public health infrastructure in devising and implementing strategies to prevent, compensate, or respond to these effects. For each of the health endpoints, the assessment addresses a number of topics, including:

- Reviewing evidence of the current burden associated with the identified health outcome;

- Characterizing the human health impacts of current climate variability and projected climate change (to the extent that the current literature allows);

- Discussing adaptation opportunities and support for effective decision making; and,

- Outlining key knowledge gaps.

Each topic chapter includes research published from 2001 through early 2007 in the United States, or in Canada, Europe, and Australia where results may provide insights for U.S. populations. As such, the health chapter serves as an update to the Health Sector Assessment conducted as part of the First National Assessment in 2001.

Chapter 3 focuses on the climate change impacts and adaptations associated with human settlements in the United States. The IPCC Third and Fourth Assessment Reports (IPCC, 2001; IPCC, 2007c) conclude that settlements are among the human systems that are the most sensitive to climate change. For example, if there are changes in climate extremes there could be serious consequences for human settlements that are vulnerable to droughts and wildfires, coastal and river floods, sea level rise and storm surge, heat waves, land slides, and windstorms. However, specific changes in these conditions in specific places cannot yet be projected with great confidence. Chapter 3 focuses on the interactions between settlement characteristics, climate, and other global stressors with a particular focus on urban areas and other densely developed population centers in the United States.

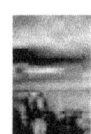

The scale and complexity of these built environments, transportation networks, energy and resource demands, and the interdependence of these systems and their populaces, suggest that urban areas are especially vulnerable to multiplying impacts in response to externally imposed environmental stresses. The collective vulnerability of American urban centers may also be determined by the disproportionate share of urban growth in areas like the Intermountain West or the Gulf Coast. The focus of Chapter 3 is on high density or rapidly growing settlements and the potential for changes over time in the vulnerabilities associated with place-based characteristics (such as their climate regime, elevation, and proximity to coasts and rivers) and spatial characteristics (such as whether development patterns are sprawling or compact).

Chapter 4 focuses on the impacts of climate change on human welfare. To examine the impacts of climate change on human welfare, this chapter reports on two relevant bodies of literature: approaches to welfare that rely on both qualitative assessment and quantitative measures, and economic approaches that monetize, or place money values, on quantitative impacts.

Finally, Chapter 5 revisits the research recommendations and data gaps of previous assessment activities and describes the progress to date and the opportunities going forward. In addition, Chapter 5 reviews the overarching themes derived from Chapters 2–4.

The remainder of this chapter is designed to provide the reader with an overview of the current state of knowledge regarding:

- Changes in climate in the united states;

- Population trends, migration patterns, and the distribution of people across settlements;

- Non-climate stressors and their interactions with climate change to realize complex impacts; and,

- A discussion of the handling of uncertainty in reporting scientific results.

1.2 CLIMATE CHANGE IN THE UNITED STATES: CONTEXT FOR AN ASSESSMENT OF IMPACTS ON HUMAN SYSTEMS

In the following chapters, the authors examine the impacts on human society of global change, especially those associated with climate change. The impact assessments in Chapters 2–4 do not rely on specific emissions or climate change scenarios, but instead rely on the existing scientific literature with respect to our understanding of climate change and its impacts on human health, settlements, and human well-being in the United States. This report does not make quantitative projections of specific impacts in specific locations based on specific projections of climate drivers of these impacts. Instead the report adopts a vulnerability perspective.

A vulnerability approach focuses on estimating *risks or opportunities* associated with possible impacts of climate change, rather than on *estimating quantitatively the impacts* themselves which would require far more detailed information about future conditions. Vulnerabilities are shaped not only by existing exposures, sensitivities, and adaptive capacities but also by responses to risks. In addition, climate change is not the only change confronting human societies: from a vulnerability perspective projected changes in populations, the economy, technology, institutions, infrastructure, and human and social capital are among the factors that also affect vulnerability to climate change. The report reviews historical trends and variability to point to vulnerabilities and then, where possible, determines the likely direction and range of potential climate-related impacts.

In the United States, we are observing the evidence of long-term changes in temperature and precipitation consistent with global warming. Changes in average conditions are being realized through rising temperatures, changes in annual and seasonal precipitation, and rising sea levels. Observations also indicate there are changes in extreme conditions, such as an increased frequency of heavy rainfall (with some increase in flooding), more heat waves, fewer very cold days, and an increase in areas affected by drought. Frequencies of tropical storms and hurricanes vary considerably from year to year and there are limitations in the quality of the data, which make it difficult to discern trends, but evidence suggests some increase in their intensity and duration since the 1970s (Christensen *et al.*, 2007).

The following sections provide a brief introduction to climate change as a context for the following chapters on impacts and adaptation. SAP 4.6 does not evaluate climate change projections as they are not used quantitatively in this assessment. The Intergovernmental Panel on Climate Change provides a comprehensive evaluation of climate change science. In their *Summary for Policy Makers* (IPCC, 2007a), the IPCC reports the following observed changes in global climate:

- "Warming of the climate system is unequivocal, as is now evident from observations of increases in global average air and ocean temperature, widespread melting of snow and ice, and rising global average sea level."

- "Eleven of the last twelve years rank among the 12 warmest years in the instrumental record of global surface temperatures (since 1850)."

- "Average temperature of the global ocean has increased to depths of at least 3000 m and that the ocean has been absorbing more than 80 percent of the heat added to the climate system. Such warming causes sea water to expand, contributing to sea level rise."

- "Mountain glaciers and snow cover have declined on average in both hemispheres."

- "The frequency of heavy precipitation events has increased over most land areas, consistent with warming and observed increases of atmospheric water vapor."

- "Widespread changes in extreme temperatures have been observed over the last 50 years… Hot days, hot nights, and heat waves have become more frequent."

- "There is observational evidence for an increase of intense tropical cyclone activity in the North Atlantic since about 1970." (IPCC, 2007a)

Note that these changes are for the entire globe: changes in the United States may be similar or different from these global changes. The following sections examine U.S. climate trends and historical records related to temperature, precipitation, sea level rise, and changes in hurricanes and other catastrophic events. Information is also drawn from the North American Chapter of the IPCC Fourth Assessment Report and the Climate Change Science Programs Synthesis and Assessment Product 3.3: Weather and Climate Extremes in a Changing Climate. Taken together, this discussion provides a context from which to assess impacts of climate change on human health, human welfare, and human settlements.

1.2.1 Rising Temperatures

Climate change is already affecting the United States. According to long-term station-based observational records such as the Historical Climatology Network (Karl *et al.*, 1990; Easterling *et al.*, 1999; Williams *et al.*, 2007), temperatures across the continental United States have been rising at a rate of 0.1°F per decade since the early 1900s. Increases in average annual temperatures over the last century now exceed 1°F (Figure 1.1a). The degree of warming has varied by region across the United States, with the West and Alaska

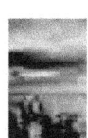

experiencing the greatest degree of warming (U.S. Environmental Protection Agency, 2007). These changes in temperature have led to an increase in the number of frost-free days, with the greatest increases occurring in the West and Southwest (Tebaldi *et al.*, 2006). The Intergovernmental Panel on Climate Change, in its most recent assessment report concluded that "Warming of the climate system is unequivocal…" (IPCC, 2007a).

The current generation of global climate models, run with IPCC SRES scenarios of future greenhouse gas emissions, simulates future changes in the earth's climate system that are greater in magnitude and scope than those already observed. According to the IPCC, by the end of the 21st century, annual surface temperature increases are projected to range from 2–3°C near the coasts in the conterminous United States to more than 5°C in northern Alaska. Nationally, annual warming in the United

States is projected to exceed 2°C, with projected increases in summertime temperatures ranging between 3 and 5°C (greatest in the Southwest). The largest warming is projected to reach 10°C for winter temperatures in the northernmost parts of Alaska. (IPCC, 2007c). For additional information about the modeling results, see the IPCC Fourth Assessment Working Group I Report, especially Chapter 11: Regional Climate Projections (Christensen *et al.*, 2007)

1.2.2 Trends in Precipitation

Shifting precipitation patterns have also been observed. Over the last century, annual precipitation across the continental United States has been increasing by an average of 0.18 inches per decade (Figure 1.1b). Broken down by season, winter precipitation around the coastal areas, including the West, Gulf, and Atlantic coasts, has been increasing by up to 30 percent while precipitation in the central part of the country (the Midwest and the Great Plains) has been decreasing by up to 20 percent. Large-scale spatial patterns in summer precipitation trends are more difficult to identify, as much of summer rainfall comes in the form of small-scale convective precipitation. However, it appears that there have been increases of 20-80 percent in summer rainfall over California and the Pacific Northwest, and decreases on the order of 20 to 40 percent across much of the south. The IPCC reports that rainfall is arriving in more intense events. (IPCC, 2007a).

El Niño events (a periodic warming of the tropical Pacific Ocean between South America and the International Date Line) are associated with increased precipitation and severe storms in some regions, such as the southeast United States and the Great Basin region of the western United States. El Niño events have also been characterized by warmer temperatures and decreased precipitation in other areas, such as the Pacific Northwest, and parts of Alaska. Historically, El Niño events occur about every 3 to 7 years and alternate with the opposite phases of below-average temperatures in the eastern tropical Pacific (La Niña). Since 1976-1977, there has been a tendency toward more prolonged and stronger El Niños (IPCC, 2007a). However, recent analyses of climate simulations indicate no consistent trends in future El Niño amplitude or frequency (Meehl *et al.*, 2007).

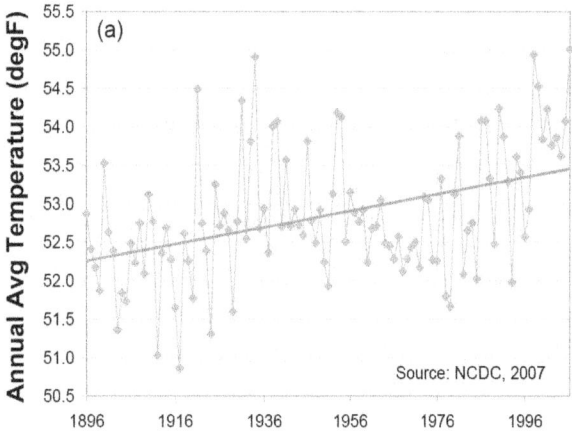

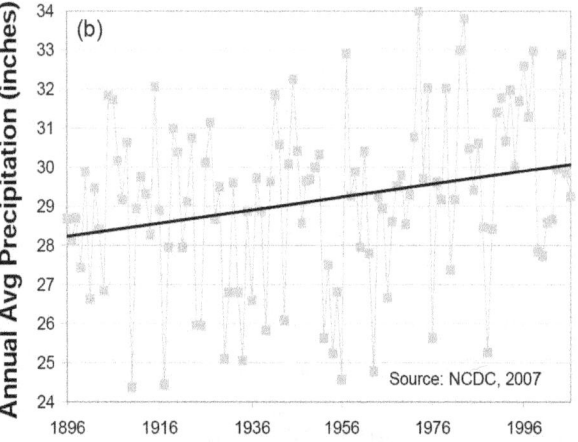

Figure 1.1 Observed trends in annual average (a) temperature (°F) and (b) precipitation (inches) across the continental United States from 1896 to 2006 (Source: NCDC, 2007)

Global model simulations summarized in the North American Chapter of the IPCC Fourth Assessment Report show moderate increases in precipitation (10 percent or less) over much of the United States over the next 100 years, except for the southwest. However, projected increases in these simulations are partially offset by increases in evaporation, resulting in greater drying in the central part of the United States. Projections for the central, eastern, and western regions of the United States show similar seasonal characteristics (*i.e.*, winter increases, summer decreases), although there is greater consensus for winter increases in the north and summer decreases in the south. However, uncertainty around the projected changes is large (IPCC, 2007b).

1.2.2.1 Changes in Snow Melt and Glacial Retreat

Warmer temperatures are melting mountain glaciers and more winter precipitation in northern states is falling as rain instead of snow (Huntington *et al.*, 2004). Snow pack is also melting faster, affecting stream flow in rivers. Over the past 50 years, changes in the timing of snow melt has shifted the schedule of snow-fed stream flow in the western part of the country earlier by 1 to 4 weeks. (Stewart *et al.*, 2005). The seasonal "center of stream flow volume" (*i.e.*, the date at which half of the expected winter-spring stream flow has occurred) also appears to be advancing by, on average, one day per decade for streams in the Northeast (Huntington *et al.*, 2003).

This trend is projected to continue, with more precipitation falling as rain rather than snow, and snow season length and snow depth are generally projected to decrease in most of the country. Such changes tend to favor increased risk of winter flooding and lower summer soil moisture and streamflows (IPCC, 2007a).

1.2.3 Rising Sea Levels and Erosion of Coastal Zones

Sea levels are rising and the IPCC concluded with high confidence that the rate of sea level rise increased from the 19th to the 20th centuries (IPCC, 2007a). The causes for observed sea level rise over the past century include thermal expansion of seawater as it warms and changes in land ice (e.g., melting

of glaciers and snow caps). Over the 20th century, sea level was rising at a rate of about 0.7 inches per decade (1.7 mm/yr ± 0.5 mm). For the period 1993 to 2003, the rate was nearly twice as fast, at 1.2 inches per decade (3.1 mm/yr ± 0.7 mm). However, there is considerably decadal variability in the tide gauge record, so it is unknown whether the higher rate in 1993 to 2003 is due to decadal variability or an increase in the longer-term trend. (Bindoff *et al.*, 2007). In the past century, global sea level rose 5–8 inches.

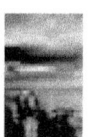

Spatially sea level change varies considerably: in some regions, rates are up to several times the global mean rise, while in other regions sea level is falling. For example, for the mid-Atlantic coast (*i.e.*, from New York to North Carolina), the "effective" or relative sea level rise rates have exceeded the global rate due to a combination of land subsidence and global sea level rise. In this region, relative sea level rise rates ranged between 3 to 4 mm per year (~1ft per century) over the 20th century. In other cases, local sea level rise is less than the global average because the land is still rising (rebounding) from when ice sheets covered the area, depressing the Earth's crust. Local sea levels can actually be falling in some cases (for example, the Pacific Northwest coast) if the land is rising more than the sea is falling (for additional details about sea level rise and its effects on U.S. coasts see Synthesis and Assessment Product 4.1 *Coastal elevations and sensitivity to sea level rise*).

Rising global temperatures are projected to accelerate the rate of sea level rise by further expanding ocean water, melting mountain glaciers, and increasing the rate at which Greenland and Antarctic ice sheets melt or discharge ice into the oceans. Estimates of sea level rise for a global temperature increase between 1.1 and 6.4°C (the IPCC estimate of likely temperature increases by 2100) are about 7 to 23 inches (0.18m to 0.59m), excluding the contribution from accelerated ice discharges from the Greenland and Antarctica ice sheets. Extrapolating the recent acceleration of ice discharges from the polar ice sheets would imply an additional contribution up to 8 inches (20cm). If melting of these ice caps increases, larger values of sea level rise cannot be excluded (IPCC, 2007a).

1.2.4 Changes in Extreme Conditions

The climatic changes described above are often referred to as changes in "average" conditions. Most observations of temperature will tend to be close to the average: days with very hot temperatures happen infrequently. Similarly, only rarely will there be days with extremely heavy precipitation. Climate change could result in a shift of the entire distribution of a meteorological variable so that a relatively small shift in the mean could be accompanied by a relatively large change in the number of relatively rare (according to today's perspective) events. For example, with an increase in average temperatures, it would be expected there would be an increase in the number of very hot days and a decrease in the number of very cold days. Other, relatively rare, extreme events of concern for human health, welfare, and settlements include hurricanes, floods and droughts.

In general, it is difficult to attribute any individual extreme event to a changing climate. Because extreme events occur infrequently, there is typically limited information to characterize these events and their trends. In addition, extreme events usually require several conditions to exist for the event to occur, so that linking a particular extreme event to a single, specific cause is problematic. For some extreme events, such as extremely hot/cold days or rainfall extremes, there is more of an observational basis for analyzing trends, increasing our understanding and ability to project future changes.

Finally, there are many different aspects to extremes. Frequency is perhaps the most often discussed but changes in other aspects of extremes such as intensity (*e.g.,* warmer hot days), time of occurrence *(e.g.,* earlier snowmelt), duration *(e.g.,* longer droughts), spatial extent, and location are also important when determining impacts on human systems.

Synthesis and Assessment Product 3.3 *Weather and Climate Extremes in a Changing Climate* (CCSP, 2008) has a much more detailed discussion of climate extremes that are only very briefly described here. The interested reader is referred to that report for additional details.

1.2.4.1 Heat and Cold Waves

Extreme temperatures (*e.g.,* temperatures in the upper 90th or 95th percentile of the distribution) often change in parallel with average temperatures. Since 1950, there are more 3-day warm spells (exceeding the 90th percentile) when averaged over all of North America (Peterson *et al.,* 2008). While the number of heat waves has increased, the heat waves of the 1930s remain the most severe in the U.S. historical record. Mirroring this shift toward more hot days is a decrease in unusually cold days during the past few decades. There has been a corresponding decrease in frost days and a lengthening of the frost-free season over the past century. The number of frost days decreased by four days per year in the United States during the 1948-1999 period, with the largest decreases, as many as 13 days per year, occurring in the western United States (Easterling, 2002). For the United States, the average length of the frost-free season over the 20th century increased by almost two weeks (Kunkel *et al.,* 2004).

Recent studies have found that there is an increased likelihood of more intense, longer-lasting, and more frequent heat waves (Meehl and Tebaldi, 2004, Schar *et al.,* 2004, Clark *et al.,* 2006). As the climate warms, the number of frost days is expected to decrease (Cubasch *et al.,* 2001) particularly along the northwest coast of North America (Meehl *et al.,* 2004). SAP 3.3, using a range of greenhouse gas emission scenarios and model simulations, found that hot days, hot nights, and heat waves are very likely to become more frequent, that cold days and cold nights are very likely to become much less frequent, and that the number of days with frost is very likely to decrease (CCSP, 2008). Growing season length is related to frost days, which is projected to increase in a warmer climate in most areas (Tebaldi *et al.*, 2006).

1.2.4.2 Heavy Precipitation Events

Over the 20th century, periods of heavy downpours became more frequent and more intense and accounted for a larger percentage of total precipitation (Karl and Knight, 1997; Groisman *et al.*, 1999, 2001, 2004, 2005; Kunkel *et al.*, 1999; Easterling *et al.*, 2000; Kunkel, 2003). These heavy rainfall events have increased in frequency by as much as 100 percent across much of the Midwest and Northeast over the past century (Kunkel *et al.*, 1999). These findings are consistent with observed warming and associated increases in atmospheric water vapor.

The intensity of precipitation events is projected to increase, particularly in high latitude areas that experience increases in mean precipitation (Meehl *et al.*, 2007). In areas where mean precipitation decreases (most subtropical and mid-latitude regions), precipitation intensity is projected to increase but there would be longer periods between rainfall events. Precipitation extremes increase more than does the mean in most tropical and mid- and high-latitude areas. Some studies project widespread increases in extreme precipitation (Christensen *et al.*, 2007), with greater risks of not only flooding from intense precipitation, but also droughts from greater temporal variability in precipitation. SAP 3.3 concluded that, over most regions, future precipitation is likely to be less frequent but more intense, and precipitation extremes are very likely to increase (CCSP, 2008).

1.2.4.3 Changes in Flooding

Heavy rainfall clearly can lead to flooding, but assessing whether observed changes in precipitation have lead to similar trends in flooding is difficult for a number of reasons. In particular, there are many human influences on streamflow (*e.g.*, dams, land-use changes, etc.) that confound climatic influences. In some cases, researchers using the same data came to opposite assessments about trends in high streamflows (Lins and Slack, 1999, 2005; Groisman *et al.*, 2001, 2004). Short duration extreme precipitation events can lead

to localized flash flooding, but for large river basins, significant flooding will not occur from these types of episodes alone; excessive precipitation must be sustained for weeks to months for flooding to occur.

1.2.4.4 Changes in Droughts

An extended period with little precipitation is the main cause of drought, but the intensity of a drought can be exacerbated by high temperatures and winds as well as a lack of cloudiness/low humidity, which result in high evaporation rates. Droughts occur on a range of geographic scales and can vary in their duration, in some cases lasting years. The 1930s and the 1950s experienced the most widespread and severe drought conditions (Andreadis *et al.*, 2005), although the early 2000s also saw severe droughts in some areas, especially in the western United States (Piechota *et al.*, 2004).

Based on observations averaged over the United States, there is no clear overall national trend in droughts (CCSP, 2008). Over the past century, the area affected by severe and extreme drought in the United States each year averaged about 14 percent: by comparison, in 1934 the area affected by drought was as high as 65 percent (CCSP, 2008). In recent years, the drought-affected area ranged between 35 and 40 percent (CCSP, 2008). These trends at the national level however mask important differences in drought conditions at regional scales: one area may be very dry while another is wet. For example, in the Southwest and parts of the interior of the West increased temperatures have led to rising drought trends (Groisman *et al.*, 2004; Andreadis and Lettenmaier, 2006).

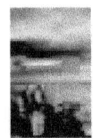

In the Southwest, the 1950s were the driest period, though droughts in the past 10 years are approaching the 1950s drought (CCSP, 2008). There are also recent regional tendencies toward more severe droughts in parts of Alaska (CCSP, 2008).

Several generations of global climate models, including the most recent, find an increase in summer drying in the mid latitudes in a future, warmer climate (Meehl *et al.*, 2007). This tendency for drying of the mid-continental areas during summer indicates a greater risk of droughts in those regions (CCSP, 2008). Analyses using several coupled global circulation models project an increased frequency of droughts lasting a month or longer in the Northeast (Hayhoe *et al.*, 2007) and greatly reduced annual water availability over the Southwest (Milly *et al.*, 2005). SAP 3.3 concluded that droughts are likely to become more frequent and severe in some regions of the country as higher air temperatures increase the potential for evaporation.

1.2.4.5 Changes in Hurricanes

Assessing changes in hurricanes is difficult: there have been large fluctuations in the number of hurricanes from year to year and from decade to decade. Furthermore, it is only since the 1960s that reliable data can be assembled for assessing trends. In general, there is increasing uncertainty in the data record the further back in time one goes but significant increases in tropical cyclone frequency are likely since 1900 (CCSP, 2008). However, the existing data and an adjusted record of tropical storms indicate no significant linear trends beginning from the mid- to late 1800s to 2005 (CCSP, 2008). Moreover, SAP 3.3 concluded that there is no evidence for a long-term increase in North American mainland land-falling hurricanes.

Evidence suggests that the intensity of Atlantic hurricanes and tropical storms has increased over the past few decades. SAP 3.3 indicates that there is evidence for a human contribution to increased sea surface temperatures in the tropical Atlantic and there is a strong correlation to Atlantic tropical storm frequency, duration, and intensity. However, a confident assessment will require further studies. An increase in extreme wave heights in the Atlantic since the 1970s has been observed that is consistent with more frequent and intense hurricanes (CCSP, 2008).

For North Atlantic hurricanes, SAP 3.3 concludes that it is likely that wind speeds and core rainfall rates will increase (Henderson-Sellers *et al.*, 1998; Knutson and Tuleya, 2004, 2008; Emanuel, 2005). However, SAP 3.3 concluded that "frequency changes are currently too uncertain for confident projection (CCSP, 2008)." SAP 3.3 also found that the spatial distribution of hurricanes will likely change. Storm surge is likely to increase due to projected sea level rise, although the degree to which this will increase has not been adequately studied (CCSP, 2008).

1.3 POPULATION TRENDS AND MIGRATION PATTERNS: A CONTEXT FOR ASSESSING CLIMATE-RELATED IMPACTS

Assessments of climate-related risk must account for the size of the population, including especially sensitive sub-populations and their geographic distribution across the landscape. The following discussion provides a basis for assessing the interactions of global change within the larger context of demographic trends. In particular, the social characteristics of a populace may interact with its spatial distribution to produce a non-linear risk. In such instances, risk assessments are shaped by questions such as:

- Which counties, states, and regions will grow most rapidly?
- How many people will live in at-risk areas, such as coastal zones, flood plains, and arid areas?
- What share of retirees will migrate and where will they move?

1.3.1 Trends in Total U.S. Population

The U.S. population numbered some 280 million individuals in 2000.[1] In 1900, the U.S. population numbered about 76 million people; fifty years later the population had roughly doubled to 151 million people.

---

1 Information on historical U.S population data and current population estimates and projections can be found at http://www.census.gov/.

Population projections are estimates of the population at future dates. They are based on assumptions about future births, deaths, international migration, and domestic migration and represent plausible scenarios of future population.

In 2000 the IPCC published a set of emission scenarios for use in the Third Assessment Report (Nakicenovic *et al.*, 2000). The SRES scenarios were constructed to explore future developments in the global environment with special reference to the production of greenhouse gases and aerosol precursor emissions. The SRES team defined four narrative storylines labeled A1, A2, B1, and B2, describing the relationships between the forces driving greenhouse gas and aerosol emissions and their evolution during the 21st century for large world regions and globally. Each storyline represents different demographic, social, economic, technological, and environmental developments that diverge in increasingly irreversible ways. (Nakicenovic *et al.*, 2000)

The U.S. Census Bureau periodically releases projections for the resident population of the United States based on Census data. The cohort-component methodology[2] is used in these projections. Alternative assumptions of fertility, life expectancy, and net immigration yield low, middle, and high projections.

Figure 1.2 displays the SRES and Census population projections[3] for the United States. The Census projections span a greater range than the SRES scenarios: by 2100 the low series projection of 282 million is below the current population while the high projection is about 1.2 billion, or about four times the current population. The Census middle series projection is relatively close to the SRES A2 scenario (570 million vs. 628 million in 2100), while the SRES A1/B1 and B2 scenarios fall below the Census middle projection.

1.3.1.1 Aging of the Population

The U.S. population has not only increased by 300 percent over the past century, it has also shifted in its demographic structure. For example, in 1900 less than 4 percent of the U.S. population was 65 years or older; currently about 12 percent of Americans are 65 or older (He *et al.*, 2005). By 2050, the US population aged 65 and older is projected to be about 86 million, or about 21 percent of the total population. Nearly 5 percent of the projected population in 2050, over 20 million people, will be 85 years or older (He *et al.*, 2005). Figure 1.3 displays the projected age distribution for the total resident population of the United States by sex for the middle projection series.

The projected increase in the elderly population is an important variable in projections of

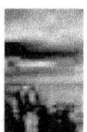

2 See Census website for additional details on the projection methodology.

3 The Census projections are based on the 1990 Census. Preliminary projections based on the 2000 Census for 2000-2050 are available.

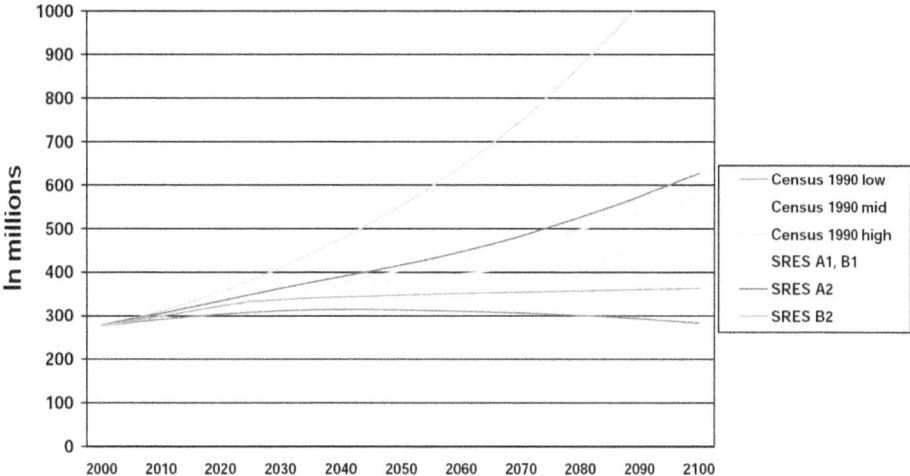

Figure 1.2 U.S. Population Projections 2000–2100

Data source: Census Population Projections http://www.census.gov/population/www/projections/natsum-T1.html
SRES Population Projections: http://sres.ciesin.columbia.edu/tgcia/

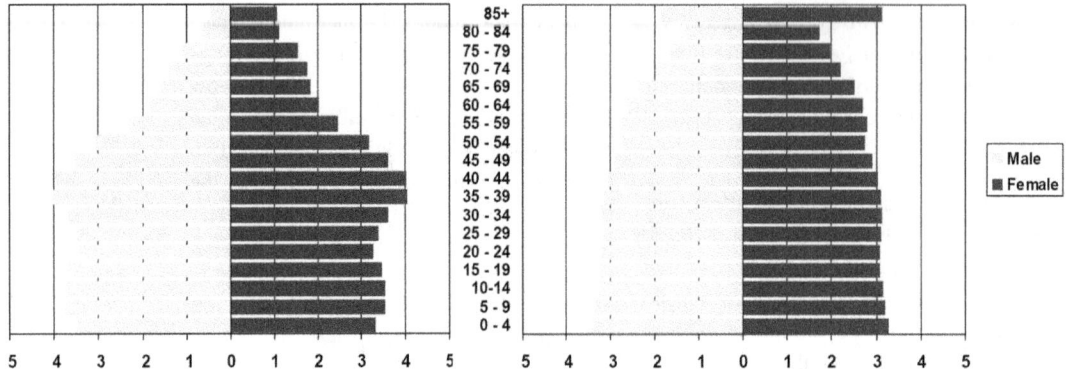

Figure 1.3 Population Pyramids of the U.S. 2000 and 2050 (Interim Projections based on 2000 Census)
Data source: Census Population Projections http://www.census.gov/ipc/www/usinterimproj/

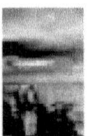

the effects of climate change. The elderly are identified in many health assessments as more vulnerable than younger age groups to a range of health outcomes associated with climate change, including injury resulting from weather extremes such as heat waves, storms, and floods (WHO, 2003; IPCC, 2007b; NAST, 2001). Aging also can be expected to be accompanied by multiple, chronic illnesses that may result in increased vulnerability to infectious disease (NAST, 2001). Chapter two in this report also identifies the elderly as a vulnerable subpopulation.

1.3.2 Migration Patterns

Although numbers produced by population projections are important, the striking relationship between potential future settlement patterns and the areas that may experience significant impacts of climate change is the critical insight. In particular, nearly all trends point to more Americans living in areas that may be especially vulnerable to the effects of climate change (see Figure 1.4). For example, many rapidly growing places in the Mountain West may also experience decreased snow pack during winter and earlier spring melting, leading to lower stream flows, particularly during the high-demand period of summer.

The continued growth of arid states in the West is therefore a critical crossroads for human settlements and climate change. These states are expected to account for one-third of all U.S. population growth over the next 25 years (U.S. Census Bureau, 2005). The combined effects of growing demand for water due to a growing population and changes in water supplies

associated with climatic change pose important challenges for these states. For example, a study commissioned by the California Energy Commission estimated that the Sierra Mountain snow pack could be reduced by 12 percent to 47 percent by 2050 (Cayan *et al.*, 2006). At the same time, state projections anticipate an additional 20 million Californians by that date (California Department of Finance, 2007).

Growth in coastal population has kept pace with population growth in other parts of the country, but given the small land area of the coasts, the density of coastal communities has been increasing (Crossett *et al.*, 2004). More than 50 percent of the U.S. population now lives in the coastal zone, and coastal areas are projected to continue to increase in population, with associated increases in population density, over the next several decades. The overlay of this migration pattern with climate change projections has several implications. Perhaps the most obvious is the increased exposure of people and property to the effects of sea level rise and hurricanes (Kunkel *et al.*, 1999). With rapidly growing communities near coastlines, property damages can be expected to increase even without any changes in storm frequency or intensity (Changnon *et al.*, 2003).

1.3.2.1 How Climate Impacts Migration Patterns

It is often said that America is a nation of movers and data collected for both the 1990 and 2000 Census support this notion. While roughly half of the U.S. population had lived in the same house for the previous five years, nearly 10 percent had recently moved from out

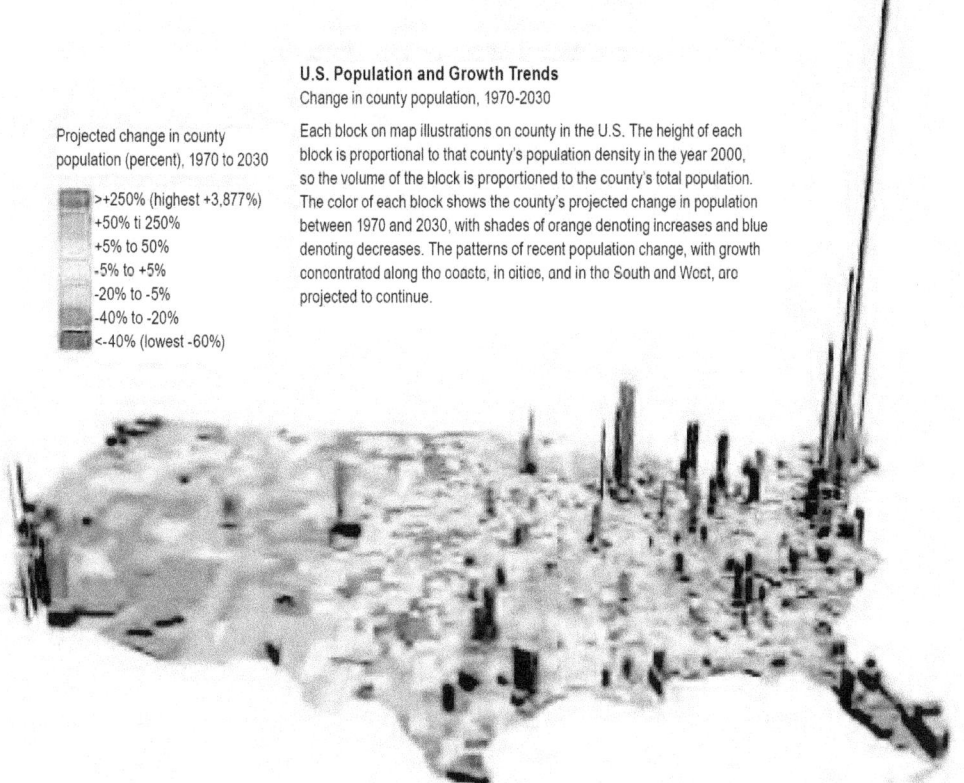

U.S. Population and Growth Trends
Change in county population, 1970-2030

Projected change in county
population (percent), 1970 to 2030

>+250% (highest +3,877%)
+50% ti 250%
+5% to 50%
-5% to +5%
-20% to -5%
-40% to -20%
<-40% (lowest -60%)

Each block on map illustrations on county in the U.S. The height of each
block is proportional to that county's population density in the year 2000,
so the volume of the block is proportioned to the county's total population.
The color of each block shows the county's projected change in population
between 1970 and 2030, with shades of orange denoting increases and blue
denoting decreases. The patterns of recent population change, with growth
concentrated along the coasts, in cities, and in the South and West, are
projected to continue.

Figure 1.4: U.S. Population and Growth Trends with evidence of more pronounced growth projected along the coasts, in urban centers, and in cities in the South and West (NAST, 2001)

of state.[4] In other words, during the five year period preceding each Census, over 20 million Americans had moved across state lines and half of those moved to different regions.

Although many forces shape domestic migration, climate is a key element of perceived quality of life. In turn, quality of life can be an important factor driving the relocation decisions of households and businesses. The popularity of the Places Rated Almanac and other publications ranking cities' livability illustrates the concept's importance. Additionally, many of the indicators in these reports are based directly on climatic conditions (average winter and summer temperature, precipitation, days of sunshine, humidity, etc.).

A range of studies have attempted to quantify how natural amenities, including a favorable climate, affect migration. While the methods vary[5] the conclusions are similar. In general:

- People move for a variety of reasons other than climate, such as: proximity to family and friends, employment opportunities, lower cost of living, and aesthetics;

- Areas with natural amenities that are close to urban centers have attracted the largest numbers of in-migrants (Serow, 2001);

- Climate's impact on migration varies by income with lower income groups also moving to colder areas in which their wages are likely to compare more favorably to the cost of living (Rebhun and Raveh, 2006);

- For retirees, weather is a far more important rationale cited for moving out of an area than moving to an area (AARP, 2006); and,

- Population growth in rural counties is strongly related to a more favorable climate and other key natural amenities (McGranahan, 1999). In addition, new information technologies may make it possible for some urban dwellers to move to and work from rural regions.

4 http://www.census.gov/Press-Release/www/2002/
   sumfile3.html
5 Study methodologies include: aggregate studies of
   population changes alongside regional characteristics, explanatory models developed from individual
   migration data and individual surveys.

1.4. COMPLEX LINKAGES: THE ROLE OF NON-CLIMATE FACTORS

Climate is only one of a number of global changes that affect human well-being. These non-climate processes and stresses interact with climate change, determining the overall severity of climate impacts. Moreover, climate change impacts can spread from directly impacted areas and sectors to other areas and sectors through extensive and complex linkages (IPCC, 2007b). Evaluating future climate change impacts therefore requires assumptions, explicit and implicit, about how future socioeconomic conditions will develop. The IPCC (1994) recommends the use of socioeconomic scenarios in impacts assessments to capture these factors in a consistent way.

Socioeconomic scenarios have tended to focus on variables such as population and measures of economic activity (*e.g.*, Gross Domestic Product) that can be quantified using well-established models or methods (for examples of economic models that have been used for long run projections, see Nakicenovic *et al.*, 2000; NAST, 2001; Yohe *et al.*, 2007). While useful as a starting point, some key socioeconomic factors may not allow this type of quantification: they could however be incorporated through a qualitative, "storyline" approach and thus yield a more fully developed socioeconomic scenario. The UNEP country study program guidance (Tol, 1998) notes the role of formal modeling in filling in (but not defining) socioeconomic scenarios but also emphasizes the role of expert judgment in blending disparate elements into coherent and plausible scenarios. Generally, socioeconomic scenarios have been developed

in situations where it is not possible to assign levels of probability to any particular future state of the world and therefore it usually is not appropriate to make confidence statements with respect to a specific socioeconomic scenario (Moss and Schneider, 2000).

Socioeconomic scenarios include non-environmental factors that influence exposures, vulnerability, and impacts. Factors that may be incorporated into a scenario include:

- Population (*e.g.*, demographics, immigration, domestic migration patterns);

- Economic status (income, prices);

- Technology (*e.g.*, pesticides, vaccines, transportation modes, wireless communications);

- Infrastructure (*e.g.*, water treatment plants, sewers, and drinking water systems; public health systems; roads, rails and bridges; flood control structures);

- Human capital and social context and behaviors (*e.g.*, skills and knowledge, social networks, lifestyles, diet); and,

- Institutions (legislative, social, managerial).

These factors are important both for characterizing potential effects of a changing climate on human health, settlements, and welfare, and for evaluating the ability of the United States to adapt to climate change.

1.4.1 Economic Status

The United States is a developed economy with GDP approaching $14 trillion and a per capita income of $38,611 in 2007 (US BEA, 2008). The U.S. economy has large private and public sectors, with strong emphasis on market mechanisms and private ownership (Christensen *et al.*, 2007). A nation's economic status clearly is important for determining vulnerability to climate change: wealthy nations have the economic resources to invest in adaptive measures and bear the costs of impacts and adaptation thereby reducing their vulnerability (WHO, 2003; IPCC, 2001). However, with the aging of the population (described in Section 1.3.1.1) the costs of health care are likely to rise over the coming decades (Christensen *et al.*, 2007). Moreover, if the trend toward globalization continues through the 21st

century, markets, primary factors of production, ownership of assets, and policies and governance will become more international in outlook (Stiglitz, 2002). Unfortunately, there has been little research to understand how these economic trends interact with climate change to affect vulnerability (*i.e.,* whether they facilitate or hinder adaptation to climate change).

1.4.2 Technology

The past half-century has seen stunning levels of technological advancement in the United States, which has done much to improve American standards of living. The availability and access to technology at varying levels, in key sectors such as energy, agriculture, water, transportation, and health is a key component to understanding vulnerability to climate change. Many technological changes, both large and small, have reduced Americans' vulnerability to climate change (NAST, 2001). Improved roads and automobiles, better weather and climate forecasting systems, computers and wireless communication, new drugs and vaccines, better building materials, more efficient energy production–the list is very long–have contributed to America's material well being while reducing vulnerability to climate. Many of the currently deployed adaptive strategies that protect human beings from climate involve technology *(e.g.,* warning systems, air conditioning and heating, pollution controls, building design, storm shelters, vector control, water treatment and sanitation) (WHO, 2003). Continued advances in technology in the 21st century can increase substantially our ability to cope with climate change (IPCC, 2007a; USGCRP, 2001).

However, it will be important to assess risks from proposed technological adaptations to avoid or mitigate adverse effects *(i.e.,* maladaptation) (Patz, 1996; Klein and Tol, 1997). For example, if new pesticides are used to control disease vectors their effects on human populations, insect predators, and insect resistance to pesticides need to be considered (Scheraga and Grambsch, 1998; Gubler *et al.,* 2001).

In addition, technological change can interact in complex ways with other socioeconomic factors *(e.g.,* migration patterns) and affect vulnerability to climate change. For example, advances in transportation technology–electric

streetcars, freight trucks, personal automobiles, and the interstate highway system–have fueled the decentralization of urban regions (Hanson and Giuliano 2004; Garreau 1991; Lang 2003). More recently, the rapid development of new information technologies, such as the internet, have made previously remote locations more accessible for work, recreation, or retirement. Whether these developments increase or decrease vulnerability is unknown, but they do indicate the need for socioeconomic scenarios to better characterize the complex linkages between climate and non-climate factors in order to evaluate vulnerability.

1.4.3 Infrastructure

Communities have reduced, and can further reduce, their vulnerability to adverse climate effects through investments in infrastructure. United States have been modified and intensively managed over the years, partly in response to climate variability (Cohan and Miller, 2001). These investments range from small, privately constructed impoundments, water diversions, and levees to major projects constructed by federal and state governments. Public health infrastructure, such as sanitation facilities, waste water treatment, and laboratory buildings reduce climate change health risks (Grambsch and Menne, 2003). Coastal communities have developed an array of systems to manage erosion and protect against flooding (see SAP 4.1 for an extensive discussion). More generally, infrastructure such as roads, rails, and bridges; water supply systems and drainage; mass transit; and buildings can reduce vulnerability (Grambsch and Menne, 2003).

However, infrastructure can increase vulnerability if its presence encourages people to locate in more vulnerable areas.

For example, increasing the density of people in coastal metropolitan areas, dependent on extensive fixed infrastructure, can increase vulnerability to extreme events such as floods, storm surges, and heat waves (NAST, 2001). In assessments of severe storms, measures of property damage are consistently higher and loss of life lower in the United States when compared with less-developed countries (Cohan and Miller, 2001). This reflects both the high level of development in coastal zones and the effectiveness of warnings and emergency preparedness (Pielke and Pielke, 1997).

Fixed infrastructure itself has the potential to be adversely impacted by climate change, which can increase vulnerability to climate change. For example, flooding can overwhelm sanitation infrastructure and lead to water-related illnesses (Grambsch and Menne, 2003). Much of the transportation infrastructure in the Gulf Coast has been constructed on land at elevations below 16.4 feet. Storm surge, therefore, poses risks of immediate flooding of infrastructure and damage caused by the force of floodwaters (see SAP 4.7 for additional information on the vulnerability of Gulf Coast transportation infrastructure to climate change). Damage to transportation infrastructure can make it more difficult to assist affected populations (Grambsch and Menne, 2003).

1.4.4 Human and Social Capital and Behaviors

While these factors are extremely difficult to quantify, much less project into the future, they are widely perceived to be important

in determining vulnerability in a number of different ways. In general, countries with higher levels of human capital (*i.e.*, the knowledge, experience, and expertise of its citizens), are considered to be less vulnerable to climate change. Effective adaptation will require individuals skilled at recognizing, reporting, and responding to climate change effects. Moreover, a number of the adaptive measures described in the literature require knowledgeable, trained, and skilled personnel to implement them. For example, skilled public health managers who understand surveillance and diagnostic information will be needed to mobilize appropriate responses. People trained in the operation, quality control, and maintenance of laboratories; communications equipment; and sanitation, wastewater, and water supply systems are also key (Grambsch and Menne, 2003). Researchers and scientists spanning a broad range of disciplines will be needed to provide a sound basis for adaptive responses.

In addition to a country's human capital, the relationships, exchange of resources, and knowledge, and the levels of trust and conflicts between individuals (*i.e.*, "social capital") are also important for understanding future vulnerability to climate change (Adger, 2003; Lehtonen, 2004; Pelling and High, 2005). Social networks can play an important role in coping and recovery from extreme weather events (Adger, 2003). For example, individuals who were socially isolated were found to be a greater risk of dying from extreme heat (Semenza *et al.*, 1996), as well as people living in neighborhoods without public gathering places and active street life (Klinenberg, 2002).

Individual behaviors and responses to changing conditions also determine vulnerability. For example, fitness, body composition, and level of activity are among the factors that determine the impact extremely hot weather will have on the human body (see Chapter 2 for additional information). Whether this trend continues or not could have important implications for determining vulnerability to climate change. Individual responses and actions to reduce exposures to extreme heat can also substantially ameliorate adverse health impacts (McGeehin and Mirabelli, 2001). Successfully motivating individuals to respond appropriately can therefore decrease vulnerability and reduce health impacts-a key goal of public health efforts (McGeehin and Mirabelli, 2001).

1.4.5 Institutions

The ability to respond to climate change and reduce vulnerability is influenced by social institutions as well as the social factors noted above. Institutions are viewed broadly in the climate change context and include a wide diversity of things such as regulations, rules, and norms that guide behavior. Examples include past development and land use patterns, existing environmental and coastal laws, building codes, and legal rights. Institutions also can determine a decision-maker's access to information and the ways in which the information can be used (Moser *et al.*, 2007).

Well-functioning institutions are essential to a modern society and provide a mechanism for stability in otherwise volatile environments (Moser *et al.*, 2007). Future options for responding to future climate impacts are thus shaped by our past and present institutions and how they evolve over time. In addition, the complex interaction of issues expected with climate change may require new arrangements and collaborations between institutions to address risks effectively, thereby enhancing adaptive capacity (Grambsch and Menne, 2003). A number of institutional changes have been identified that improve adaptive capacity and reduce vulnerability (see Chapter 3 for additional details). While the importance of institutions is clear, there are few scenarios that incorporate an explicit representation of them.

1.4.6 Interacting Effects

The same social and economic systems that bear the stress of climate change also bear the stress of non-climate factors, including: air and water pollution, the influx of immigrants, and an aging and over-burdened infrastructure in rapidly growing metropolitan centers and coastal zones. While non-climate stressors are currently more pronounced than climate impacts, one cannot assume that this trend will

persist. Understanding the impacts of climate change and variability on health and quality of life assumes knowledge of how these dynamics might vary by location and across time and socioeconomic group. The effects of climate change often spread from directly affected areas and sectors to other areas and sectors through complex linkages. The relative importance of climate change depends on the directness of each climate impact and on demographic, social, economic, institutional, and political factors, including the degree of emergency preparedness.

Consider the damage left by Hurricanes Katrina and Rita in 2005. Damage was measured not only in terms of lives and property lost, but also in terms of the devastating impacts on infrastructure, neighborhoods, businesses, schools, and hospitals as well as in the disruption to families and friends in established communities, with lost lives and lost livelihoods, challenges to psychological well-being, and exacerbation of chronic illnesses. While the aftermath of a single hurricane is not the measure of climate change, such an event demonstrates the disruptive power of climate impacts and the resulting tangle of climate and non-climate stressors that complicate efforts to respond and to adapt. The impacts following these hurricanes reveal that socioeconomic factors and failures in human systems may be as damaging as the storms themselves.

Another trend of significance for climate change is the suburbanization of poverty. A recent study noted that by 2005 the number of low income households living in suburban communities had for the first time surpassed the number living in central cities (Berube and Kneebone, 2006). Although the poverty rate in

cities was still double the suburban rate, there were 1 million more people living in poverty in America's suburbs. Many of these people live in older inner-ring suburbs developed in the 1950's and 60's. The climate adaptation challenge for these places is captured succinctly by a recent study: "Neither fully urban nor completely suburban, America's older, inner-ring, "first" suburbs have a unique set of challenges—such as concentrations of elderly and immigrant populations as well as outmoded housing and commercial buildings—very different from those of the center city and fast growing newer places. Yet first suburbs exist in a policy blind spot with little in the way of state or federal tools to help them adapt to their new realities" (Puentes and Warren, 2006).

1.5 REPORTING UNCERTAINTY IN SAP 4.6

Uncertainty can be traced to a variety of sources: (1) a misspecification of the cause(s), such as the omission of a causal factor resulting in spurious correlations; (2) mischaracterization of the effect(s), such as a model that predicts cooling rather than warming; (3) absence of or imprecise measurement or calibration (such as devices that fail to detect minute causal agents); (4) fundamental stochastic (chance) processes; (5) ambiguity over the temporal ordering of cause and effect; (6) time delays in cause and effect; and, (7) complexity where cause and effect between certain factors are camouflaged

by a context with multiple causes and effects, feedback loops, and considerable noise.

A new perspective on the treatment of uncertainty has emerged from the IPCC Third and Fourth Assessment processes.[6] This new perspective suggests that uncertainties about projections of climate changes, impacts, and responses include two fundamentally different dimensions. One dimension recognizes that most processes and systems being observed are characterized by inherent variability in outcomes: the more variable the process or system, the greater the uncertainty associated with any attempt to project an outcome. A second dimension recognizes limitations in our knowledge about processes and systems.

This report is a summary of the state of the science on the impacts of climate change on human health, human settlements, and human welfare. With this focus, the assessment of uncertainty in this report is based on the literature and the author team's expert judgment. The considerations in determining confidence include the degree of belief within the scientific community that available understanding, models, and analyses are accurate, expressed by

---

6 SAP 4.6 follows *the Guidance Notes for Lead Authors of the IPCC Fourth Assessment Report on Addressing Uncertainties*, produced by the IPCC in July 2005. See http://www.ipcc.ch/pdf/supporting-material/uncertainty-guidance-note.pdf for more details.

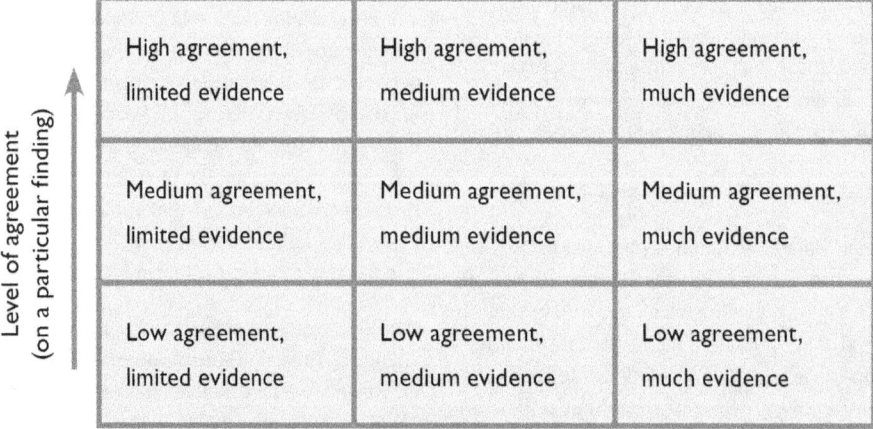

Figure 1.5 Considerations in determining confidence

Source: IPCC Guidance Notes on risk and uncertainty (2005)

the degree of consensus in the available evidence and its interpretation. This can be thought of using two different dimensions related to consensus. Figure 1.5 represents the qualitatively defined levels of understanding. It considers both the amount of evidence available in support of findings and the degree of consensus among experts on its interpretation.

In this report, each chapter author team assigned likelihood judgments that reflect their assessments of the current consensus of the science and the quality and amount of evidence. This represents their expert judgment that the given likelihood impact statement is true given a specified climatic change. The likelihood terminology and corresponding values used in this report are shown in Table 1.1. As the focus of this report is on impacts, it is important to note that these likelihood statements refer to the impact, not the underlying climatic changes *(i.e.,* the report does not address whether the specific climatic change is likely to occur). Moreover,

the authors do not attempt an assessment that takes into account a probabilistic accounting of both the likelihood of the climatic change and the impact. The terms defined in Table 1.1 are intended to be used in a relative sense to summarize judgments of the scientific understanding relevant to an issue, or to express uncertainty in a finding where there is no basis for making more quantitative statements.

The application of this approach to likelihood estimates demonstrates some variability across each of the three core chapters (Chapters 2–4). This variability in reporting uncertainty is based on the degree of richness of their respective knowledge bases. A relatively more extensive and specific application of likelihood and state of the knowledge estimates is possible for health impacts, only a more general approach is warranted for conclusions about human settlements, and uncertainty statements about human welfare conclusions are necessarily the least explicit.

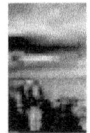

Table 1.1 Description of likelihood: probabilistic assessment of outcome having occurred or occurring in the future based on quantitative analysis or elicitation of expert views.

Likelihood Terminology	Likelihood of the Occurrence/Outcome
Virtually certain	> 99 percent probability
Very likely	> 90 percent probability
Likely	> 66 percent probability
About as likely as not	33 - 66 percent probability
Unlikely	< 33 percent probability
Very unlikely	< 10 percent probability
Exceptionally unlikely	< 1 percent probability

1.6 REFERENCES

AARP, 2006: *Aging, Migration, and Local Communities: The Views of 60+ Residents and Community Leaders*. Washington, DC.

Adger, W. N., 2003: Social capital, collective action and adaptation to climate change. *Journal of Economic Geography*, **79(4)**, 387-404.

Andreadis, K.M., E.A. Clark, A.W. Wood, A.F. Hamlet, and D.P. Lettenmaier, 2005: 20th century drought in the conterminous United States. *Journal of Hydrometeorology*, **6(6)**, 985-1001.

Andreadis, K.M. and D.P. Lettenmaier, 2006: Trends in 20th century drought over the continental United States. Geophysical Research Letters, 33, DOI:10.1029/2006GL025711.

Arctic Climate Impact Assessment, 2004: *Impacts of a Warming Arctic*. Cambridge U n i v e r s i t y Press, Cambridge, UK.

Bernard, S.M. and M.A. McGeehin, 2004: Municipal heat wave response plans. *American Journal of Public Health*, **94(9),** 1520-1522.

Bernard, S.M., J.M. Samet, A. Grambsch, K.L. Ebi, I. Romieu, 2001: The potential impacts of climate variability and change on air pollution-related health effects in the United States. Environmental Health Perspectives 109, **Supplement 2**, 199-209.

Berube, A. and E. Kneebone, 2006: *Two Steps Back: City and Suburban Poverty Trends 1999–2005*. Metropolitan Policy Program, Brookings Institution, Washington DC.

Bindoff, N.L., J. Willebrand, V. Artale, A. Cazenave, J. Gregory, S. Gulev, K. Hanawa, C. Le Quéré, S. Levitus, Y. Nojiri, C.K. Shum, L.D. Talley and A. Unnikrishnan, 2007: Observations: Oceanic Climate Change and Sea Level. In: Climate Change 2007: The Physical Science Basis. Contribution of Working Group I to the Fourth Assessment Report of the Intergovernmental Panel on Climate Change [Solomon, S., D. Qin, M. Manning, Z. Chen, M. Marquis, K.B. Averyt, M. Tignor and H.L. Miller (eds.)]. Cambridge University Press, Cambridge, UK and New York, USA, pp. 385-432.

Borrell, C., M. Marí-Dell'Olmo, M. Rodríguez-Sanz, P. Garcia-Olalla, J. Caylà, J. Benach, and C. Muntaner, 2006: Socioeconomic position and excess mortality during the heat wave of 2003 in Barcelona. *European Journal of Epidemiology*, **21(9)**, 633-640.

California Climate Change Center, 2003. *Our Changing Climate: Assessing the Risks to California*. A Summary Report from the California Climate Change Center.

Carruthers, J. I., A.C. Vias, 2005: Urban, suburban, and exurban sprawl in the Rocky Mountain West: evidence from regional adjustment models. *Journal of Regional Science*, **45(1)**, 21–48.

Cayan, D., A. L. Luers, M. Hanemann, G. Franco and B. Croes, 2006: *Scenarios of Climate Change in California: An Overview*. A Report by the California Climate Change Center, CEC-500-2005-186-SF.

CCSP, 2003: *Strategic Plan for the U.S. Climate Change Science Program*. A Report by the U.S. Climate Change Science Program and the Subcommittee on Global Change Research.

CCSP, 2008: *Weather and Climate Extremes in a Changing Climate. Regions of Focus: North America, Hawaii, Caribbean, and U.S. Pacific Islands*. A Report by the U.S. Climate Change Science Program and the Subcommittee on Global Change Research. [Karl, T.R., G.A. Meehl, C.D. Miller, S.J. Hassol, A.M. Waple, and W.L. Murray (eds.)]. Department of Commerce and National Oceanic and Atmospheric Administration, National Climatic Data Center, Washington, DC.

Centers for Disease Control and Prevention, 2001: Heat-related deaths—Los Angeles County, California, 1999-2000, and United States 1979-1998. *Morbidity and Mortality Weekly Report*, **50(29)**, 623-626.

Chagnon, S., 2003: Shifting economic impacts from weather extremes in the United States: result of societal changes, not global warming. *Natural Hazards*, **29**, 273-290.

Christensen, J.H., B. Hewitson, A. Busuioc, A. Chen, X. Gao, I. Held, R. Jones, R.K. Kolli, W.-T. Kwon, R. Laprise, V. Magaña Rueda, L. Mearns, C.G. Menéndez, J. Räisänen, A. Rinke, A. Sarr and P. Whetton, 2007: Regional climate projections. In: *Climate Change 2007: The Physical Science Basis. Contribution of Working Group I to the Fourth Assessment Report of the Intergovernmental Panel on Climate Change* [Solomon, S., D. Qin, M. Manning, Z. Chen, M. Marquis, K.B. Averyt, M. Tignor and H.L. Miller (eds.)]. Cambridge University Press, Cambridge, UK and New York, USA, pp. 848-940.

Clark, R.T., S. Brown, and J.M. Murphy, 2006: Modeling northern hemisphere summer heat extreme changes and their uncertainties using a physics ensemble of climate sensitivity experiments. *Journal of Climate*, **19(17)**, 4,418-4,435.

Cohen, S., K. Miller, K. Duncan, E. Gregorich, P. Groffman, P. Kovacs, V. Magana, D. McKnight, E. Mills, and D. Schimel, 2001: North America. In: *Climate Change 2001: Impacts, Adaptation, and Vulnerability. Contribution of Working Group II to the Third Assessment Report of the Intergovernmental Panel on Climate Change* [McCarthy, J.J., O.F. Canziani, N.A. Leary, D.J. Dokken, and K.S. White (eds.)]. Cambridge University Press, Cambridge, UK and New York, USA, pp. 735-800.

Cromartie, J.B., 1998: Net migration in the Great Plains increasingly linked to natural amenities and suburbanization. *Rural Development Perspectives*, **13(1)**, 27-34.

Crossett, K.M., T.J. Culliton, P.C. Wiley, T.R. Goodspeed, 2004: Population trends along the coastal United States: 1980-2008. National Oceanographic and Atmospheric Administration, Washington, DC.

Cubasch, U., G.A. Meehl, G.J. Boer, R.J. Stouffer, M. Dix, A. Noda, C.A. Senior, S.Raper, and K.S. Yap, 2001: Projections of future climate. In: *Climate Change 2001: The Scientific Basis. Contribution of Working Group 1 to the Third Assessment Report of the Intergovernmental Panel on Climate Change* [Houghton, J.T.,Y. Ding, D.J. Griggs, M. Noguer, P.J. van der Linden, X. Dai, et al. (eds.)].

Easterling, D.R., 2002: Recent changes in frost days and the frost-free season in the United States. *Bulletin of the American Meteorological Society*, **83(9)**, 1327-1332.

Easterling, D.R., T.R. Karl, J.H. Lawrimore, and S.A. Del Greco, 1999: *United States Historical Climatology Network Daily Temperature, Precipitation, and Snow Data for 1871-1997.* ORNL/CDIAC-118, NDP-070. Carbon Dioxide Information Analysis Center, Oak Ridge National Laboratory, Oak Ridge, Tennessee.

Easterling, D.R., G.A. Meehl, C. Parmesan, S.A. Changnon, T.R. Karl, and L.O. Mearns, 2000: Climate extremes: observations, modeling, and impacts. *Science*, **289**, 2068-2074.

Edwards, L. and P. Torcellini, 2002: *A literature review of the effects of natural light on building occupants*, TP-550-30769, Natural Renewable Energy Laboratory, Golden, Colorado.

Emanuel, K, 2005: Increasing destructiveness of tropical cyclones over the past 30 years. *Nature* **436**, 686-688.

EPA, 2000: Guidelines for Preparing Economic Analyses.EPA240-R-00-003,U.S.Environmental Protection Agency, Washington, DC.

Field, C.B., L.D. Mortsch, M. Brklacich, D.L. Forbes, P. Kovacs, J.A. Patz, S.W. Running and M.J. Scott, 2007: North America. *Climate Change 2007: Impacts, Adaptation and Vulnerability. Contribution of Working Group II to the Fourth Assessment Report of the Intergovernmental Panel on Climate Change* [M.L. Parry, O.F. Canziani, J.P. Palutikof, P.J. van der Linden and C.E. Hanson (eds.)]. Cambridge University Press, Cambridge, UK, pp. 617-652.

Flegal, K.M., M.D. Carroll, R.J. Kuczmarski, C.L. Johnson, 1998: Overweight and obesity in the United States: prevalence and trends, 1960–1994. *International Journal of Obesity*, 22(1), 39-47.

Frich, P., L.V. Alexander, P. Della-Marta, B. Gleason, M. Haylock, A. Tank, and T. Peterson, 2002: Observed coherent changes in climatic extremes during the second half of the twentieth century. *Climate Research*, **19**, 193–212.

Garreau, J., 1991: *Edge Cities: Life on the New Frontier.* Doubleday, New York.

Glantz, M.H., 2004: *Usable science 8: early warning systems: do's and don'ts.* Report of Workshop held 20-23 October 2003 in Shanghai, China. Boulder, Colorado.

Graham,S.andS.Marvin,1996:Telecommunications and the city: electronic spaces, urban places. Routledge Press, London.

Grambsch, A. and B. Menne, 2003: Adaptation and adaptive capacity in the public health context. In: *Climate Change and Human Health: Risks and Responses* [McMichael, A.J., D.H. Campbell-Lendrum, C.F. Corvalan, K.L. Ebi, A. Githeko, J.D. Scheraga, *et al.*, (eds.)]. World Health Organization, Geneva, Switzerland, pp. 220-236.

Greenough, G., M. McGeehin, S.M. Bernard, J. Trtanj, J. Riad, D. Engelberg, 2001: The potential impacts of climate variability and change on health impacts of extreme weather events in the United States. *Environmental Health Perspectives.* **109 (Supplement 2)**, 191-198.

Groisman P.Y., T.R. Karl, D.R. Easterling, R.W. Knight,P.B. Jamason, K.J. Hennessy, *et al.*, 1999: Changes in the probability of heavy precipitation: important indicators of climatic change. *Climatic Change*, **42**, 243–283.

Groisman, P.Y., R.W. Knight, D.R. Easterling, T.R. Karl, G.C. Hegerl, V.N. Razuvaez, 2004: Trends in precipitation intensity in the climate record. Abstract #A52A-06, American Geophysical Union Spring Meeting 2004.

Groisman, P.Y., R.W. Knight, D.R. Easterling, D. Levinson,R.R.Heim Jr.,T.R.Karl,P.H.Whitfield, G.C. Hegerl, V.N. Razuvaez, B.G. Sherstyukov, J.G. Enloe, and N.S. Stroumentova, 2005: Changes in precipitation distribution spectra and contemporary warming of the extratropics: implications for intense rainfall, droughts, and potential forest fire danger. Sixteenth Conference on Climate Variability and Change, January 2005, American Meteorological Society.

Groisman, P.Y., R.W. Knight, T.R. Karl., 2001: Heavy precipitation and high streamflow in the contiguous United States: trends in the twentieth century. *Bulletin of the American Meteorological Society*, **82(2)**, 219-246.

Gubler, D.J., P. Reiter, K.L. Ebi, W. Yap, R. Nasci, J.A. Patz, 2001: Climate variability and change in the United States: potential impacts on vector- and rodent-borne diseases. *Environmental Health Perspectives*, **109 (Supplement 2)**, 223-233.

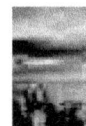

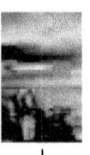

Hall, M.H.P. and D.B. Fagre, 2003: Modeled climate-induced glacier change in Glacier National Park, 1850-2100. *Bioscience*, **53**, 131-140.

Hanson, S. and G. Giuliano (eds.), 2004: *The Geography of Urban Transportation*. Guilford Press, New York, 3rd edition.

Hayhoe, K., C.P. Wake, T.G. Huntington, L. Lifeng, M.D. Schwartz, J. Sheffield, E. Wood, B. Anderson, J. Bradbury, A. Degaetano, T.J. Troy, D. Wolfe, 2007: Past and future changes in climate and hydrological indicators in the US Northeast. *Climate Dynamics*, **28(4)**, 381-407.

He, W., M. Sengupta, V.A. Velkoff, and K.A. DeBarros, 2005: *65+ in the United States: 2005*. U.S. Census Bureau, Current Population Reports, P23-209, U.S. Government Printing Office, Washington, DC.

Houghton, J. T., Y. Ding, D.J. Griggs, M. Noguer, P.J. van der Linden, X. Dai, K. Maskell, and C.A. Johnson (eds.), 2001: *Climate Change 2001: The Scientific Basis. Contribution of Working Group I to the Third Assessment Report of the Intergovernmental Panel on Climate Change*. Cambridge University Press, Cambridge, UK and New York, NY, USA, 881 pp.

Huntington, T. G., G.A. Hodgkins and R.W. Dudley, 2003: Historical trend in river ice thickness and coherence in hydroclimatological trends in Maine. *Climatic Change*, **61**, 217-236.

Huntington, T. G., G.A. Hodgkins, B.D. Keim, and R.W. Dudley, 2004: Changes in the proportion of precipitation occurring as snow in New England (1949 to 2000). *Journal of Climate*, **17**, 2626-2636.

IPCC, 1995. *Climate Change 1994: Radiative Forcing of Climate Change and an Evaluation of the IPCC IS92 Emission Scenarios* [Houghton, J.T., L.G. Meira Filho, J.P. Bruce, H. Lee, B.T. Callander, E.F. Haites, N. Harris, and K. Maskell (eds.)]. Cambridge University Press, Cambridge, UK, 339 pp.

IPCC, 2001: *Climate Change 2001: Synthesis Report. A Contribution of Working Groups I, II, and III to the Third Assessment Report of the Intergovernmental Panel on Climate Change* [Watson, R.T. and the Core Writing Team (eds.)]. Cambridge University Press, Cambridge, UK, 398 pp.

IPCC, 2007a. *Climate Change 2007: The Physical Science Basis. Contribution of Working Group I to the Fourth Assessment Report of the Intergovernmental Panel on Climate Change* [Solomon, S., D. Qin, M. Manning, Z. Chen, M. Marquis, K.B. Averyt, M. Tignor and H.L. Miller (eds.)]. Cambridge University Press, Cambridge, UK and New York, USA, 996 pp.

IPCC, 2007b: *Climate Change 2007: Impacts, Adaptation and Vulnerability. Contribution of Working Group II to the Fourth Assessment Report of the Intergovernmental Panel on Climate Change* [Parry, M.L., O.F. Canziani, J.P. Palutikof, P.J. van der Linden and C.E. Hanson (eds.)]. Cambridge University Press, Cambridge, UK, 976 pp.

IPCC, 2007c. *Climate Change 2007: Synthesis Report. Contribution of Working Groups I, II and III to the Fourth Assessment Report of the Intergovernmental Panel on Climate Change* [Core Writing Team, R.K. Pachauri and A. Reisinger (eds.)]. Intergovernmental Panel on Climate Change, Geneva, Switzerland, 104 pp.

Karl, T.R. and R.W. Knight, 1997: The 1995 Chicago heat wave: how likely is a recurrence? *Bulletin of the American Meteorological Society*, **78**, 1107-1119.

Karl, T.R., C. Williams, F. Quinlan and T. Boden, 1990: *United States Historical Climatology Network (HCN) Serial Temperature and Precipitation Data*. Environmental Science Division, Publication No. 3404, Carbon Dioxide Information and Analysis Center, Oak Ridge National Laboratory, Oak Ridge, Tennessee.

Klein, R.J.T. and R.S.J. Tol, 1997: Adaptation to climate change: options and technologies, an overview paper. Technical Paper FCCC/TP/1997/3, United Nations Framework Convention on Climate Change Secretariat, Bonn, Germany.

Klinenberg, E., 2002: *Heat wave: A Social Autopsy of Disaster in Chicago*. The University of Chicago Press, Chicago, Illinois.

Knutson, T.R. and R.E. Tuleya, 2004: Impact of CO_2-induced warming on simulated hurricane intensity and precipitation: sensitivity to the choice of climate model and convective parameterization. *Journal of Climate*, **17**, 3477-3495.

Kunkel, K. E., 2003: North American trends in extreme precipitation. *Natural Hazards*, **29**, 291–305.

Kunkel, K.E., D.R. Easterling, K. Hubbard, and K. Redmond, 2004: Temporal variations in frost-free season in the United States: 1895–2000. *Geophysical Research Letters*, **31**, L03201, DOI:10.1029/2003GL018624.

Kunkel, K.E., R. Pielke Jr., and S.A. Changnon, 1999: Temporal fluctuations in weather and climate extremes that cause economic and human health impacts: a review. *Bulletin of the American Meteorological Society*, **80**, 1077–1098.

Landsea, C.W. and R.D. Knabb, 2007: *Tropical Cyclone Wind Probabilities: Better Defining Uncertainty at the National Hurricane Center*. 19th Conference on Climate Variability and Change.

Lang, R., 2003: *Edgeless Cities: Exploring the Elusive Metropolis*. Brookings Institution Press, Washington, DC.

Lang, R.E. and D. Chavale, 2004: *Micropolitan America: A Brand New Geography*. Metropolitan Institute at Virginia Tech Census Note 05:01 May 2004.

Lehtonen, M., 2004: The environmental–social interface of sustainable development: capabilities, social capital, institutions. *Ecological Economics*, **49(2)**, 199-214.

Lins, H.F. and J.R. Slack, 1999: Streamflow trends in the United States. *Geophysical Research Letters*, **26(2)**, 227-230.

Lins, H.F. and J.R. Slack, 2005: Seasonal and regional characteristics of U.S. streamflow trends in the United States from 1940 to 1999. *Physical Geography*, **26(6)**, 489-501.

McGheehin, M.A. and M. Mirabelli, 2001: The potential impacts of climate variability and change on temperature-related morbidity and mortality in the United States. *Environmental Health Perspectives*, **109(2)**, 185-190.

McGranahan, D., 1999: *Natural Amenities Drive Rural Population Change*. U.S. Department of Agriculture Economic Research Service Report No. 781.

Meehl, G.A. and C. Tebaldi, 2004: More intense, more frequent, and longer lasting heat waves in the 21st century. *Science*, **305**, 994–997.

Meehl, G.A., C. Tebaldi, H. Teng, T.C. Peterson, 2007: Current and future U.S. weather extremes and El Niño. *Geophysical Research Letters*, **34**, L20704, DOI: 10.1029/2007GL031027.

Meehl, G.A., W.M. Washington, W.D. Collins, J.M. Arblaster, A. Hu, L.E. Buja, W.G. Strand, and H. Teng, 2005: How much more global warming and sea level rise. *Science*, **307**, 1769-1772.

Mendelsohn, R. and J.E. Neumann (eds.), 1999: *The Impact of Climate Change on the United States Economy*. Cambridge University Press, UK.

Milly, P.C.D., K.A. Dunne, and A.V. Vecchia, 2005: Global pattern of trends in streamflow and water availability in a changing climate. *Nature*, **438(7066)**, 347-350.

Moser, S.C., R.E. Kasperson, G. Yohe, and J. Agyeman, 2008: Adaptation to climate change in the Northeast United States: opportunities, processes, constraints. *Mitigation and Adaptation Strategies for Global Change*, **13(5-6)**, 643-659.

Moss, R.H. and S.H. Schneider, 2000: Uncertainties in the IPCC TAR: recommendations to lead authors for more consistent assessment and reporting. In: *Guidance Papers on the Cross Cutting Issues of the Third Assessment Report of the IPCC* [R. Pachauri, T. Taniguchi and K. Tanaka (eds.)]. World Meteorological Organization, Geneva, Switzerland, pp. 33-51.

Nakicenovic, N., R. Swart (eds.), 2000: *Special Report on Emissions Scenarios: A Special Report of Working Group III of the Intergovernmental Panel on Climate Change*. Cambridge University Press, Cambridge, UK, 599 pp.

National Assessment Synthesis Team, 2001: *Climate Change Impacts on the United States: The Potential Consequences of Climate Variability and Change*. [Melilo, J.M., A.C. Jacentos, T.R. Karl, and the National Assessment Synthesis Team (eds.)]. U.S. Climate Change Research Program, Washington, DC.

National Center for Atmospheric Research, 2005: Most of Arctic's near-surface permafrost may thaw by 2100. *Geophysical Research Letters*, **32**, L24401, DOI: 10.1029/2005GL025080.

National Research Council Division on Earth and Life Studies, 2001: *Under the Weather: Climate, Ecosystems, and Infectious Disease*. National Academy Press, Washington, DC.

Natural Resources Conservation Service, 2007: *National Resource Inventory*. U.S. Department of Agriculture.

Naughton, M., A. Henderson, M.C. Mirabelli, R. Kaiser, J.L. Wilhelm, S.M. Kieszak, C.H. Rubin, and M.A. McGeehin, 2002: Heat-related mortality during a 1999 heat wave in Chicago. *American Journal of Preventive Medicine*, **22(4)**, 221-227.

National Center for Climatic Data (NCDC), 2007: U.S. climate at a glance. Retrieved May 28, 2008, from http://www.ncdc.noaa.gov/oa/climate/research/cag3/cag3.html.

North Dakota State University. U.S. Geological Survey, North Dakota Water Science Center, 1997. Available online at: http://www.ndsu.edu/fargogeology/whyflood.htm

Patz, J.A., 1996: Health adaptation to climate change: need for far-sighted, integrated approaches. In: *Adapting to Climate Change: An International Perspective*. [Smith, J., *et al.* (eds.)]. Springer-Verlag, New York, pp. 450–464.

Patz, J., S. Khoury and C. Parker, 2005: *Climate Change and Health in California: A Pier Research Roadmap*. A Report Prepared for the California Energy Commission. CEC-500-2005-093.

Pelling, M. and C. High, 2005: Understanding adaptation: what can social capital offer assessments of adaptive capacity? *Global Environmental Change, Part A: Human and Policy Dimensions*, **5(4)**, 308-319.

Piechota, T., J. Timilsena, G. Tootle, and H. Hidalgo, 2004: The western U.S. drought: how bad is it? *EOS Transactions and American Geophysical Union*, **85(342)**, 301-308.

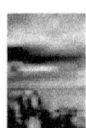

Pielke, Jr., R.A., and R.A. Pielke Sr., 1997: *Hurricanes: Their Nature and Impacts on Society.* John Wiley and Sons, England.

Public Law 101-606, 104 Stat. 3096-3104. Global Change Research Act of 1990.

Puentes, R. and D. Warren, 2006: *One Fifth of America, A Comprehensive Guide to America's First Suburbs.* Brookings Metropolitan Policy Program, Washington, DC.

Rebhun, U. and A. Raveh, 2006: The spatial distribution of quality of life in the United States and interstate migration, 1965–1970 and 1985–1990. *Social Indicators Research*, **78**, 137–178.

Rignot, E., 2006: Changes in ice dynamics and mass balance of the Antarctic ice sheet. *Philosophical Transactions of the Royal Society*, **364**, 1637-1655.

Rose, J.B., P.R. Epstein, E.K. Lipp, B.H. Sherman, S.M. Bernard, and J.A. Patz, 2001: Climate variability and change in the United States: potential impacts on water and foodborne diseases by microbiological agents. *Environmental Health Perspectives*, **109(2)**, 211-220.

Rosenthal, J.K. and P.W. Brandt-Rauf. 2006: Damage and low income households. *Environmental Planning and Urban Health Annals Academy of Medicine*, **35(8)**, 517-522.

Rosenzweig, C. and W.D. Solecki (eds.), 2001: *Climate Change and a Global City: The Potential Consequences of Climate Variability and Change: Metro East Coast.* Columbia Earth Institute. Columbia University, New York.

Ross, T. and N. Lott. 2000: *A Climatology of Recent Extreme Weather and Climate Events.* National Climatic Data Center, Technical Report 2000-02.

Schar, C., P.L. Vidale, D. Luthi, C. Frei, C. Haberli, M.A. Liniger, and C. Appenzeller, 2004: The role of increasing temperature variability in European summer heat waves. *Nature*, **427**, 332-336.

Scheraga, J.D., K.L. Ebi, A.R. Moreno, and J. Furlow, 2003: From science to policy: developing responses to climate change. In: *Climate Change and Human Health: Risks and Responses* [McMichael, A.J., D.H. Campbell-Lendrum, C.F. Corvalan, K.L. Ebi, A. Githeko, J.D. Scheraga, et al., (eds.)]. World Health Organization, World Meteorological Organization, and the United Nations Environment Program.

Scheraga, J.D. and A.E. Grambsch, 1998: Risks, opportunities and adaptation to climate change. *Climate Research*, **10**, 85-95.

Seager, R., M.F. Ting, I.M. Held, Y. Kushnir, J. Lu, G. Vecchi, H.P. Huang, N. Harnik, A. Leetmaa, N.C. Lau, C. Li, J. Velez, and N. Naik, 2007: Model projections of an imminent transition to a more arid climate in southwestern North America. *Science*, DOI: 10.1126/science.1139601

Semenza, J.C., C.H. Rubin, K.H. Falter, J.D. Selanikio, W.D. Flanders, H.L. Howe, and J.L. Wilhelm, 1996: Heat-related deaths during the July 1995 heat wave in Chicago. *New England Journal of Medicine*, **335**, 84-90.

Serow, W.J., 2001: Retirement migration counties in the southeastern United States: geographic, demographic, and economic correlates. *The Gerontologist*, **41(2)**, 220–222.

State of California, Department of Finance, 2004: *Population Projections by Race/Ethnicity, Gender and Age for California and Its Counties 2000-2050*, Sacramento, California.

Stewart, I.T., D.R. Cayan, and M.D. Dettinger, 2005: Changes toward earlier stream flow timing across western North America. *Journal of Climate*, **18**, 1136-1155.

Stott, P.A. 2004: Human contribution to the European heat wave of 2003. *Nature*, **432**, 610-614.

Tebaldi, C., K. Hayhoe, J.M. Arblaster, G.A. Meehl. 2006: Going to the extremes: an intercomparison of model-simulated historical and future changes in extreme events. *Climatic Change*, **79(3-4)**, 185-211.

Tol, R. 1998: Socio-economic scenarios. In: *UNEP Handbook on Methods for Climate Change Impact Assessment and Adaptation Studies.* [Burton, I., J.F. Feenstra, J.B. Smith, and R.S.J. Tol (eds.)]. Version 2.0, United Nations Environment Programme and Institute for Environmental Studies, Vrije Universiteit, Amsterdam, The Netherlands, 1-19.

U.S. BEA, 2008: *Survey of Current Business.* **88(3)**.

U.S. Census Bureau, 2000: *Methodology and Assumptions for the Population Projections of the United States: 1999 to 2100.* Population Division Working Paper No. 38, Washington, DC.

U.S. Census Bureau, 2000: *Population Projections of the United States by Age, Sex, Race, Hispanic Origin, and Nativity: 1999 to 2100.* Washington, DC.

U.S. Census Bureau, 2002: *Demographic Trends in the 20th Century.* Census 2000 Special Reports, CENSR-4, U.S. Government Printing Office, Washington, DC.

U.S. Census Bureau, 2005: Florida, California and Texas to dominate future population growth. *Census Bureau Reports.* Retrieved May 28, 2008, from: http://www.census.gov/Press-Release/www/releases/archives/population/004704.html.

U.S. Census Bureau, 2006: *Statistical Abstract of the United States: 2007.* (126th Edition) Washington, DC.

U.S. Climate Change Science Program and the Subcommittee on Global Change Research, 2003: *Strategic Plan for the U.S. Climate Change Science Program*. A Report by the U.S. Climate Change Science Program and the Subcommittee on Global Change Research, 202 pp.

United States Department of Agriculture Natural Resources Conservation Service, *National Water and Climate Center* (reservoir data). Retrieved May 28, 2008, from http://www.wcc. nrcs.usda.gov/cgibin/rs.pl.

U.S. Department of Energy, 2000: Trend in residential air-conditioning usage from 1978 to 1997. Retrieved May, 28, 2008, from http:// www.eia.doe.gov/emeu/consumptionbriefs/recs/ actrends/recs_ac_trends.html

U.S. National Research Council Committee on the Analysis of Global Change Assessments, 2007: *Analysis of Global Change Assessments: Lessons Learned*. National Academy Press, Washington, DC.

Webster, P.J., G.J. Holland, J.A Curry, and H.R. Chang, 2005: Changes in tropical cyclone number, duration, and intensity in a warming environment. *Science, 309*, 1844-1846.

Weisler, R.H., J.G. Barbee IV, and M.H. Townsend, 2006: Mental health and recovery in the Gulf Coast after Hurricanes Katrina and Rita. *Journal of the American Medical Association*, **296(5)**, 585-588.

Westerling, A.L., H.G. Hidalgo, D.R. Cayan, and T.W. Swetnam, 2006: Warming and earlier spring increase western U.S. forest wildfire activity. *Science*, **313(5789)**, 940-943.

Whitman, S., G. Good, E.R. Donoghue, N. Benbow, W.Y. Shou, and S.X. Mou, 1997: Mortality in Chicago attributed to the July 1995 heat wave. *American Journal of Public Health,* 87, 1515- 1519.

WHO, 2003: *Climate Change and Human Health - Risks and Responses*. [McMichael A.J., D.H. Campbell-Lendrum, C.F. Corvalán, K.L. Ebi, A. Githeko, J.D. Scheraga and A. Woodward (eds.)]. World Health Organization, World Meteorological Organization, and the United Nations Environment Program.

Williams, C. N., M. J. Menne, R.S. Vose, and D.R. Easterling, 2007: *United States Historical Climatology Network Monthly Temperature and Precipitation Data*. ORNL/CDIAC-87, NDP- 019. Carbon Dioxide Information Analysis Center, Oak Ridge National Laboratory, Oak Ridge, Tennessee.

Yohe, G.W., R.D. Lasco, Q.K. Ahmad, N.W. Arnell, S.J. Cohen, C. Hope, A.C. Janetos and R.T. Perez, 2007: Perspectives on climate change and sustainability. In: *Climate Change 2007: Impacts, Adaptation and Vulnerability. Contribution of Working Group II to the Fourth Assessment Report of the Intergovernmental Panel on Climate Change* [M.L. Parry, O.F. Canziani, J.P. Palutikof, P.J. van der Linden and C.E. Hanson, (eds.)]. Cambridge University Press, Cambridge, UK, 811-841.

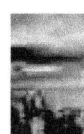

Effects of Global Change on Human Health

Lead Author: Kristie L. Ebi, ESS, LLC

Contributing Authors: John Balbus, Environmental Defense; Patrick L. Kinney, Columbia University; Erin Lipp, University of Georgia; David Mills, Stratus Consulting; Marie S. O'Neill, University of Michigan; Mark Wilson, University of Michigan

2.1 INTRODUCTION

Climate change can affect health directly and indirectly. Directly, extreme weather events (floods, droughts, windstorms, fires, and heat waves) can affect the health of Americans and cause significant economic impacts. Indirectly, climate change can alter or disrupt natural systems, making it possible for vector-, water-, and food-borne diseases to spread or emerge in areas where they had been limited or not existed, or for such diseases to disappear by making areas less hospitable to the vector or pathogen (NRC, 2001). Climate change can also affect the incidence of diseases associated with air pollutants and aeroallergens (Bernard *et al.,* 2001).[1] The cause-and-effect chain from climate change to changing patterns of health outcomes is complex and includes factors such as initial health status, financial resources, effectiveness of public health programs, and access to medical care. Therefore, the severity of future impacts will be determined by changes in climate as well as by concurrent changes in nonclimatic factors and by adaptations implemented to reduce negative impacts.

A comprehensive assessment of the potential impacts of climate change on human health in the United States was published in 2000. This First National Assessment was undertaken by the U.S. Global Change Research Program. The Health Sector Assessment examined potential impacts and identified research and data gaps to be addressed in future research.

---

1 Any of various air-borne substances, such as pollen or spores, that can cause an allergic response.

The results appeared in a special issue of *Environmental Health Perspectives* (May 2001). The Health Sector Assessment's conclusions on the potential health impacts of climate change in the United States included:

- Populations in northeastern and midwestern U.S. cities are likely to experience the greatest number of illnesses and deaths in response to changes in summer temperatures (McGeehin and Mirabelli, 2001).

- The health impacts of extreme weather events hinge on the vulnerabilities and recovery capabilities of the natural environment and the local population (Greenough *et al.,* 2001).

- If the climate becomes warmer and more variable, air quality is likely to be affected (Bernard *et al.,* 2001). However, uncertainties in climate models make the direction and degree of change speculative (Bernard and Ebi, 2001).

- Federal and state laws and regulatory programs protect much of the U.S. population from water-borne disease. However, if climate variability increases, current and future deficiencies in areas such as watershed protection, infrastructure, and storm drainage systems will probably increase the risk of contamination events (Rose *et al.,* 2000).

- It is unlikely that vector- and rodent-borne diseases will cause major epidemics in the United States if the public health infrastructure is maintained and improved (Gubler *et al.,* 2001).

- Multiple uncertainties preclude any definitive statement on the direction of potential future change for each of the health outcomes assessed (Patz *et al.*, 2000).

The assessment further concluded that much of the U.S. population is protected against adverse health outcomes associated with weather and/or climate by existing public health and medical care systems, although certain populations are at increased risk.

This chapter of SAP 4.6 updates the 2000 Health Sector Assessment. It also examines adaptation strategies that have been or are expected to be developed by the public health community in response to the challenges and opportunities posed by climate change. The first section of this chapter focuses on climate-related impacts on human morbidity and mortality from extreme weather, vector-, water- and food-borne diseases, and changes in air quality. For each health endpoint, the assessment addresses the potential impacts, populations that are particularly vulnerable, and research and data gaps that, if bridged, would allow significant advances in future assessments of the health impacts of global change. The assessment includes research published from 2001 through early 2007 in the United States or in Canada, Europe, and Australia, where results may provide insights for U.S. populations.

This chapter summarizes the current burden of climate-sensitive health determinants and outcomes for the United States before assessing the potential health impacts of climate change.

Two types of studies are assessed: (1) studies that increase our understanding of the associations between weather variables and health outcomes raise possible concerns about the impacts of a changing climate, and (2) studies that project the burden of health outcomes using scenarios of socioeconomic and climate change.

It is important to note that this assessment focuses on how climate change could affect the future health of Americans. However, the net impact of any changes will depend on many other factors, including demographics; population and regional vulnerabilities; the future social, economic, and cultural context; availability of resources and technological options; built and natural environments; public health infrastructure; and the availability and quality of health and social services.

The chapter then turns to adaptation to the potential health impacts of environmental change in the United States. It also considers public health interventions (including prevention, response, and treatment strategies) that could be revised, supplemented, or implemented to protect human health in response to the challenges and opportunities posed by global change, and considers how much adaptation could achieve.

2.2 OBSERVED CLIMATE-SENSITIVE HEALTH OUTCOMES IN THE UNITED STATES

2.2.1 Thermal Extremes: Heat Waves

Excess deaths occur during heat waves, on days with higher-than-average temperatures, and in places where summer temperatures vary more or where extreme heat is rare (Braga *et al.*, 2001). Figure 2.1 illustrates that the relation between temperature and mortality is nonlinear, typically J- or U-shaped, and that increases in mortality occur even below temperatures considered to be extremely hot. This figure was created using log-linear regression to analyze 22 years of data on daily mortality and outdoor temperature in 11 U.S. cities (Curriero *et al.*, 2002). Exposure to excessive natural heat caused a reported 4,780 deaths during the period 1979 to 2002, and an additional 1,203 deaths had hyperthermia reported as a contributing factor (CDC, 2005). These numbers are underestimates of the total mortality associated with heat waves because the person filling out the death certificate may not always list heat as a cause. Furthermore, heat can exacerbate chronic health conditions, and several analyses have reported associations with cause-specific mortality, including cardiovascular, renal, and respiratory diseases; diabetes; nervous system disorders; and other causes not specifically described as heat-related (Conti *et al.*, 2007; Fouillet *et al.*, 2006; Medina-Ramon *et al.*, 2006). Among the most well-documented heat waves in the United States are those that occurred in 1980 (St. Louis and Kansas City, Missouri), 1995 (Chicago, Illinois), and 1999 (Cincinnati, Ohio; Philadelphia, Pennsylvania; and Chicago, Illinois). The highest death rates in these heat waves occurred in people over 65 years of age.

Less information exists on temperature-related morbidity, and those studies that have examined hospital admissions and temperature have not seen consistent effects, either by cause or by demonstrated coherence with mortality effects where both deaths and hospitalizations were examined simultaneously (Kovats *et al.*, 2004; Michelozzi *et al.*, 2006; Schwartz *et al.*, 2004; Semenza *et al.*, 1999).

Age, fitness, body composition, and level of activity are important determinants of how the human body responds to exposure to thermal extremes (DeGroot *et al.*, 2006; Havenith *et al.*, 1995; Havenith *et al.*, 1998; Havenith, 2001). Groups particularly vulnerable to heat-related mortality include the elderly, very young, city-dwellers, those with less education, people on medications such as diuretics, the socially isolated, the mentally ill, those lacking access to air conditioning, and outdoor laborers (Diaz *et al.*, 2002; Klinenberg, 2002; McGeehin and Mirabelli, 2001; Semenza *et al.*, 1996; Whitman *et al.*, 1997; Basu *et al.*, 2005; Gouveia *et al.*, 2003; Greenberg *et al.*, 1983; O'Neill *et al.*, 2003; Schwartz, 2005; Jones *et al.*, 1982; Kovats *et al.*, 2004; Schwartz *et al.*, 2004; Semenza *et al.*, 1999; Watkins *et al.*, 2001). A sociological analysis of the 1995 Chicago heat wave found that people living in neighborhoods without public gathering places and active street

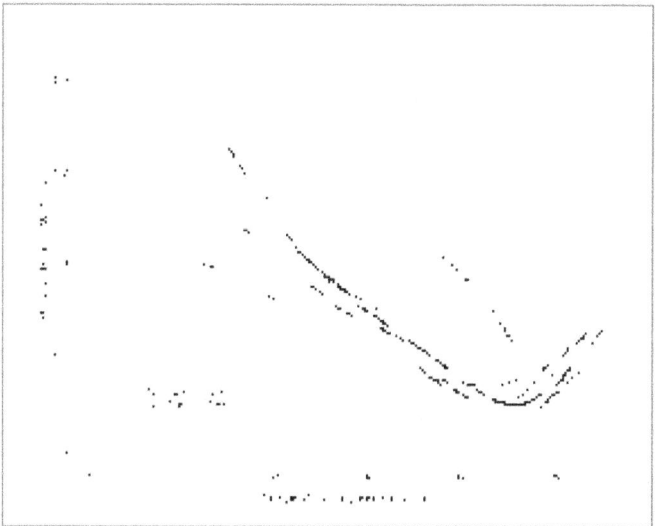

Figure 2.1 Temperature-mortality relative risk functions for 11 U.S. cities, 1973–1994. Northern cities: Boston, Massachusetts; Chicago, Illinois; New York, New York; Philadelphia, Pennsylvania; Baltimore, Maryland; and Washington, DC. Southern cities: Charlotte, North Carolina; Atlanta, Georgia; Jacksonville, Florida; Tampa, Florida; and Miami, Florida. Relative risk is defined as the risk of an event such as mortality relative to exposure, such that the relative risk is a ratio of the probability of the event occurring in the exposed group versus the probability of occurrence in the control (non-exposed) group. (Curriero *et al.*, 2002)

life were at higher risk, highlighting the important role that community and societal characteristics can play in determining vulnerability (Klinenberg, 2002).

Urban heat islands may increase heat-related health impacts by raising air temperatures in cities 2-10°F over the surrounding suburban and rural areas due to absorption of heat by dark paved surfaces and buildings; lack of vegetation and trees; heat emitted from buildings, vehicles, and air conditioners; and reduced air flow around buildings (EPA, 2005; Pinho and Orgaz, 2000; Vose *et al.*, 2004; Xu and Chen, 2004). However, in some regions, urban areas may not experience greater heat-related mortality than in rural areas (Sheridan and Dolney, 2003); few comparisons of this nature have been published.

The health impacts of high temperatures and high air pollution can interact, with the extent of interaction varying by location (Bates, 2005; Goodman *et al.*, 2004; Goodman *et al.*, 2004; Keatinge and Donaldson, 2001; O'Neill *et al.*, 2005; Ren *et al.*, 2006).

2.2.2 Thermal Extremes: Cold Waves

From 1979 to 2002, an average of 689 reported deaths per year (range 417-1,021), totaling 16,555 over the period, were attributed to exposure to excessive cold temperatures (Fallico *et al.*, 2005). Cold also contributes to deaths caused by respiratory and cardiovascular diseases, so the overall mortality burden is likely underestimated. Factors associated with increased vulnerability to cold include African American race (Fallico *et al.*, 2005); living in Alaska, New Mexico, North Dakota, and Montana, or living in milder states that experience rapid temperature changes (North and South Carolina) and western states with greater ranges in nighttime temperatures (*e.g.*, Arizona) (Fallico *et al.*, 2005); having less education (O'Neill *et al.*, 2003); being female or having pre-existing respiratory illness (Wilkinson *et al.*, 2004); lack of protective clothing (Donaldson *et al.*, 2001); income inequality, fuel poverty, and low residential thermal standards (Healy, 2003); and living in nursing homes (Hajat *et al.*, 2007).

Because climate change is projected to reduce the severity and length of the winter season (IPCC, 2007a), there is considerable speculation concerning the balance of climate change-related decreases in winter mortality compared with increases in summer mortality. Net changes in mortality are difficult to estimate because, in part, much depends on complexities in the relationship between mortality and the changes

associated with climate change. Few studies have attempted to link the epidemiological findings to climate scenarios for the United States, and studies that have done so have focused on the effects of changes in average temperature, with results dependent on climate scenarios and assumptions of future adaptation. Moreover, many factors contribute to winter mortality, making the question of how climate change could affect mortality highly uncertain. No projections have been published for the United States that incorporate critical factors such as the influence of influenza outbreaks.

2.2.3 Extreme Events: Hurricanes, Floods, and Wildfires

The United States experiences a wide range of extreme weather events, including hurricanes, floods, tornadoes, blizzards, windstorms, and drought. Other extreme events, such as wildfires, are strongly influenced by meteorological conditions. Direct morbidity and mortality due to an event increase with the intensity and duration of the event, and can decrease with advance warning and preparation. Health also can be affected indirectly. Examples include carbon monoxide poisonings from portable electric generator use following hurricanes (CDC, 2006b) and an increase in gastroenteritis cases among hurricane evacuees (CDC, 2005a). The mental health impacts (*e.g.*, post-traumatic stress disorder [PTSD], depression) of these events are likely to be especially important but are difficult to assess (Middleton *et al.*, 2002; Russoniello *et al.*, 2002; Verger *et al.*, 2003; North *et al.*, 2004; Fried *et al.*, 2005; Weisler *et al.*, 2006). However, failure to fully account for direct and indirect health impacts may result in inadequate preparation for and response to future extreme weather events.

Figure 2.2 shows the annual number of deaths attributable to hurricanes in the United States from the 1900 Galveston storm, (NOAA, 2006), records for the years 1940-2004 (NOAA, 2005a), and a summary of a subset of the 2005 hurricanes (NOAA, 2007). The data shown are dominated by the 1900 Galveston storm and a subset of 2005 hurricanes, particularly Katrina and Rita, which together accounted for 1,833 of the 2,002 lives lost to hurricanes in 2005 (NOAA, 2007b). While Katrina was a Category

3 hurricane and its path was forecast well in advance, there was a secondary failure of the levee system in Louisianna. This illustrates that multiple factors contribute to making a disaster and that adaptation measures may not fully avert adverse consequences.

From 1940 through 2005 roughly 4,300 lives were lost in the United States to hurricanes. The impact of the 2005 hurricane season is especially notable as it doubled the estimate of the average number of lives lost to hurricanes in the United States over the previous 65 years.

Figure 2.3 shows the annual number of deaths attributed to flooding in the United States from 1940-2005 (NOAA, 2007a). Over this period roughly 7,000 lives were lost.

A wildfire's health risk is largely a function of the population in the affected area and the speed and intensity with which the wildfire moves through those areas. Wildfires can increase eye and respiratory illnesses due to fire-related air pollution. Climate conditions affect wildfire incidence and severity in the West (Westerling *et al.*, 2003; Gedalof *et al.*, 2005; Sibold and Veblen, 2006). Between 1987-2003 and 1970-1986, there was a nearly fourfold increase in the incidence of large Western wildfires (*i.e.*, fires that burned at least 400 hectares) (Westerling *et al.*, 2006). The key driver of this increase was an average increase in springtime temperature of 0.87°C that affected spring snowmelt, subsequent potential for evapotranspiration, loss of soil moisture, and drying of fuels (Running, 2006; Westerling *et al.*, 2006). Data providing a time-series summary of deaths similar to the data in Figures 2.2 and 2.3 were not identified.

There is a rich body of literature detailing the mental health impacts of extreme weather events. Anxiety and depression, the most common mental health disorders, can be directly attributable to the experience of the event (*i.e.*, being flooded) or indirectly during the recovery process (*e.g.*, Gerrity and Flynn, 1997). These psychological effects tend to be much longer lasting and can be worse than the physical effects experienced during an event and its immediate aftermath.

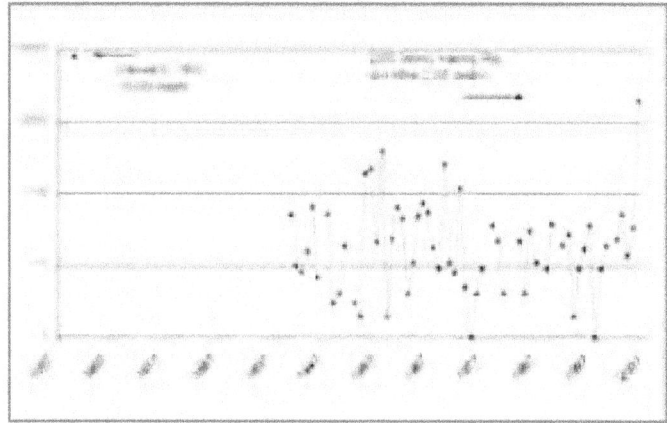

Figure 2.2 Annual Deaths Attributed to Hurricanes in the United States, 1900 and 1940-2005

Source: NOAA, 2007

Extreme events are often multi-strike stressors, with stress associated with the event itself; the disruption and problems of the recovery period; and the worry or anxiety about the risk of recurrence of the event (Tapsell *et al.*, 2002). During the recovery period, mental health problems can arise from the challenges associated with geographic displacement, damage to the home or loss of familiar possessions, and stress involved with the process of repairing. The full impact often is not appreciated until after people's homes have been put back in order. For instance, in the aftermath of Hurricane Katrina in 2005, mental health services in New Orleans were challenged by an increased incidence of serious mental illness, including anxiety, major depression, and PTSD. Shortly after Katrina, a Centers for Disease Control and Prevention poll found that

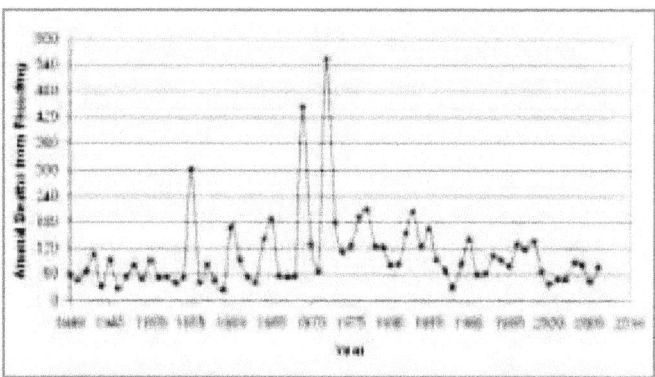

Figure 2.3 Annual Deaths Attributed to Flooding in the United States, 1940-2005

Source: NOAA, 2007a

nearly half of all survey respondents indicated a need for mental health care, yet less than 2 percent were receiving professional attention (Weisler *et al.*, 2006).

2.2.4 Indirect Health Impacts of Climate Change

The observation that most vector-, water- or food-borne and/or animal-associated diseases exhibit a distinct seasonal pattern suggests *a priori* that weather and/or climate influence their distribution and incidence. The following sections differentiate between zoonotic and water- and food-borne diseases, although many water- and food-borne diseases are zoonotic.

2.2.4.1 Vector-borne and Zoonotic (VBZ) Diseases

Transmission of infectious agents by blood-feeding arthropods (particular insect or tick species) and/or by non-human vertebrates (certain rodents, canids, and other mammals) has changed significantly in the United States during the past century. Diseases such as rabies and cholera have become less widespread and diseases such as typhus, malaria, yellow fever, and dengue fever have largely disappeared, primarily because of environmental modification and/or socioeconomic development (Philip and Bozeboom, 1973; Beneson, 1995; Reiter, 1996). While increasing average temperatures may allow the permissive range for *Aedes aegypti*, the mosquito vector of dengue virus, to move further north in the United States, it is unlikely that more cases of dengue fever will

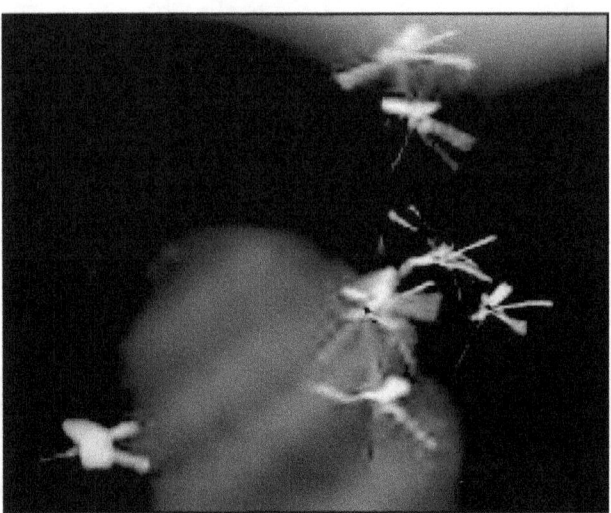

be observed because most people are protected living indoors due to quality housing. Indeed, a recent epidemic of dengue in southern Texas and northern Mexico produced many cases among the relatively poor Mexicans, and very few cases among Texans (Reiter *et al.*, 1999). At the same time, the distubution of other diseases changed either because of suitable environmental conditions (including climate) or enhanced detection (examples include Lyme disease, ehrlichioses, and Hantavirus pulmonary syndrome), or have been introduced and are expanding their range due to appropriate climatic and ecosystem conditions (West Nile Virus; *e.g.*, Reisen *et al.*, 2006). Still others are associated with non-human vertebrates that have complex associations with climate variability and human disease (*e.g.*, plague, influenza). The burden of VBZ diseases in the United States is not negligible and may grow in the future because the forces underlying VBZ disease risk involve weather/climate, ecosystem change, social and behavioral factors simultaneously, and larger political-economic forces that are part of globalization. In addition, introduction of pathogens from other regions of the world is a very real threat.

Few original research articles on climate and VBZ diseases have been published in the United States and in other developed temperate countries since the First National Assessment. Overall, these studies provide evidence that climate affects the abundance and distributions of vectors that may carry West Nile virus, Western Equine encephalitis, Eastern Equine encephalitis, Bluetongue virus, and Lyme disease. Climate also may affect disease risk, but sometimes in counterintuitive ways that do not necessarily translate to increased disease incidence (Wegbreit and Reisen, 2000; Subak, 2003; McCabe and Bunnell, 2004; DeGaetano, 2005; Purse *et al.*, 2005; Kunkel *et al.*, 2006; Ostfeld *et al.*, 2006; Shone *et al.*, 2006). Changes in other factors such as hosts, habitats, and human behavior also are important.

2.2.4.2 Water-borne and Food-borne Diseases

Water- and food-borne diseases continue to cause significant morbidity in the United States. In 2002, there were 1,330 food-related disease outbreaks (Lynch *et al.*, 2006), 34 outbreaks from recreational water (2004), and 30 outbreaks from drinking water (2004) (Dziuban *et al.*, 2006; Liang *et al.*, 2006). For outbreaks of food-borne disease with known etiology, bacteria (*Salmonella*) accounted for 55 percent and viruses accounted for 33 percent (Lynch *et al.*, 2006). Viral associated outbreaks rose from 16 percent in 1998 to 42 percent in 2002, primarily due to increases in norovirus (Lynch *et al.*, 2006). In recreational water, bacteria accounted for 32 percent of outbreaks, parasites (primarily *Cryptosporidium*) for 24 percent, and viruses 10 percent (Dziuban *et al.*, 2006). Similarly in drinking water outbreaks of known etiology, bacteria were the most commonly identified agent (29 percent, primarily *Campylobacter*), followed by parasites and viruses (each identified 5 percent of the time) (2003—2004; Liang *et al.*, 2006). Gastroenteritis continues to be the primary disease associated with food and water exposure. In 2003 and 2004, gastroenteritis was noted in 48 percent and 68 percent of reported recreational and drinking water outbreaks, respectively (Dziuban *et al.*, 2006; Liang *et al.*, 2006).

Water- and food-borne disease remain highly underreported (*e.g.*, Mead *et al.*, 1999). Few people seek medical attention and of those that do, few cases are diagnosed (many pathogens are difficult to detect and identify in stool samples) or reported. Using a combination of underreporting estimates, passive and active surveillance data, and hospital discharge data, Mead *et al.* (1999) estimated that more than 210 million cases of gastroenteritis occur annually in the United States, including more than 900,000 hospitalizations and more than 6,000 deaths. More recently, Herikstad *et al.* (2002) estimated as many as 375 million episodes of diarrhea occur annually in the United States, based on a self-reporting study. These numbers far exceed previous estimates. Of the total estimated annual cases, just over 39 million can be attributed to a specific pathogen and approximately 14

million are transmitted by food (Mead *et al.*, 1999). While bacteria continue to cause the majority of documented food- and water-borne outbreaks (Lynch *et al.*, 2006; Liang *et al.*, 2006), the majority of sporadic (non-outbreak) cases of disease are caused by viruses (67 percent; primarily noroviruses), followed by bacteria (30 percent, primarily *Campylobacter* and *Salmonella*) and parasites (3 percent, primarily *Giardia* and *Cryptosporidium*). While the outcome of many gastrointestinal diseases is mild and self-limiting, they can be fatal or significantly decrease fitness in vulnerable populations, including young children, the immunocompromised, and the elderly. Children ages 1-4 and older adults (>80 years) each make up more than 25 percent of hospitalizations involving gastroenteritis, but older adults contributed to 85 percent of the associated deaths (Gangarosa *et al.*, 1992). As the U.S. population ages, the economic and public health burden of diarrheal disease will increase proportionally without appropriate interventions.

Most pathogens of concern for food- and water-borne exposure are enteric and transmitted by the fecal-oral route. Climate may affect the pathogen directly by influencing its growth, survival, persistence, transmission, or virulence. In addition, there may be important interactions between land-use practices and climate variability. For example, incidence of food-borne disease associated with fresh produce is growing (FDA, 2001; Powell and Chapman, 2007). Storm events and flooding

may result in the contamination of food crops (especially produce such as leafy greens and tomatoes) with feces from nearby livestock or feral animals. Therefore, changing climate or environments may alter the transmission of pathogens or affect the ecology and/or habitat of zoonotic reservoirs (NAS, 2001).

Studies in North America (United States and Canada) (Fleury *et al.*, 2006; Naumova *et al.*, 2006), Australia (D'Souza *et al.*, 2004), and several countries across Europe (Kovats *et al.*, 2004a) report striking similarities in correlations between peak ambient temperatures (controlled for season) and peak in clinical cases of salmonellosis. Over this broad geographic range, yearly peaks in salmonellosis cases occur within 1 to 6 weeks of the highest reported ambient temperatures. Mechanisms suggested include replication in food products at various stages of processing (D'Souza *et al.*, 2004; Naumova *et al.*, 2006) and changes in eating habits during warm summer months (*i.e.*, outdoor eating) (Fleury *et al.*, 2006). Additionally, because *Salmonella* are well adapted to both host conditions and the environment, they can grow readily even under low nutrient conditions at warm temperatures (*e.g.*, in water and associated with fruits and vegetables) (Zhuang *et al.*, 1995; Mouslim *et al.*, 2002). Evidence supports the notion that increasing global temperatures will likely increase rates of salmonellosis. However, additional research is needed to determine the critical drivers behind this trend (*i.e.*, intrinsic properties of the pathogen or extrinsic factors related to human behavior).

The possible effects of increasing temperatures on *Campylobacter* infection rates and patterns cannot be reliably projected. The apparent seasonality of

campylobacteriosis incidence is more variable than salmonellosis, and temperature models are less consistent in their ability to account for the observed infection patterns. In the northeastern United States, Canada, and the U.K., *Camplyobacter* infection peaks coincide with high annual daily or weekly temperatures (Louis *et al.*, 2005; Fleury *et al.*, 2006; Naumova *et al.*, 2006). However, in several other European countries, campylobacteriosis rates peak earlier, before high annual temperatures, and in those cases temperature accounts for only 4 percent of the interannual variability (Kovats, *et al.*, 2005). Pathogenic species of *Campylobacter* cannot replicate in the environment and will not persist long under non-microaerophilic conditions, suggesting that high ambient temperatures would not contribute to increased replication in water or in food products.

Leptospirosis is a re-emerging disease in the United States and, given its wide case distribution, high number of pathogenic strains, and wide array of hosts, it is often cited as one of the most widespread zoonotic disease in the world (Meites *et al.*, 2004; WHO, 1999). While it has not been a reportable disease nationally since 1995, several states continue to collect passive surveillance data and cases continue to be reported (Katz *et al.*, 2002; Meites *et al.*, 2004). Because increased disease rates are linked to warm temperatures, epidemiological evidence suggest that climate change may increase the number of cases.

Pathogenic species of *Vibrio* (primarily *V. vulnificus*) account for 20 percent of sporadic shellfish-related illnesses and over 95 percent of deaths (Lipp and Rose 1997; Morris, 2003). While the overall incidence of illness from *Vibrio* infections remains low, the rate of infection increased 41 percent since 1996 (Vugia *et al.*, 2006). *Vibrio* species are more frequently associated with warm climates (*e.g.*, Janda *et al.*, 1988; Lipp *et al.*, 2002). Coincident with proliferation in the environment, human cases also occur during warm temperatures. In the United States, the highest case rates occur in the summer months (Dziuban *et al.*, 2006). Given the close association between temperature, the pathogen, and disease, increasing temperatures may increase the geographic range and disease burdens of *Vibrio* pathogens (*e.g.,* Lipp *et al.,* 2002). For example, increasing prevalence and diversity of *Vibrio* species has been noted in northern Atlantic waters of the United States coincident with warm water (Thompson *et al.,* 2004). Additionally, although most cases of *V. vulnificus* infection are attributed to Gulf Coast states, this species recently has been isolated from northern waters in the United States (Pfeffer *et al.,* 2003; Randa *et al.,* 2004).

The most striking example of an increased range in pathogen distribution and incidence was documented in 2004, when an outbreak of shellfish-associated *V. parahaemolyticus* was reported from Prince William Sound in Alaska (McLaughlin *et al.,* 2005). *V. parahaemolyticus* had never been isolated from Alaskan shellfish before and it was thought that Alaskan waters were too cold to support the species (McLaughlin *et al.,* 2005). In the period preceding the July 2004 outbreak, water temperatures in the harvesting area consistently exceeded 15° C and the mean daily water temperatures were significantly higher than in the prior six years (McLaughlin *et al.,* 2005). This outbreak extended the northern range of oysters known to contain *V. parahaemolyticus* and cause illness by 1,000 km. Given the well-documented association between increasing sea surface temperatures and proliferation of many *Vibrio* species, evidence suggests that increasing global temperatures will lead to an increased burden of disease associated with certain *Vibrio* species in the United States, especially *V. vulnificus* and *V. parahaemolyticus.*

Protozoan parasites, particularly *Cryptosporidium* and *Giardia*, contribute significantly to water-borne and to a lesser extent food-borne disease burdens in the United States. Both parasites are zoonotic and form environmentally resistant infective stages, with only 10-12 oocysts or cysts required to cause disease. In 1998, 1.2 cases of cryptosporidiosis per 100,000 people were reported in the United States (Dietz and Roberts, 2000); the immunocompromised are at particularly high risk (Casman *et al.,* 2001; King and Monis, 2006). Between 2003 and 2004, of the 30 reported outbreaks of gastroenteritis from recreational water, 78.6 percent were due to

Cryptosporidium and 14.3 percent were due to *Giardia* (Dzuiban *et al.,* 2006). *Giardia* has historically been the most commonly diagnosed parasite in the United States. Between 1992 and 1997 there were 9.5 cases of *Giardia* per 100,000 people (Furness *et al.,* 2000). Both *Cryptosporidium* and *Giardia* case reports peak in late summer and early fall, particularly among younger age groups (Dietz and Roberts, 2000; Furness *et al.,* 2000). For both parasites, peak rates of reported infection in Massachusetts occurred approximately one month after the annual temperature peak (Naumova *et al.,* 2006). The lagged association between peak annual temperatures and peaks in reported cases in late summer has been attributed to increased exposure during the summer bathing season, especially in the younger age groups, and to a slight lag in reporting (Dietz and Roberts, 2000; Furness *et al.,* 2000; Casman *et al.,* 2001). With increasing global temperatures, an increase in recreational use of water can be reasonably expected and could lead to increased exposure among certain groups, especially children.

Naegleria fowleri is a free-living amboeboflagellate found in lakes and ponds at warm temperatures, either naturally or in thermally polluted bodies of water. While relatively rare, infections are almost always fatal (Lee *et al.,* 2002). *N. fowleri* can be detected in environmental waters at rates up to 50 percent (Wellings *et al.,* 1977) at water temperatures above 25°C (Cabanes *et al.,* 2001). Cases are consistently reported in the United States. Between 1999 and 2000, four cases (all fatal) were reported. While *N. fowleri* continues to be a rare disease, it remains more common in the United States than elsewhere in the world (Marciano-Cabral *et al.,* 2003). Given its association with warm water, elevated temperatures could increase this pathogen's range.

Epidemiologically significant viruses for food and water exposure include enteroviruses, rotaviruses, hepatitis A virus, and norovirus. Viruses account for 67 percent of food-borne disease, and the vast majority of these are due to norovirus (Mead *et al.,* 1999). Rotavirus accounts for a much smaller fraction of viral food-borne disease (Mead *et al.,* 1999), but is a significant cause of diarrheal disease among

infants and young children (Charles *et al.*, 2006). Enteroviruses are not reportable and therefore incidence rates are poorly reflected in surveillance summaries (Khetsuriani *et al.*, 2006). With the exception of hepatitis A (Naumova *et al.*, 2006), enteric viral infection patterns follow consistent year to year trends. Enteroviruses are characterized by peaks in cases in the early to late summer (Khetsuriani *et al.*, 2006), while rotavirus and norovirus infections typically peak in the winter (Cook *et al.*, 1990; Lynch *et al.*, 2006). No studies have been able to identify a clear role for temperature in viral infection patterns.

An analysis of water-borne outbreaks associated with drinking water in the United States between 1948 and 1994 found that 51 percent of outbreaks occurred following a daily precipitation event in the 90th percentile and 68 percent occurred when precipitation levels reached the 80th percentile (Curriero *et al.*, 2001) (Figure 2.4). Similarly, Thomas *et al.* (2006) found that the risk of water-borne disease doubled when rainfall amounts surpassed the 93rd percentile. Rose *et al.* (2000) found that the relationship between rainfall and disease was stronger for surface water outbreaks, but the association was significant for both surface and groundwater sources. In 2000, groundwater used for drinking water in Walkerton, Ontario was contaminated with *E. coli* O157:H7 and *Campylobacter* during rains that surpassed the 60-year event mark for the region and the 100-year event mark in

local areas (Auld *et al.*, 2004). In combination with preceding record high temperatures, 2,300 people in a community of 4,800 residents became ill (Hrudey *et al.*, 2003; Auld *et al.*, 2004).

Floodwaters may increase the likelihood of contaminated drinking water and lead to incidental exposure to standing floodwaters. In 1999, Hurricane Floyd hit North Carolina and resulted in severe flooding of much of the eastern portion of the state, including extensive hog farming operations. Residents in the affected areas experienced more than twice the rate of gastrointestinal illness following the flood as before it (Setzer and Domino, 2004). Following the severe floods of 2001 in the Midwest, contact with floodwater was shown to increase the rate and risk of gastrointestinal illness, especially among children (Wade *et al.*, 2004); however, consumption of tap water was not a risk factor as drinking water continued to meet all regulatory standards (Wade *et al.*, 2004).

2.2.4.3 Influenza

Influenza may be considered a zoonosis in that pigs, ducks, etc. serve as non-human hosts to the influenza viruses (*e.g.*, H3N2, H1N1) that normally infect humans (not H5N1). A number of recent studies evaluated the influence of weather and climate variability on the timing and intensity of the annual influenza season in the United States and Europe. Results indicated that cold winters alone do not predict pneumonia and influenza (P&I)-related winter deaths, even though cold spells may serve as a short-term trigger (Dushoff *et al.*, 2005), and that regional differences in P&I mortality burden may be attributed to climate patterns and to the dominant circulating virus subtype (Greene *et al.*, 2006). Studies in France and the United States demonstrated that the magnitude of seasonal transmission (whether measured as mortality or morbidity) during winter seasons is significantly higher during years with cold El Niño Southern Oscillation (ENSO) conditions than during warm ENSO years (Flahault *et al.*, 2004; Viboud *et al.*, 2004), whereas a study in California concluded that higher temperatures and El Niño years increased hospital admissions for viral pneumonia (Ebi *et al.*, 2001). In an attempt to better understand the spatio-temporal patterns of ENSO and influenza, Choi *et al.*,

Figure 2.4. Drinking Water-borne Disease Outbreaks and 90th percentile Precipitation Events (a two month lag precedes outbreaks); 1948–1994.

Source: Curriero *et al.*, 2001

(2006) used stochastic models (mathematical models that take into account the presence of randomness) to analyze California county-specific influenza mortality and produced maps that showed different risks during the warm and cool phases. In general, these studies of influenza further support the importance of climate drivers at a global and regional scale, but have not advanced our understanding of underlying mechanisms.

2.2.4.4 Valley Fever

Valley fever (Coccidioidomycosis) is an infectious disease caused by inhalation of the spores of a soil-inhabiting fungus that thrives during wet periods following droughts. The disease is of public health importance in the Desert Southwest. In the early 1990s, California experienced an epidemic of Valley Fever following five years of drought (Kolivras and Comrie, 2003). Its incidence varies seasonally and annually, which may be due partly to climatic variations (Kolivras and Comrie, 2003; Zender and Talamantes, 2006). If so, climate change could affect its incidence and geographic range.

2.2.4.5 Morbidity and Mortality Due to Changes in Air Quality

Millions of Americans continue to live in areas that do not meet the health-based National Ambient Air Quality Standards for ozone and fine particulate matter (PM2.5). Both ozone and PM2.5 have well-documented health effects, and levels of these two pollutants have the potential to be influenced by climate change in a variety of ways.

Ground-level ozone is formed mainly by reactions that occur in polluted air in the presence of sunlight. Nitrogen oxides (emitted mainly by burning of fossil fuels) and volatile organic compounds (VOCs) (emitted both by burning of fossil fuels and by evaporation from vegetation and stored fuels, solvents, and other chemicals) are the key precursor pollutants for ozone formation. Ozone formation increases with greater sunlight and higher temperatures; it reaches peak concentrations during the warm half of the year, and then mostly in the late afternoon and early evening. Cloud cover and mixing height are two additional meteorological

factors that influence ozone concentrations. It has been firmly established that breathing ozone results in short-term, reversible decreases in lung function (Folinsbee *et al.*, 1988) as well as inflammation deep in the lungs (Devlin *et al.*, 1991). In addition, epidemiologic studies of people living in polluted areas have suggested that ozone may increase the risk of asthma-related hospital visits (Schwartz, 1995), premature mortality (Kinney and Ozkaynak, 1991; Bell *et al.*, 2004), and possibly the development of asthma (McConnell *et al.*, 2002). Vulnerability to ozone health effects is greater for persons who spend time outdoors during episode periods, especially with physical exertion, because this results in a higher cumulative dose to the lung. Thus, children, outdoor laborers, and athletes may be at greater risk than people who spend more time indoors and who are less active. At a given lung dose, little has been firmly established about vulnerability as a function of age, race, and/or existing health status. However, because their lungs are inflamed, asthmatics are potentially more vulnerable than non-asthmatics.

PM2.5 is a far more complex pollutant than ozone, consisting of all air-borne solid or liquid particles that share the property of being less than 2.5 micrometers in aerodynamic diameter.[2] All such particles are included, regardless of their size, composition, and biological reactivity. PM2.5 has complex origins, including primary particles directly emitted from sources and secondary particles that form via atmospheric reactions of precursor gases. Most of the particles captured as PM2.5 arise from burning of fuels, including primary particles such as diesel soot and secondary particles such as sulfates and nitrates. Epidemiologic studies have demonstrated associations between both short-term and long-term average ambient concentrations and a variety of adverse health outcomes including respiratory symptoms such as coughing and difficulty breathing, decreased lung function, aggravated asthma,

2 Aerodynamic diameter is defined in a complex way to adjust for variations in shape and density of various particles, and is based on the physical diameter of a water droplet that would settle to the ground at the same rate as the particle in question. For a spherical water particle, the aerodynamic and physical diameters are identical.

development of chronic bronchitis, heart attack, and arrhythmias (Dockery *et al.*, 1993; Samet *et al.*, 2000; Pope *et al.*, 1995, 2002, 2004; Pope and Dockery, 2006; Dominici *et al*, 2006; Laden *et al.*, 2006). Associations have also been reported for increased school absences, hospital admissions, emergency room visits, and premature mortality. Susceptible individuals include people with existing heart and lung disease, and diabetics, children, and older adults. Because the mortality risks of PM2.5 appear to be mediated through narrowing of arteries and resultant heart impacts (Künzli *et al.*, 2005), persons or populations with high blood pressure and/or pre-existing heart conditions may be at increased risk. In a study of mortality in relation to long-term PM2.5 concentrations in 50 U.S. cities, individuals without a high school education demonstrated higher concentration/response functions than those with more education (Pope *et al.*, 2002). This result suggests that low education was a proxy for increased likelihood of engaging in outdoor labor with an associated increase in exposure to ambient air.

Using a coupled climate-air pollution three-dimensional model, Jacobson (2008) compared the health effects of pre-industrial vs. present day atmospheric concentrations of CO_2. The results suggest that increasing concentrations of CO_2 increased tropospheric ozone and PM2.5, which increased mortality by about 1.1 percent per degree temperature increase over the baseline rate. Jacobson estimated that about 40 percent of the increase was due to ozone and the rest to particulate matter. The estimated

mortality increase was higher in locations with poorer air quality.

2.2.4.6 Aeroallergens and Allergenic Diseases

Climate change has caused an earlier onset of the spring pollen season for several species in North America (Casassa *et al.*, 2007). Although data are limited, it is reasonable to infer that allergenic diseases caused by pollen, such as allergic rhinitis, also have experienced concomitant changes in seasonality (Emberlin *et al.*, 2002; Burr *et al.*, 2003). Several laboratory studies suggest that increasing CO_2 concentrations and temperatures could increase ragweed pollen production and prolong the ragweed pollen season (Wan *et al.*, 2002; Wayne *et al.*, 2002; Singer *et al.*, 2005; Ziska *et al.*, 2005; Rogers *et al.*, 2006) and increase some plant metabolites that can affect human health (Ziska *et al.*, 2005; Mohan *et al.*, 2006). Although there are suggestions that the abundance of a few species of air-borne pollens has increased due to climate change, it is unclear whether the allergenic content of these pollen types has changed (Huynen and Menne, 2003; Beggs and Bambrick, 2005). The introduction of regionally new invasive species associated with climatic and other changes, such as ragweed and poison ivy, may increase current health risks. There are no projections of the possible impacts of climate change on allergenic diseases.

2.3 PROJECTED HEALTH IMPACTS OF CLIMATE CHANGE IN THE UNITED STATES

2.3.1 Heat-Related Mortality

Determinants of how climate change could alter heat-related mortality include actual changes in the mean and variance of future temperatures; factors affecting temperature variability at the local scale; demographic and health characteristics of the population; and policies that affect the social and economic structure of communities, including urban design, energy policy, water use, and transportation planning. Barring an unexpected and catastrophic economic decline, residential and industrial development will increase over the coming decades, which could increase urban heat islands in the absence of urban design and new technologies to reduce heat loads.

The U.S. population is aging. The portion of the population over age 65 is projected to be 13 percent by 2010 and 20 percent by 2030 (over 50 million people) (Day, 1996). Older adults are physiologically and socially vulnerable (Khosla and Guntupalli, 1999; Klinenberg, 2002) to hot weather and heat waves, suggesting that heat-related mortality could increase. Evidence that diabetics are at greater risk of heat-related mortality (Schwartz 2005), along with the increasing prevalence of obesity and diabetes (Seidell, 2000; Visscher and Seidell, 2001), suggests that reduced fitness and higher-fat body composition may contribute to increased mortality.

Table 2.1 summarizes projections of temperature-related mortality either in the United States or in temperate countries whose experience is relevant to the United States (Dessai, 2003) (Woodruff et al., 2005) (Knowlton et al., 2007) (CLIMB, 2004; Hayhoe et al., 2004). Similar studies are underway in Europe (Kosatsky et al., 2006; Lachowsky and Kovats, 2006). All studies used downscaled projections of future temperature distributions in the geographic region of interest. The studies used different approaches to incorporate likely future adaptation, addressing such issues as increased availability of air conditioning, heat wave early warning systems, demographic changes, and enhanced services such as cooling shelters and physiological adaptation.

Time-series studies also can shed light on potential future mortality during temperature extremes. Heat-related mortality has declined over the past decades (Davis et al., 2002; Davis et al., 2003a; Davis et al., 2003b). A similar trend, for cold- and heat-related mortality, was observed in London over the past century (Carson et al., 2006). The authors speculate that these declines are due to increasing prevalence of air-conditioning (in the United States), improved health care, and other factors. These results do not necessarily mean that future increases in heat-related mortality may not occur in the United States, as some have claimed (Davis et al., 2004), because the percentage of the population with access to air conditioning is high in most regions (thus with limited possibilities for increasing access). Further, population level declines may obscure persistent mortality impacts in vulnerable groups.

Table 2.1. Projections of Impacts of Climate Change on Heat-Related Mortality

Location	Period	Adaptation considered	Projected Impact on Heat- Related Deaths
Lisbon, Portugal[3]	2020s, 2050s compared to 1980–1998	yes	Increase of 57 percent–113 percent in 2020s, 97 percent–255 percent in 2050s, depending on adaption
8 Australian cites[4]	2100 compared to 1900s	no	Increase of 1700 to 3200 deaths, depending on policy approach followed and age structure of population
New York, NY[5]	2050s compared to 1990s	yes	Increase 47 percent to 95 percent; reduced by 25 percent with adaptation
California[6]	2090s compared to 1990s	yes	Depending on emissions, mortality increases 2–7fold from 1990 levels, reduced 20–25 percent with adaption
Boston, MA[7]	projections to 2100 compared to 1970–92	yes	Decrease after 2010 due to adaptation

3 Dessai, 2003
4 Woodruff, 2005
5 Knowlton, in press
6 Hayhoe, 2004
7 CLIMB, 2004

The impacts projected for Lisbon were more sensitive to the choice of regional climate model than the method used to calculate excess deaths, and the author described the challenge of extrapolating health effects at the high end of the temperature distribution, for which data are sparse or nonexistent (Dessai, 2003).

In summary, given the projections of increases in the frequency, intensity, and duration of heat waves and projected demographic changes, the at-risk population will increase (highly likely). The extent to which mortality increases will depend on the effective implementation of a range of adaptation options, including heat wave early warning systems, urban design to reduce heat loads, and enhanced services during heat waves.

2.3.2 Hurricanes, Floods, Wildfires, and Health Impacts

No studies have projected the future health burdens of extreme weather events. There is concern that climate change could increase the frequency and/or severity of extreme events, including hurricanes, floods, and wildfires.

Theoretically, climate change could increase the frequency and severity of hurricanes by warming tropical seas where hurricanes first emerge and gain most of their energy (Pielke *et al.*, 2005; Trenberth, 2005; Halverson, 2006). Controversy over whether hurricane intensity increased over recent decades stems less from the conceptual arguments than from the limitations of available hurricane incidence data (Halverson, 2006; Landsea, 2005; Pielke *et al.*, 2005; Trenberth, 2005). Even if climate change increases the frequency and severity of hurricanes, it will be difficult to definitively identify this trend for some time because of the relatively short and highly variable historical data available as a baseline for

comparison. Adding to the uncertainty, some research has projected that climate change could produce future conditions that might hinder the development of Atlantic hurricanes despite the warming of tropical seas (NOAA, 2007c).

Evidence suggests that the intensity of Atlantic hurricanes and tropical storms has increased over the past few decades. SAP 3.3 indicates that there is evidence for a human contribution to increased sea surface temperatures in the tropical Atlantic and there is a strong correlation to Atlantic tropical storm frequency, duration, and intensity. However, a confident assessment will require further studies. An increase in extreme wave heights in the Atlantic since the 1970s has been observed, consistent with more frequent and intense hurricanes (CCSP, 2008).

For North Atlantic hurricanes, SAP 3.3 concludes that it is likely that wind speeds and core rainfall rates will increase (Henderson-Sellers *et al.*, 1998; Knutson and Tuleya, 2004, 2008; Emanuel, 2005). However, SAP 3.3 concludes that "frequency changes are currently too uncertain for confident projection" (CCSP, 2008). SAP 3.3 also found that the spatial distribution of hurricanes will likely change. Storm surge is likely to increase due to projected sea level rise, though the degree to which storm surges will increase has not been adequately studied (CCSP, 2008).

Theoretical arguments for increases in extreme precipitation and flooding are based on the principles of the hydrological cycle where increasing average temperature will intensify evaporation and subsequently increase precipitation (Bronstert, 2003; Kunkel, 2003, Senior *et al.*, 2002). Looking at the available data for evidence of a climate change signal, evidence suggests that the number of extreme precipitation events in the United States has increased (Balling Jr. and Cerveny, 2003; Groisman *et al.*, 2004; Kunkel, 2003). However, these results are not as consistent when evaluated by season or region (Groisman *et al.*, 2004).

Projections of changes in the future incidence of extreme precipitation and flooding rely on the results from general circulation models (GCMs). These models project increases in mean precipitation with a disproportionate increase in the frequency of extreme precipitation events

(Senior *et al.*, 2002). Kim (2003) used a regional climate model to project that a doubling of CO_2 concentrations in roughly 70 years could increase the number of days with at least 0.5 mm of precipitation by roughly 33 percent across the study's defined elevation gradients in the western United States. Furthermore, the IPCC concluded that it is very likely (>90 percent certainty) that trends in extreme precipitation will continue in the 21st century (IPCC, 2007a).

Studies modeling future wildfire incidence in the western United States using GCM outputs project increasingly severe wildfires, measured both in terms of energy released and the number of fires that avoid initial containment in areas that GCMs project will be increasingly dry (Brown *et al.*, 2004; Fried *et al.*, 2004). In general, these results suggest much of the western United States could face an increasing wildfire risk from climate change. The apparent exception could be the Pacific Northwest, including northern California, where GCMs generally project a wetter future.

Factors independent of the impacts of and responses to climate change will affect vulnerability to extreme events, including population growth, continued urban sprawl, population shifts to coastal areas, and differences in the degree of community preparation for extreme events (U.S. Census Bureau, 2004).

All else equal, the anticipated demographic changes will increase the size of the U.S. population at risk for future extreme weather events (very likely). This raises the potential for increasing total numbers of adverse health impacts from these events, even if the rate at which these impacts are experienced decreases (where the rate reflects the number of impacts per some standard population size among those actually experiencing the events).

2.3.3 Vector-borne and Zoonotic Diseases

Modeling the possible impacts of climate change on VBZ diseases is complex, and few studies have made projections for diseases of concern in the United States. Studies suggest that temperature influences the distributions of

Ixodes spp. ticks that transmit pathogens causing Lyme disease in the United States (Brownstein *et al.*, 2003) and Canada (Ogden *et al.*, 2006), and tick-borne encephalitis (TBE) in Sweden (Lindgren *et al.*, 2000). Higher minimum temperatures generally were favorable to the potential of expanding tick distributions and greater local abundance of these vectors. However, changing patterns of tick-borne encephalitis in Europe are not consistently related to changing climate (Randolph, 2004a). Climate change is projected to decrease the geographic range of TBE in areas of lower latitude and elevation as transmission expands northward (Randolph and Rogers, 2000).

2.3.4 Water- and Food-borne Diseases

Several important pathogens that are commonly transmitted by food or water may be susceptible to changes in replication, survival, persistence, habitat range, and transmission under changing climatic and environmental conditions (Table 2.2). Many of these agents show seasonal infection patterns (indicating potential underlying environmental or weather control), are capable of survival or growth in the environment, or are capable of water-borne transport. Factors that may affect these pathogens include changes in temperature, precipitation, extreme weather events (*i.e.*, storms), and ecological shifts. While the United States has successful programs to protect water quality under the Safe Drinking Water Act and the Clean Water Act, some contamination pathways and routes of exposure do not fall under regulatory programs (*e.g.*, dermal absorption from floodwaters, swimming in lakes and ponds with elevated pathogen levels, etc.).

2.3.5 Air Quality Morbidity and Mortality

The sources and conditions that give rise to elevated ozone and PM2.5 in outdoor air in the United States have been and will continue to be affected by global environmental changes related to land use, economic development, and climate change. Conversions of farmland and forests into housing developments and the infrastructure of schools and businesses that support them change the spatial patterns

Table 2.2. Possible Influence of Climate Change on Climate-Susceptible Pathogens and/or Disease, Based on Observational Models or Empirical Evidence

Pathogen	Climate-Related Driver	Possible Influence of Climate Change	Likelihood of Change[a]	Basis for Assessment	References
Bacteria					
Salmonella	Rising Temperature	Increasing temperature associated with increasing clinical cases	Likely	Likelihood of climate event is high and published research supports disease trend	D'Souza et al., 2004; Kovats et al., 2004a; Fleury et al., 2006; Naumova et al., 2006
	Changes in Precipitation	Precipitation and runoff associated with increased likelihood of contamination of surface waters used for recreation, drinking, or irrigation	Likely	Likelihood of climate event is probable but more research is needed to confirm disease trend	Haley 2006; Holley et al., 2006
	Shifts in Reservoir Host Ranges	Shifts in habitat and range of reservoir hosts may influence exposure routes and/or rate of contact with humans	More likely than not	Likelihood of climate event is probable but there is insufficient research on this relationship	Srikantiah et al., 2003
Campylobacter	Rising Temperature	Increasing temperatures may expand typical peak season of clinical infection, or result in earlier peak (commonly spring and summer)	More likely than not	Likelihood of climate event is high and published research supports disease trend, but mechanisms are not understood	Skelly & Weinstein, 2003; Louis et al., 2005; Kovats et al., 2005
		Increasing temperatures may result in shorter developmental times for flies, contributing to increased transmission by this proposed vector	About as likely as not	Likelihood of climate event and fly development trend is high but additional research is needed to confirm disease association	Nichols, 2005
	Changes in Precipitation	Increasing precipitation and runoff associated with increased likelihood of contamination of surface waters used for recreation or drinking	More likely than not	Likelihood of climate event is probable but more research is needed to confirm disease trend	Auld et al., 2004; Vereen et al., 2007

Pathogen	Climate-Related Driver	Possible Influence of Climate Change	Likelihood of Change[a]	Basis for Assessment	References
	Shifts in Reservoir Host Ranges or Behavior	Shifts in habitat and range of reservoir hosts (geographically or temporally) may influence exposure routes and/or rate of contact with humans	More likely than not	Likelihood of climate event is probable but there is insufficient research on this relationship	Stanley et al., 1998; Lacey, 1993; Southern et al., 1990
Vibrio species	Rising Temperature	Increasing ambient temperatures associated with growth in pre-harvest and post-harvest shellfish (in absence of appropriate post-harvest controls) and increasing disease	Very likely	Likelihood of climate event is high and evidence supports growth trend in ambient waters; adaptive (control) measures (refrigeration) would reduce this effect for post-harvest oysters	Cook, 1994
		Increasing temperature associated with higher environmental prevalence and disease	Extremely likely	Likelihood of climate event is high and evidence supports environmental growth trend	Janda et al., 1988; Lipp et al., 2002; McLaughlin et al., 2005; Dziuban et al., 2006
		Increasing temperature associated with range expansion	Very likely	Likelihood of climate event is high and evidence collected to date supports trend; more data needed to confirm	McLaughlin et al., 2005
	Changes in Precipitation	Increasing precipitation and fresh water runoff leads to depressed estuarine salinities and increase in some Vibrio species	About as likely as not	Likelihood of climate event is probable but additional research is needed to confirm pathogen distribution patterns	Lipp et al., 2001b; Louis et al., 2003
	Sea Level Changes	Rising sea level and or storm surge increase range and human exposure	Likely	Likelihood of climate event is probable but confirmatory research is needed on disease patterns	Lobitz et al., 2000
Leptospira	Rising Temperature	Increasing temperatures may increase range of pathogen (temporally and geographically)	Likely	Likelihood of climate event is high but additional research is needed to confirm pathogen distribution patterns	Bharti et al., 2003; Howell and Cole, 2006
	Changes in Precipitation	Increasing precipitation and run off precedes outbreaks	Likely	Likelihood of climate event in probable and research supports this pattern	Meites et al., 2004
Viruses					
Enteroviruses	Rising Temperature	Increasing temperature associated with increased or expanded peak clinical season (summer)	Unlikely	Likelihood of climate event is high but no mechanistic studies are available to explain the underlying cause of this seasonality	Khetsuriani et al., 2006

Pathogen	Climate-Related Driver	Possible Influence of Climate Change	Likelihood of Change[a]	Basis for Assessment	References
		Increasing temperature associated with increased decay and inactivation of viruses in the environment	About as likely as not	Likelihood of climate event is high and research demonstrates decreased persistence under increasing temperatures but little data are available to relate this with disease	Gantzer et al., 1998; Wetz et al., 2004
	Changes in Precipitation	Increasing precipitation associated with increased loading of viruses to water and increased exposure or disease	Likely	Likelihood of climate is probable and research supports this pattern	Lipp et al., 2001a; Frost et al., 2002; Fong et al., 2005
Norovirus	Rising Temperature	Increasing temperature leads to decreased retention of virus in shellfish	Unlikely	Likelihood of climate event is high and research indicates seasonally high shellfish loading in winter but there is no evidence for direct control of temperature on seasonality of infection	Burkhardt and Calci, 2000
		Increasing temperature associated with shorter peak clinical season (winter)	Unlikely	Likelihood of climate event is high and research indicates seasonal disease peak in winter but there is no evidence for direct control of temperature on seasonality of infection	Mounts et al., 2000
		Increasing temperature associated with increased decay and inactivation of viruses in the environment	About as likely as not	Likelihood of climate event is high and research demonstrates decreased persistence under increasing temperatures but little data are available to relate this with disease	Griffin et al., 2003
	Changes in Precipitation	Increasing precipitation associated with increased loading of viruses to crops and fresh produce	More likely than not	Likelihood of climate event is probable but there is insufficient research on this relationship	Miossec et al., 2000
		Increasing precipitation associated with increased loading of viruses to water and increased exposure or disease	Likely	Likelihood of climate is probable and research supports this pattern	Goodman et al., 1982
Rotavirus	Rising Temperature	Increasing temperature associated with increased decay and inactivation of viruses in the environment	About as likely as not	Likelihood of climate event is high and research demonstrates decreased persistence under increasing temperatures but little data are available to relate this with disease	Rzezutka and Cook, 2004

Pathogen	Climate-Related Driver	Possible Influence of Climate Change	Likelihood of Change[a]	Basis for Assessment	References
		Dampening of winter seasonal peak in temperate latitudes	About as likely as not	Likelihood of climate event is high and research indicates seasonal disease peak in winter but there is no evidence for direct control of temperature on seasonality of infection; note that tropical countries do not exhibit a seasonal peak	Cook et al., 1990
Parasites					
Naegleria fowleri	Rising Temperature	Increasing temperature associated with expanded range and conversion to flagellated form (infective)	More likely than not	Likelihood of climate event is high but more research is needed to confirm disease trend	Cabanes et al., 2001
Cryptosporidium	Rising Temperature	Expanding recreational (swimming) season may increase likelihood of exposure and disease	About as likely as not	Likelihood of climate event is high but there is insufficient research on this relationship	Naumova et al., 2006
	Changes in Precipitation	Increasing precipitation associated with increased loading of parasite to water and increased exposure and disease	Very likely	Likelihood of climate event is probable and research supports this pattern but adaptive measures (water treatment and infrastructure) would reduce this effect	Curriero et al., 2001; Davies et al., 2004
Giardia	Rising Temperature	Expanding recreational (swimming) season may increase likelihood of exposure and disease	About as likely as not	Likelihood of climate event is high but there is insufficient research on this relationship	Naumova et al., 2006
	Changes in Precipitation	Increasing precipitation associated with increased loading of parasite to water and increased disease	Very likely	Likelihood of climate event is probable and research supports this pattern but adaptive measures (water treatment and infrastructure) would reduce this effect	Kistemann et al., 2002
	Shifts in Reservoir Host Ranges or Behavior	Increasing temperature associated with shifting range in reservoir species (carriers) and expanded disease range	About as likely as not	Likelihood of climate event is probable but there is insufficient research on this relationship	Parkinson and Butler, 2005

[a] Likelihood was based on expert judgment of the strength of the research and the likelihood of the event. See Chapter 1 for a discussion of likelihood (section 1.5).

and absolute amounts of emissions from fuel combustion related to transportation, space heating, energy production, and other activities. Resulting vegetation patterns affect biogenic (VOC) emissions that influence ozone production. Conversion of land cover from natural to man-made also changes the degree to which surfaces absorb solar energy (mostly in the form of light) and later re-radiate that energy as heat, which contributes to urban heat islands. In addition to their potential for increasing heat-related health effects, heat islands also can influence local production and dispersion of air pollutants such as ozone and PM2.5.

It is important to recognize that U.S. Environmental Protection Agency administers a well-developed and successful national regulatory program for ozone, PM2.5, and other criteria pollutants. Although many areas of the United States remain out of compliance with the ozone and PM2.5 standards, there is evidence for gradual improvements in recent years, and this progress can be expected to continue with more stringent emissions controls going forward in time. Thus, the influence of climate change on air quality will play out against a backdrop of ongoing regulatory control of both ozone and PM2.5 that will shift the baseline concentrations of these two important air pollutants. On the other hand, most of the studies that have examined potential future climate impacts on air quality reviewed below have tried to isolate the climate effect by holding precursor emissions constant over future decades. Thus, the focus has been on examining the sensitivity of ozone concentrations to alternative future climates rather than on attempting to project actual future ozone concentrations.

The influence of meteorology on air quality is substantial and well-established (EPRI, 2005), raising the possibility that changes in climate could alter patterns of air pollution concentrations. Temperature and cloud cover affect the chemical reactions that lead to ozone and secondary particle formation. Winds, vertical mixing, and rainfall patterns influence the movement and dispersion of anthropogenic pollutant emissions in the atmosphere, with generally improved air quality at higher winds, mixing heights, and rainfall. The most severe U.S. air pollution episodes occur with

atmospheric conditions that limit both vertical and horizontal dispersion over multi-day periods. Methods used to study the influence of climatic factors on air quality range from statistical analyses of empirical relationships to integrated modeling of future air quality resulting from climate change. To date, most studies have been limited to climatic effects on ozone. Additional research is needed on the impacts of climate change on anthropogenic particulate matter concentrations.

Leung and Gustafson (2005) used regional climate simulations for temperature, solar radiation, precipitation, and stagnation/ventilation, and projected worse air quality in Texas and better air quality in the Midwest in 2045-2055 compared with 1995-2005. Aw and Kleeman (2003) simulated an episode of high air pollution in southern California in 1996 with observed meteorology and then with higher temperatures. Ozone concentrations increased up to 16 percent with higher temperatures, while the PM2.5 response was more variable due to opposing forces of increased secondary particle formation and more evaporative losses from nitrate particles. Bell and Ellis (2004) showed greater sensitivity of ozone concentrations in the Mid-Atlantic to changes in biogenic than to changes in anthropogenic emissions. Ozone's sensitivity to changing temperatures, absolute humidity, biogenic VOC emissions, and pollution boundary conditions on a fine-scale (4 km grid resolution) varied in different regions of California (Steiner *et al.*, 2006).

Several studies explored the impacts of climate change alone on future ozone projections. In a coarse-scale analysis of pollution over the continental United States, Mickley *et al.*, (2004) used the GISS (NASA Goddard Institute for Space Studies) 4x5° model to project that, due to climate change alone (A1b emission scenario), air pollution could increase in the upper Midwest due to decreases between 2000 and 2052 in the frequency of Canadian frontal passages that clear away stagnating air pollution episodes. The 2.8x2.8° Mozart global chemistry/climate model was used to explore global background and urban ozone changes over the 21st century in response to climate change, with ozone precursor emissions kept constant at 1990s levels (Murazaki and Hess,

2006). While global background decreased slightly, the urban concentrations due to U.S. emissions increased.

As part of the New York Climate and Health Study, Hogrefe and colleagues conducted local-scale analyses of air pollution impacts of future climate changes using integrated modeling (Hogrefe *et al.*, 2004a,b,c; 2005a,b) to examine the impacts of climate and land use changes on heat- and ozone-related health impacts in the NYC metropolitan area (Knowlton *et al.*, 2004; Kinney *et al.*, 2006; Bell *et al.*, 2007; Civerolo *et al.*, 2006). The GISS 4x5⁰ model was used to simulate hourly meteorological data from the 1990s through the 2080s based on the A2 and B2 SRES scenarios. The A2 scenario assumes roughly double the CO_2 emissions of B2. The global climate outputs were downscaled to a 36 km grid over the eastern United States using the MM5 regional climate model. The MM5 results were used in turn as inputs to the Congestion Mitigation and Air Quality Improvement Program regional-scale air quality model. Five summers (June, July, and August) in each of four decades (1990s, 2020s, 2050s, and 2080s) were simulated at the 36 km scale. Pollution precursor emissions over the eastern United States were based on U.S. EPA estimates at the county level for 1996. Compared with observations from ozone monitoring stations, initial projections were consistent with ozone spatial and temporal patterns over the eastern United States in the 1990s (Hogrefe *et al.*, 2004a). Average daily maximum 8-hour concentrations were projected to increase by 2.7, 4.2, and 5.0 ppb in the 2020s, 2050s, and 2080s, respectively, due to climate change (Figure 2.5) (Hogrefe *et al.*, 2004c). The influence of climate on mean ozone values was similar in magnitude to the influence of rising global background concentrations by the 2050s, but climate had a much greater impact on extreme values than did the global background. When biogenic VOC emissions were allowed to increase in response to warming, an additional increase in ozone concentrations was projected that was similar in magnitude to that of climate alone (Hogrefe *et al.*, 2004b). Climate change shifted the distribution of ozone concentrations toward higher values, with larger relative increases in future decades (Figure 2.6).

Projections in Germany also found larger climate impacts on extreme ozone values (Forkel and Knoche, 2006). Using the IS92a business-as-usual scenario, the ECHAM4 GCM projected changes for the 2030s compared with the 1990s; the output was downscaled to a 20 km grid using a modification of the MM5 regional model, which was in turn linked to the RADM2 ozone chemistry model. Both biogenic VOC emissions and soil nitric oxide emissions were projected to increase as temperatures rose. Daily maximum ozone concentrations increased by between 2 and 6 ppb (6-10 percent) across the study region. The number of cases where daily maximum ozone exceeded 90 ppb increased by nearly four-fold, from 99 to 384.

Using the New York Climate & Health Project (NYCHP) integrated model, PM2.5 concentrations were projected to increase with climate change, with the effects differing by component species, with sulfates and primary PM increasing markedly and with organic and nitrated components decreasing, mainly due to movement of these volatile species from the particulate to the gaseous phase (Hogrefe *et al.*, 2005b; 2006).

Hogrefe *et al.*, (2005b) noted that "the simulated changes in pollutant concentrations stemming from climate change are the result of a complex interaction between changes in transport, mixing, and chemistry that cannot be parameterized by spatially uniform linear regression relationships." Additional uncertainties include how population

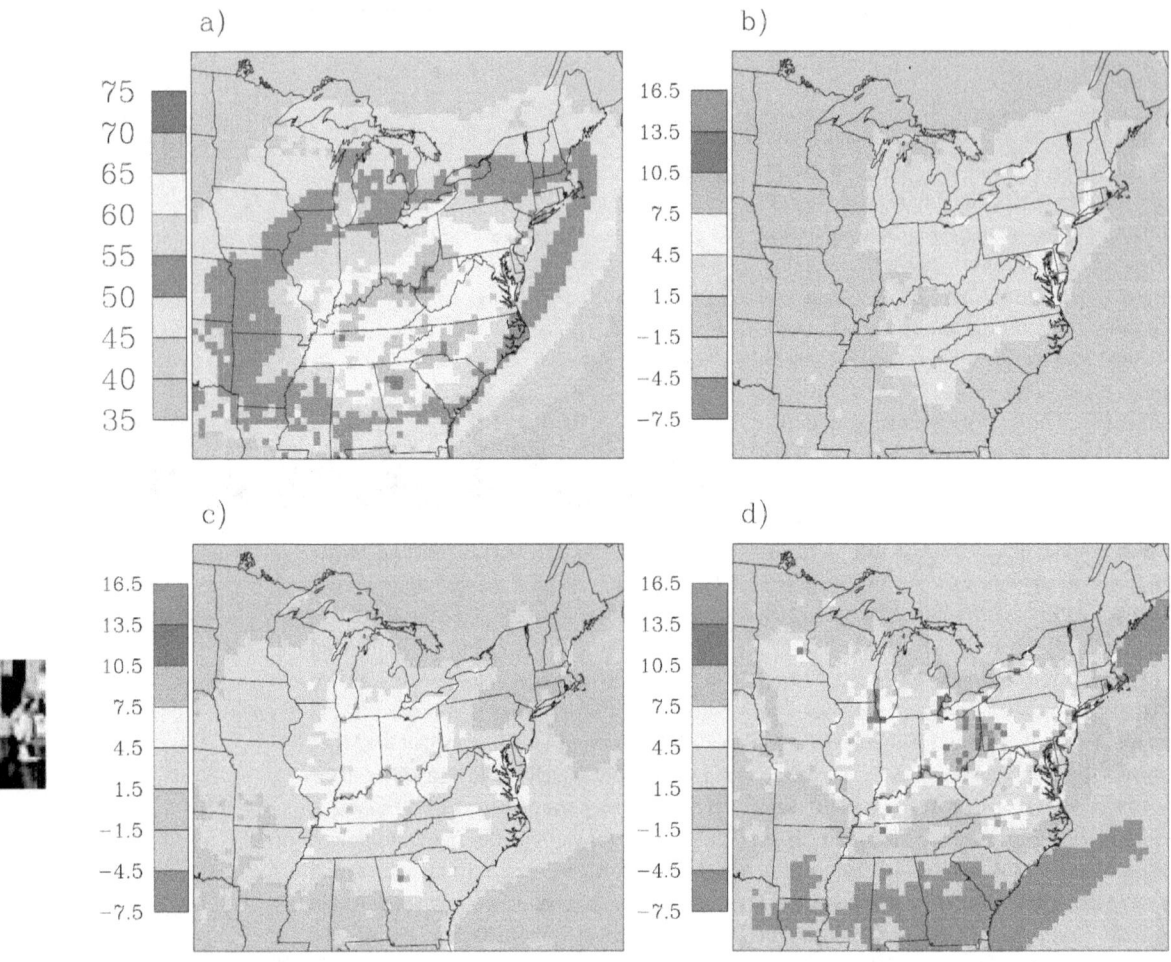

Figure 2.5 (a) Summertime Average Daily Maximum 8-hour Ozone Concentrations (ppb) for the 1990s and Changes for the (b) 2020s relative to the 1990s, (c) 2050s relative to the 1990s, and (d) 2080s relative to the 1990s. All are based on the A2 Scenario relative to the 1990s. Five consecutive summer seasons were simulated in each decade.

Source: Hogrefe *et al.*, 2004c.

vulnerability, mix of pollutants, housing characteristics, and activity patterns may differ in the future. For example, in a warmer world, more people may stay indoors with air conditioners in the summer when ozone levels are highest, decreasing personal exposures (albeit with potential increases in pollution emissions from power plants). Baseline mortality rates may change due to medical advances, changes in other risk factors such as smoking and diet, and aging of the population.

The NYCHP examined the marginal sensitivity of health to changes in climate to project the potential health impacts of ozone in the eastern United States (Knowlton *et al.*, 2004; Bell *et al.*, 2007). Knowlton and colleagues computed absolute and percentage increases in ozone-related daily summer-season deaths in the NYC metropolitan region in the 2050s compared with the 1990s using a downscaled GCM/RCM/air quality model (Knowlton *et al.*, 2004; Kinney *et al.*, 2006). The availability of county-scale ozone projections made it possible to compare impacts in the urban core with those in outlying areas. Projected increases in ozone-related mortality due to climate change ranged from 0.4 to 7.0 percent across 31 counties. Bell and colleagues expanded the analysis to 50 eastern cities and examined both mortality and hospital admissions (Bell *et al.*, 2007). Average ozone concentrations were projected to increase by 4.4 ppb (7.4 percent) in the 2050s; the range was 0.8 percent to 13.7 percent. In addition, ozone red alert days could increase by 68 percent. Changes in health impacts were of corresponding magnitude.

Based on the new research findings published since the previous assessment, the following summary statements can be made:

- There is an established but incomplete level of knowledge suggesting that both ozone and fine particle concentrations may be affected by climate change.

- A substantial body of new evidence on ozone supports the interpretation that ozone concentrations would be more likely to increase than decrease in the United States as a result of climate change, holding precursor emissions constant.

- Too few data yet exist for PM to draw firm conclusions about the direction or magnitude of climate impacts

2.4 VULNERABLE REGIONS AND SUBPOPULATIONS

In adapting the IPCC (1996) definitions to public health, "vulnerability" can be defined as the summation of all risk and protective factors that ultimately determine whether an individual or subpopulation experiences adverse health outcomes, and "sensitivity" can be defined as an individual's or subpopulation's increased responsiveness, primarily for biological reasons, to a given exposure. Thus, specific subpopulations may experience heightened vulnerability for climate-related health effects for a wide variety of reasons. Biological sensitivity may be related to the developmental stage, presence of pre-existing chronic medical conditions (such as the sensitivity of people with chronic heart conditions to heat-related illness), acquired factors (such as immunity), and genetic factors (such as metabolic enzyme subtypes that play a role in sensitivity to air pollution effects). Socioeconomic factors also play a critical role in altering vulnerability and sensitivity to environmentally mediated factors. They may alter the likelihood of exposure to

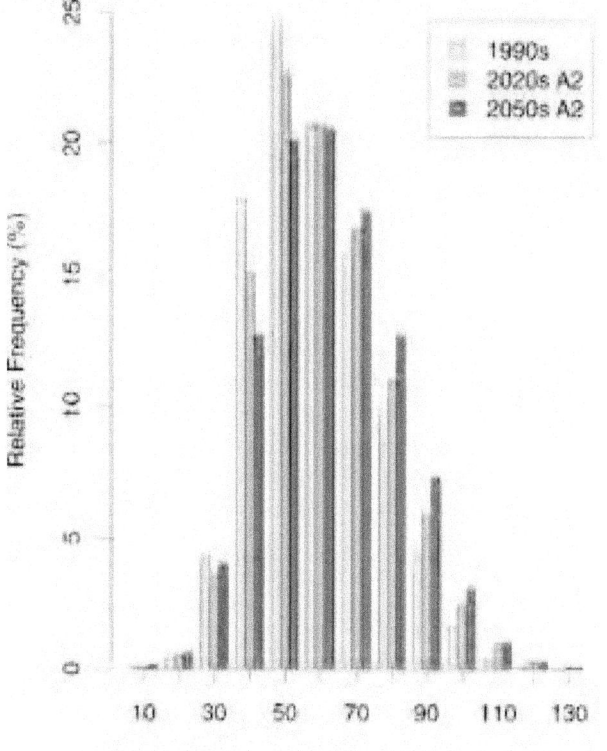

Figure 2.6 Frequency Distributions of Summertime Daily Maximum 8-hr Ozone Concentrations over the eastern United States in the 1990s, 2020s, and 2050s based on the A2 Scenario.

Source: From Hogrefe *et al.*, 2005a

harmful agents, interact with biological factors that mediate risk (such as nutritional status), and/or lead to differences in the ability to adapt or respond to exposures or early phases of illness and injury. For public health planning, it is critical to recognize populations that may experience synergistic effects of multiple risk factors for health problems related to climate change and to other temporal trends.

2.4.1 Vulnerable Regions

Populations living in certain regions of the United States may experience altered risks for specific climate-sensitive health outcomes due to their regions' baseline climate, abundance of natural resources such as fertile soil and fresh water supplies, elevation, dependence on private wells for drinking water, or vulnerability to coastal surges or riverine flooding. Some regions' populations may in fact experience

multiple climate-sensitive health problems simultaneously. One approach to identifying such areas is to map regions currently experiencing increased rates of climate-sensitive health outcomes or other indicators of increased climate risk, as illustrated in Figure 2.7a-2.7d.

Residents of low-lying coastal regions, which are common locations for hurricane landfalls and flooding, are particularly vulnerable to the health impacts of climate change. Those who live in the Gulf Coast region, for example, are likely to experience increased human health burdens due to the constellation of more intense storms, greater sea level rise, coastal erosion, and damage to freshwater resources and infrastructure. Other coastal areas may also experience the combination of sea level rise chronically threatening water supplies and periodic infrastructure damage

from more intense storms. Populations in the Southwest and Great Lakes regions may experience increased strain on water resources and availability due to climate change. More intense heat waves and heat-related illnesses may take place in regions where extreme heat events (EHE) already occur, such as interior continental zones of the United States. High-density urban populations will experience heightened health risks, in part due to the heat-island effect. In addition, increased demand for electricity during summers may lead to greater air pollution levels (IPCC, 2007b).

2.4.2 Specific Subpopulations at Risk

Vulnerable subpopulations may be categorized according to specific health endpoints. (Table 2.3). While this is typically the way the scientific literature reports risk factors for adverse health

Geographic Vulnerability of US Residents to Selected Climate Health Impacts

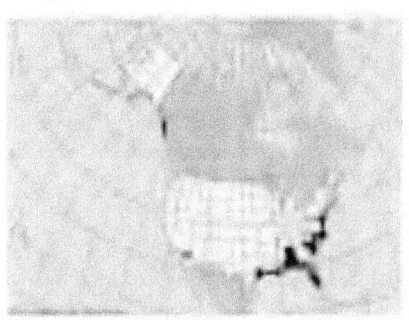

Location of Hurricane Landfalls, 1995-2000

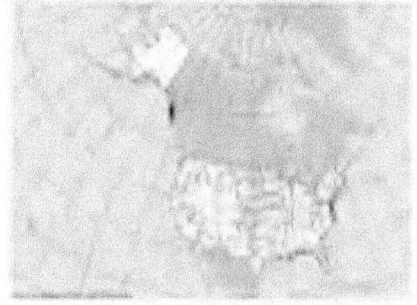

Percentage of US Population 65 or older, 2000

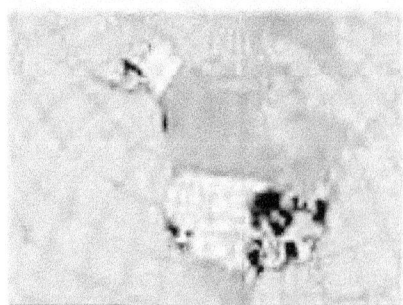

Locations of Extreme Heat Events, 1995-2000

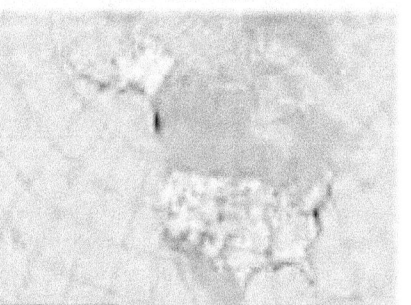

West Nile Virus Cases, 2004

Figure 2.7 a-d U.S. maps indicating counties with existing vulnerability to climate sensitive health outcomes: (a) location of hurricane landfalls; (b) EHEs, defined by CDC as temperatures 10 or more degrees Fahrenheit above the average high temperature for the region and lasting for several weeks; (c) percentage of population over age 65; (d) West Nile Virus cases reported in 2004. Historical disease activity, especially in the case of WNV, is not necessarily predictive of future vulnerability. Maps were generated using NationalAtlas.gov™ Map Maker (2008).

Climate-Sensitive Health Outcome	Particularly Vulnerable Groups
Heat-Related Illnesses and Deaths	Elderly, chronic medical conditions, infants and children, pregnant women, urban and rural poor, outdoor workers
Diseases and Deaths Related to Air Quality	Children, pre-existing heart or lung disease, diabetes, athletes, outdoor workers
Illnesses and Deaths Due to Extreme Weather Events	Poor, pregnant women, chronic medical conditions, mobility and cognitive constraints
Water- and Food-borne Illness	Immunocompromised, elderly, infants; specific risks for specific consequences (e.g., Campylobacter and Guillain-Barre syndrome, E. coli O157:H7)
Vector-borne Illnesses	
Lyme Disease	Children, outdoor workers
Hantavirus	Rural poor, occupational groups
Dengue	Infants, elderly
Malaria	Children, immunocompromised, pregnant women, genetic (e.g., G6PD status)

Table 2.3. Climate-Sensitive Health Outcomes and Particularly Vulnerable Groups

effects, this section discusses vulnerability for a variety of climate-sensitive health endpoints one subpopulation at a time.

2.4.2.1 Children

Children's small body mass to surface area ratio and other factors make them more vulnerable to heat-related morbidity and mortality (AAP, 2000), while their increased breathing rates relative to body size, time spent outdoors, and developing respiratory tracts heighten their sensitivity to harm from ozone air pollution (AAP, 2004). In addition, children's relatively naive immune systems increase the risk of serious consequences from water- and food-borne diseases. Specific developmental factors make them more vulnerable to complications from specific severe infections such as *E coli* O157:H7. Children's lack of immunity also plays a role in higher risk of mortality from malaria (CDC, 2004b). Conversely, maternal antibodies to dengue in infants convey increased risk of developing dengue hemorrhagic syndromes. A second peak of greater risk of complications from dengue appears in children between the ages of 3 and 5 (Guzman and Khouri, 2002).

Children may also be more vulnerable to psychological complications of extreme weather events related to climate change.

Following two floods in Europe in the 1990s, children demonstrated moderate to severe stress symptoms (Becht *et al.*, 1998; cited in Hajat *et al.*, 2003) and long-term PTSD, depression, and dissatisfaction with ongoing life (Bokszanin, 2000; cited in Hajat *et al.*, 2003).

2.4.2.2 Older Adults

Health effects associated with climate change pose significant risks for the elderly, who often have frail health and limited mobility. Older adults are more sensitive to temperature extremes, particularly heat (Semenza *et al.*, 1996; Medina-Ramon *et al.*, 2006); individuals 65 years of age and older comprised 72 percent of the heat-related deaths in the 1995 Chicago heat wave (Whitman *et al.*, 1997). The elderly are also more likely to have preexisting medical conditions, including cardiovascular and respiratory illnesses, which may put them at greater risk of exacerbated illness by climate-related events or conditions. For example, a 2004 rapid needs assessment of older adults in Florida found that Hurricane Charley exacerbated preexisting, physician-diagnosed medical conditions in 24-32 percent of elderly households (CDC, 2004a). Also, effects of ambient particulate matter on daily mortality tend to be greatest in older age groups (Schwartz, 1995).

2.4.2.3 Impoverished Populations

Even in the United States, the greatest health burdens related to climate change are likely to fall on those with the lowest socioeconomic status (O'Neill *et al.*, 2003a). Most affected are individuals with inadequate shelter or resources to find alternative shelter in the event their community is disrupted. While quantitative methods to assess the increase in risk related to these social and economic factors are not well-developed, qualitative insights can be gained by examining risk factors for mortality and morbidity from recent weather-related extreme events such as the 1995 heat wave in Chicago and Hurricane Katrina in 2005 (Box 2.1).

Studies of heat waves identify poor housing conditions, including lack of access to air conditioning and living spaces with fewer rooms, as significant risk factors for heat-related mortality (Kalkstein, 1993; Semeza *et al.*, 1996). Higher heat-related mortality has been associated with socioeconomic indicators, such as lacking a high school education and living in poverty (Curriero *et al.*, 2002). Financial stress plays a role, as one study of the 1995 Chicago heat wave found that concern about the affordability of utility bills influenced individuals to limit air conditioning use (Klinenberg, 2002). The risk for exposure

and sensitivity to air pollution is also elevated among groups in a lower socioeconomic position (O'Neill *et al.*, 2003a).

Air conditioning is an important short-term method for protecting health, but is not a sustainable long-term adaptation technology because the electricity use is often associated with greenhouse gas emissions and during heat waves can overload the grid and contribute to outages (O'Neill, 2003c). Furthermore, the elderly with limited budgets and racial minorities are less likely to have access to air conditioning or to use it during hot weather (O'Neill *et al.*, 2005b, Sheridan, 2006). Incentives for and availability of high-efficiency, low energy-demand residential cooling systems, especially among disadvantaged populations, can advance health equity and minimize some of the negative aspects of air conditioning.

Another area of concern for impoverished populations is the impact that climate change may have on food systems and food supply. In the United States, food insecurity is a prevalent health risk among the poor, particularly poor children (Cook *et al.*, 2007). On a global scale, studies suggest that climate change is likely to contribute to food insecurity by reducing crop yield, most significantly at lower latitudes, due to shortened growing periods and decreases in water availability (Parry *et al.*, 2005). In

Box 2.1 Vulnerable Populations and Hurricane Katrina

In 2005, Hurricane Katrina caused more than 1,500 deaths along the Gulf Coast. Many of these victims were members of vulnerable subpopulations, such as hospital and nursing-home patients, older adults who required care within their homes, and individuals with disabilities (U.S. CHSGA, 2006). The hurricane was complicated by a catastrophic failure of the levee system that was intended to shield those areas in New Orleans that lie at or below sea level. According to the Louisiana Department of Health and Hospitals, more than 45 percent of the state's identified victims were 75 years of age or older; 69 percent were above age 60 (LDHH, 2006). In Mississippi, 67 percent of the victims whose deaths were directly, indirectly, or possibly related to Katrina were 55 years of age or older (MSDH, 2005).

At hurricane evacuation centers in Louisiana, Mississippi, Arkansas, and Texas, chronic illness was the most commonly reported health problem, accounting for 33 percent or 4,786 of 14,531 visits (CDC, 2006a). Six of the fifteen deaths indirectly related to the hurricane and its immediate aftermath in Alabama were associated with preexisting cardiovascular disease (CDC, 2006c), and the storm disrupted an estimated 100,000 diabetic evacuees across the region from obtaining appropriate care and medication (Cefalu *et al.*, 2006). One study suggested that the hurricane had a negative effect on reproductive outcomes among pregnant women and infants, who experienced exposure to environmental toxins, limited access to safe food and water, psychological stress, and disrupted health care (Callaghan *et al.*, 2007). Other vulnerable individuals included those without personal means of transportation and poor residents in Louisiana and Mississippi who were unable to evacuate in time (U.S. CHSGA, 2006).

the United States, changes in the price of food would likely contribute to food insecurity to a greater degree than overall scarcity.

The tragic loss of life that occurred after Hurricane Katrina underscores the increased vulnerability of special populations and demonstrates that, in the wake of extreme weather events, particularly those that disrupt medical infrastructure and require large-scale evacuation, treating individuals with chronic diseases is of critical concern (Ford *et al.*, 2006).

2.4.2.4 People with Chronic Conditions and Mobility and Cognitive Constraints

People with chronic medical conditions have an especially heightened vulnerability for the health impacts of climate change. Extreme heat poses a great risk for individuals with diabetes (Schwartz, 2005), and extreme cold has an increased effect on individuals with chronic obstructive pulmonary disease (Schwartz, 2005). People with mobility and cognitive constraints may be at particular risk during heat waves and other extreme weather events (EPA, 2006). As noted above, those with chronic medical conditions are also at risk of worsened status as the result of climate-related stressors and limited access to medical care during extreme events.

2.4.2.5 Occupational Groups

Certain occupational groups, primarily by virtue of spending their working hours outdoors, are at greater risk of climate-related health outcomes. Outdoor workers in rural or suburban areas, such as electricity and pipeline utility workers, are at increased risk of infection with Lyme Disease, although evidence is lacking for greater risk of clinical illness (Schwartz and Goldstein, 1990; Piacentino and Schwartz, 2002). They and other outdoor workers have increased exposures to ozone air pollution and heat stress, especially if work tasks involve heavy exertion.

2.4.2.6 Recent Migrants and Immigrants

Residential mobility, migration, and immigration may increase vulnerability. For example, new residents in an area may not be acclimated to the weather patterns, have lower awareness of risks posed by local vector-borne diseases, and have fewer social networks to provide support during an extreme weather event. U.S. immigrants returning to their countries of origin to visit friends and relatives have also been shown to suffer increased risks of severe travel-associated diseases (Bacaner *et al.*, 2004, Angell and Cetron, 2005). This vulnerability may become more significant if such diseases, which include malaria, viral hepatitis, and typhoid fever, become more prevalent in immigrants' countries of origin because of climate change.

2.5 ADAPTATION

Realistically assessing the potential health effects of climate change must include consideration of the capacity to manage new and changing climatic conditions. Individuals, communities, governments, and other organizations currently engage in a wide range of actions to identify and prevent adverse health outcomes associated with weather and climate. Although these actions have been largely successful, recent extreme events and outbreaks of vector-borne diseases highlight areas for improvement (Confalonieri *et al.*, 2007). Climate change is likely to further challenge the ability of current programs and activities to control climate-sensitive health determinants and outcomes. Preventing additional morbidity and mortality requires consideration of all upstream drivers of adverse health outcomes, including developing and deploying adaptation policies and measures that consider the full range of health risks that are likely to arise with climate change.

In public health, prevention is the term analogous to adaptation, acknowledging that adaptation implies a set of continuous or evolving practices and not just upfront investments. Public health prevention is classified as primary, secondary, or tertiary. Primary prevention aims to prevent the onset of disease in an otherwise unaffected population (such as regulations to reduce harmful exposures to ozone). Secondary prevention entails preventive action in response to early evidence of health effects (including strengthening disease surveillance programs to provide early intelligence on the emergence

or re-emergence of health risks at specific locations, and responding effectively to disease outbreaks, such as West Nile virus). Tertiary prevention consists of measures (often treatment) to reduce long-term impairment and disability and to minimize suffering caused by existing disease. In general, primary prevention is more effective and less expensive than secondary and tertiary prevention. For every health outcome, there are multiple possible primary, secondary, and tertiary preventions.

The degree to which programs and measures will need to be modified to address the additional pressures due to climate change will depend on factors such as the current burden of climate-sensitive health outcomes, the effectiveness of current interventions, projections of where, when, and how quickly the health burdens could change with changes in climate and climate variability (which depends on the rate and magnitude of climate change), the feasibility of implementing additional cost-effective interventions, other stressors that could increase or decrease resilience to impacts, and the social, economic, and political context within which interventions are implemented (Ebi *et al.*, 2006a). Failure to invest in adaptation may leave communities poorly prepared and increase the probability of severe adverse consequences (Haines *et al.*, 2006a,b).

Adaptation to climate change is basically a risk management issue. Adaptation and mitigation are the primary responses to manage current and projected risks. Mitigation and adaptation are not mutually exclusive. Co-benefits to human health can result concurrently with implementation of mitigation and adaptation actions. A dialogue is needed on prioritizing the costs of mitigation actions designed to limit future climate change and the potential costs of continually trying to adapt to its impacts. This dialogue should explicitly recognize that there is no guarantee that future changes in climate will not present a threshold that poses technological or physical limits to which adaptation is not possible.

Adaptation policies and measures should address both projected risks and the regions and populations that currently are not well adapted to climate-related health risks. Because the degree and rate of climate change are projected to increase over time, adaptation will be a continual process of designing and implementing policies and programs to prevent adverse impacts from changing exposures and vulnerabilities (Ebi *et al.*, 2006). Clearly, the extent to which effective proactive adaptations are developed and deployed will be a key determinant of future morbidity and mortality attributable to climate change.

Regional vulnerabilities to the health impacts of climate change are influenced by physical, social, demographic, economic, and other factors. Adaptation activities take place within the context of slowly changing factors that are specific to a region or population, including specific population and regional vulnerabilities, social and cultural factors, the built and natural environment, the status of the public health infrastructure, and health and social services. Because these factors vary across geographic and temporal scales, adaptation policies and measures generally are more successful when focused on a specific population and location. Additional important factors include the degree of risk perceived, the human and financial resources available for adaptation, the available technological options, and the political will to undertake adaptation.

2.5.1 Actors and Their Roles and Responsibilities for Adaptation

Responsibility for the prevention of climate-sensitive health risks rests with individuals, community and state governments, national agencies, and others. The roles and responsibilities vary by health outcome. For example, individuals are responsible for taking appropriate action on days with declared poor air quality, with health care providers and others responsible for providing the relevant information, and government agencies providing the regulatory framework. Community governments play a central role in preparedness and response for extreme events because of their jurisdiction over police, fire, and emergency medical services. Early warning systems for extreme events such as heat waves (Box 2.2) and outbreaks of infectious diseases may be developed at the community or state level. The federal government funds research and development to increase the range of decision support planning and response tools. Medical and nursing schools are responsible for ensuring that health professionals are trained in the identification and treatment of climate-sensitive diseases. The Red Cross and other nongovernmental organizations (NGOs) often play critical roles in disaster response.

Ensuring that surveillance systems account for and anticipate the potential effects of climate change will be beneficial. For example, surveillance systems in locations where changes in weather and climate may foster the spread of climate-sensitive pathogens and vectors into new regions would help advance our understanding of the associations between disease patterns and environmental variables. This knowledge could be used to develop early warning systems that warn of outbreaks before most cases have occurred. Increased understanding is needed of how to design these systems where there is limited knowledge of the interactions of climate, ecosystems, and infectious diseases (NAS, 2001).

There are no inventories in the United States of the various actors taking action to cope with climate change-related health

impacts. However, the growing numbers of city and state actions on climate change show increasing awareness of the potential risks. As of 1 November 2007, more than 700 cities have signed the U.S. Mayors' Climate Protection Agreement (http://www.seattle.gov/mayor/climate/cpaText.htm). Although this agreement focuses on mitigation through increased energy efficiency, one strategy, planting trees, can both sequester CO_2 and reduce urban heat islands. The New England Governors and Eastern Canadian Premiers developed a Climate Change Action Plan because of concerns about public health associated with degradation in air quality, public health risks, the magnitude and frequency of extreme climatic phenomena, and availability of water. (NEG/ECP, 2001). One action item focuses on the reduction and/or adaptation of negative social, economic, and environmental impacts. Activities being undertaken include a long-term phenology study and studies on temperature increases and related potential impacts.

Strategies, policies, and measures implemented by community and state governments, federal agencies, NGOs, and other actors can change the

---

Box 2.2 Heat Wave Early Warning Systems

Projections for increases in the frequency, intensity, and duration of heat waves suggest that more cities need heat wave early warning systems, including forecasts coupled with effective response options, to warn the public about the risks during such events (Meehl and Tebaldi, 2004). Prevention programs designed to reduce the toll of hot weather on the public have been instituted in several cities, and guidance has been developed to further aid communities seeking to plan such interventions, including buddy systems, cooling centers, and community preparedness (EPA, 2006b). Although these systems appear to reduce the toll of hot weather (Ebi et al., 2004; Ebi and Schmier, 2005; Weisskopf et al., 2002), and enhance preparedness following events such as the 1995 heat waves in Chicago and elsewhere, a survey of individuals 65 or older in four North American cities (Dayton, OH; Philadelphia, PA; Phoenix, AZ; and Toronto, Ontario, Canada) found that the public was unaware of appropriate preventive actions to take during heat waves (Sheridan, 2006). Although respondents were aware of the heat warnings, the majority did not consider they were vulnerable to the heat, or did not consider hot weather to pose a significant danger to their health. Only 46 percent modified their behavior on the heat advisory days. Although many individuals surveyed had access to home air conditioning, their use of it was influenced by concerns about energy costs. Precautionary steps recommended during hot weather, such as increasing intake of liquids, were taken by very few respondents (Sheridan, 2006). Some respondents reported using a fan indoors with windows closed and no air conditioning, a situation that can increase heat exposure and be potentially deadly. Further, simultaneous heat warnings and ozone alerts were a source of confusion, because recommendations not to drive conflicted with the suggestion to seek cooler locations if the residence was too warm. Critical evaluation of heat wave early warning systems is needed, including a determination of which components are effective and why (Kovats and Ebi, 2006; NOAA, 2005).

context for adaptation by conducting research to assess vulnerability and to identify technological options available for adaptation, implementing programs and activities to reduce vulnerability, and shifting human and financial resources to address the health impacts of climate change. State and federal governments also can provide guidance for vulnerability assessments that consider a range of plausible future scenarios. The results of these assessments can be used to identify priority health risks (over time), particularly vulnerable populations and regions, effectiveness of current adaptation activities, and modifications to current activities or new activities to address current and future climate change-related risks.

Table 2.4 summarizes the roles and responsibilities of various actors for adapting to climate change. Note that viewing adaptation from a public health perspective results in similar activities being classified as primary rather than secondary prevention under different health outcomes. It is not possible to prevent the occurrence of a heat wave, so primary prevention focuses on actions such as developing and enforcing appropriate infrastructure standards, while secondary prevention focuses on implementing early warning systems and other activities. For vector-borne diseases, primary prevention refers to preventing exposure to infected vectors. In this case, early warning systems can be considered primary prevention. For most vector-borne diseases, there are few options for preventing disease onset once an individual has been bitten.

A key activity not included in this framework is research on the associations between weather / climate and various health outcomes, taking into consideration other drivers of those outcomes (*e.g.*, taking a systems-based approach), and projecting how those risks may change with changing weather patterns. Increased understanding of the human health risks posed by climate change is needed for the design of effective, efficient, and timely adaptation options.

2.5.2 Adaptation Measures to Manage Climate Change-Related Health Risks

Determining where populations are not effectively coping with current climate variability and extremes facilitates identification of the additional interventions that are needed now. However, given uncertainties in climate change projections, identifying current adaptation deficits is not sufficient to protect against projected health risks.

Adaptation measures can be categorized into legislative policies, decision support tools, technology development, surveillance and monitoring of health data, infrastructure development, and other measures. Table 2.5 lists some adaptation measures for health impacts from heat waves, extreme weather events, vector-borne diseases, water-borne diseases, and air quality. These measures are generic because the local context, including vulnerabilities and adaptive capacity, needs to be considered in the design of programs and activities to be implemented.

An additional category of measures includes public education and outreach to provide information to the general public and specific vulnerable groups on climate risks to which they may be exposed and appropriate actions to take. Messages need to be specific to the region and group. For example, warnings to senior citizens of an impending heat wave should focus on keeping cool and drinking lots of water. Box 2.3 provides tips for dealing with extreme heat waves developed by U.S. EPA with assistance from federal, state, local, and academic partners (U.S. EPA, 2006).

2.6 CONCLUSIONS

The conclusions from this assessment are consistent with those of the First National Assessment: climate change poses a risk for U.S. populations, with uncertainties limiting quantitative projections of the number of increased injuries, illnesses, and deaths attributable to climate change. However, the strength and consistency of projections for climatic changes for some exposures of concern to human health suggest that implementation

Table 2.4: Actors and Their Roles and Responsibilities for Adaptation to Climate Change Health Risks

Actor	Reduce Exposures	Prevent Onset of Adverse Health Outcomes	Reduce Morbidity and Mortality
Extreme Temperature and Weather Events			
Individuals	Stay informed about impending weather events Follow guidance for emergency preparedness	Follow guidance for conduct during and following an extreme weather event (such as seeking cooling centers during a heat wave or evacuation during a hurricane)	Seek treatment when needed
Community, State, and National Agencies	Provide scientific and technical guidance for building and infrastructure standards Enforce building and infrastructure standards, including identification of restricted building zones where necessary	Develop scientific and technical guidance and decisions support tools for development of early warning systems and emergency response plans, including appropriate individual behavior Implement early warning systems and emergency response plans Conduct tests of early warning systems and response plans before events Conduct education and outreach on emergency preparedness	Ensure that emergency preparedness plans include medical services Improve programs to monitor the air, water, and soil for hazardous exposures Improve surveillance programs to collect, analyze, and disseminate data on the health consequences of extreme events and heat waves Monitor and evaluate the effectiveness of systems
NGOs and Other Actors		NGOs and other actors play critical roles in emergency preparedness and disaster relief	Educate and train health professionals on risks from extreme weather events
Vector-borne and Zoonotic Diseases			
Individuals	Take appropriate actions to reduce exposure to infected vectors, including eliminating vector breeding sites around residence	Vaccinate for diseases to which one would likely be exposed	Seek treatment when needed
Community, State, and National Agencies	Provide scientific and technical guidance and decision support tools for development of early warning systems Conduct effective vector (and pathogen) surveillance and control programs (including consideration of land use policies that affect vector distribution and habitats) Develop early warning systems for disease outbreaks, such as West Nile virus Develop and disseminate information on appropriate individual behavior to avoid exposure to vectors	Conduct research on vaccines and other preventive measures Conduct research and development on rapid diagnostic tools Provide vaccinations to those likely to be exposed	Conduct research on treatment options Develop and disseminate information on signs and symptoms of disease to guide individuals on when to seek treatment

Table 2.4. Actors and Their Roles and Responsibilities for Adaptation to Climate Change Health Risks

Actor	Reduce Exposures	Prevent Onset of Adverse Health Outcomes	Reduce Morbidity and Mortality
Water-borne and Food-borne Diseases			
Individuals	Follow proper food-handling guidelines Follow guidelines on drinking water from outdoor sources		Seek treatment when needed
Community, State, and National Agencies	Improve surveillance and control programs for early detection of disease outbreaks Develop methods to ensure watershed protection and safe water and food handling (e.g., Clean Water Act)	Sponsor research and development on rapid diagnostic tools for food- and water-borne pathogens	Sponsor research and development on treatment options Develop and disseminate information on signs and symptoms of disease to guide individuals on when to seek treatment
Diseases Related to Air Quality			
Individuals	Follow advice on appropriate behavior on high ozone days	For individuals with certain respiratory diseases, follow medical advice during periods of high air pollution	Seek treatment when needed
Community, State, and National Agencies	Develop and enforce regulations of air pollutants (e.g., Clean Air Act)	Develop decision support tools for early warning systems Conduct education and outreach on the risks of exposure to air pollutants	Conduct research on treatment options

	Heat waves	Extreme Weather Events	Vector-borne Diseases	Waterborne Diseases	Air Quality
Decision Support Tools	Enhance early warning systems	Enhance early warning systems and emergency response plans	Enhance early warning systems based on climate and environmental data for selected diseases	Develop early warning systems based on climate and environmental data for conditions that may increase selected diseases	Enhance alert systems for high air pollution days
Technology Development	Improve building design to reduce heat loads during summer months		Develop vaccines for West Nile virus and other vector-borne diseases; Develop more rapid diagnostic tests	Develop more rapid diagnostic tests	
Surveillance and Monitoring	Alter health data collection systems to monitor for increased morbidity and mortality during a heat wave	Alter health data collection systems to monitor for disease outbreaks during and after an extreme event	Enhance vector surveillance and control programs; Monitor disease occurrence	Enhance surveillance and monitoring programs for water-borne diseases	Enhance health data collection systems to monitor for health outcomes due to air pollution
Infrastructure Development	Improve urban design to reduce urban heat islands by planting trees, increasing green spaces, etc.	Design infrastructure to withstand projected extreme events	Consider possible impacts of infrastructure development, such as water storage tanks, on vector-borne diseases	Consider possible impacts of placement of sources of water- and food-borne pathogens (e.g., cattle near drinking water sources)	Improve public transit systems to reduce traffic emissions
Other	Conduct research on effective approaches to encourage appropriate behavior during a heat wave	Conduct research on effective approaches to encourage appropriate behavior during an extreme event			

of adaptation actions should commence now (Confalonieri *et al.*, 2007). Further, trends in factors that affect vulnerability, such as a larger and older U.S. population, will increase overall vulnerability to health risks. At the same time, the capacity of the United States to implement effective and timely adaptation measures is assumed to remain high throughout this century, thus reducing the likelihood of severe health impacts if appropriate programs and activities are implemented. However, the nature of the risks posed by climate change means that some adverse health outcomes might not be avoidable, even with attempts at adaptation. Severe health impacts will not be evenly distributed across populations and

Box 2.3: Quick Tips for Responding to Excessive Heat Waves

For the Public

Do

- Use air conditioners or spend time in air-conditioned locations such as malls and libraries

- Use portable electric fans to exhaust hot air from rooms or draw in cooler air

- Take a cool bath or shower

- Minimize direct exposure to the sun

- Stay hydrated: regularly drink water or other nonalcoholic fluids

- Eat light, cool, easy-to-digest foods such as fruit or salads

- Wear loose-fitting, light-colored clothes

- Check on older, sick, or frail people who may need help responding to the heat

- Know the symptoms of excessive heat exposure and the appropriate responses.

Don't

- Direct the flow of portable electric fans toward yourself when room temperature is hotter than 90°F

- Leave children and pets alone in cars for any amount of time

- Drink alcohol to try to stay cool

- Eat heavy, hot, or hard-to-digest foods

- Wear heavy, dark clothing.

Useful Community Interventions

For Public Officials

Send a clear public message

- Communicate that EHEs are dangerous and conditions can be life-*threatening. In the event of conflicting environmental safety recommendations, emphasize that health protection should be the first priority.*

Inform the public of anticipated EHE conditions

- When will EHE conditions be dangerous?

- How long will EHE conditions last?

- How hot will it feel at specific times during the day (e.g., 8 a.m., 12 p.m., 4 p.m., 8 p.m.)?

Assist those at greatest risk

- Assess locations with vulnerable populations, such as nursing homes and public housing

- Staff additional emergency medical personnel to address the anticipated increase in demand

- Shift/expand homeless intervention services to cover daytime hours

- Open cooling centers to offer relief for people without air conditioning and urge the public to use them.

Provide access to additional sources of information

- Provide toll-free numbers and Website addresses for heat exposure symptoms and responses

- Open hotlines to report concerns about individuals who may be at risk

- Coordinate broadcasts of EHE response information in newspapers and on television and radio.

Source: U.S. EPA, 2006

regions, but will be concentrated in the most vulnerable groups.

Proactive policies and measures should be identified that improve the context for adaptation, reduce exposures related to climate variability and change, prevent the onset of climate-sensitive health outcomes, and increase treatment options. Future community, state, and national assessments of the health impacts of climate variability and change should identify gaps in adaptive capacity, including where barriers and constraints to implementation, such as governance mechanisms, need to be addressed.

Because of regional variability in the types of health stressors attributable to climate change and their associated responses, it is difficult to summarize adaptation at the national level. Planning for adaptation is hindered by the fact that downscaled climate projections, as well as other climate information and tools, are generally not available to local governments. Such data and tools are essential for sectors potentially affected by climate change to assess their vulnerability and possible adaptation options, and to catalogue, evaluate, and disseminate adaptation measures. Explicit consideration of climate change is needed in the many programs and research activities within federal, state, and local agencies that are relevant to adaptation to ensure that they have maximum effectiveness and timeliness in reducing future vulnerability. In addition, collaboration and coordination are needed across agencies and sectors to ensure protection of the American population from the current and projected impacts of climate change.

2.7 EXPANDING THE KNOWLEDGE BASE

Few research and data gaps have been filled since the First National Assessment. An important shift in perspective that occurred since the First National Assessment is a greater appreciation of the complex pathways and relationships through which weather and climate affect health, and the understanding that many social and behavioral factors will influence disease risks and patterns (NRC,

2001). Several research gaps identified in the First National Assessment have been partially filled by studies that address the differential effects of temperature extremes by community, demographic, and biological characteristics; that improve our understanding of exposure-response relationships for extreme heat; and that project the public health burden posed by climate-related changes in heat waves and air quality. Despite these advances, the body of literature remains small, limiting quantitative projections of future impacts.

Improving our understanding of the linkages between climate change and health in the United States, may require a wide range of activities.

- Improve characterization of exposure-response relationships, particularly at regional and local levels, including identifying thresholds and particularly vulnerable groups.

- Collect data on the early effects of changing weather patterns on climate-sensitive health outcomes.

- Collect and enhance long-term surveillance data on health issues of potential concern, including VBZ diseases, air quality, pollen and mold counts, reporting of food- and water-borne diseases, morbidity due to temperature extremes, and mental health impacts from extreme weather events.

- Develop quantitative models of possible health impacts of climate change that can be used to explore the consequences of a range of socioeconomic and climate scenarios.

- Increase understanding of the processes of adaptation, including social and behavioral dimensions, as well as the costs and benefits of interventions.

- Evaluate the implementation of adaptation measures. For example, evaluation of heat wave warning systems, especially as they become implemented on a wider scale (NOAA, 2005), is needed to understand how to motivate appropriate behavior.

- Understand local- and regional-scale vulnerability and adaptive capacity to characterize the potential risks and the time horizon over which climate risks might arise. These assessments should include stakeholders to ensure their needs are identified and addressed in subsequent research and adaptation activities.

- Improve comprehensive estimates of the co-benefits of adaptation and mitigation policies in order to clarify trade-offs and synergies.

- Improve collaboration across the multiple agencies and organizations with responsibility and research related to climate change-related health impacts, such as weather forecasting, air and water quality regulations, vector control programs, and disaster preparation and response.

- Anticipate infrastructure requirements that will be needed to protect against extreme events such as heat waves, and food- and water-borne diseases, or to alter urban design to decrease heat islands, and to maintain drinking and wastewater treatment standards and source water and watershed protection.

- Develop downscaled climate projections at the local and regional scale in order to conduct the types of vulnerability and adaptation assessments that will enable adequate response to climate change and to determine the potential for interactions between climate and other risk factors, including societal, environmental, and economic. The growing concern over impacts from extreme events demonstrates the importance of climate models that allow for stochastic generation of possible future events, assessing not only how disease and pathogen population dynamics might respond, but also to assess whether levels of preparedness are likely to be adequate.

2.8 REFERENCES

American Academy of Pediatrics (AAP) Committee on Sports Medicine and Fitness, 2000: Climatic heat stress and the exercising child and adolescent. *Pediatrics*, **106(1)**, 158-159.

American Academy of Pediatrics (AAP), 2004: Ambient air pollution: health hazards to children. *Pediatrics*, **114(6)**, 1699-1707.

Angell, S.Y., and M.S. Cetron, 2005: Health disparities among travelers visiting friends and relatives abroad. *Annals of Internal Medicine*, **142**, 67-72.

Auld, H., D. MacIver, and J. Klaassen, 2004: Heavy rainfall and waterborne disease outbreaks: the Walkerton example. *Journal of Toxicology and Environmental Health, Part A*, **67**, 1879-1887.

Aw, J. and M.J. Kleeman, 2003: Evaluating the first-order effect of inter-annual temperature variability on urban air pollution. Journal of Geophysical Research, **108**, 7-1—7-18.

Bacaner N., B. Stauffer, D.R. Boulware, P.F. Walker, and J.S. Keystone, 2004: Travel medicine considerations for North American immigrants visiting friends and relatives. *Journal of the American Medical Association*, **291**, 2856-64.

Balling, Jr., R.C., and R.S. Cerveny, 2003: Compilation and discussion of trends in severe storms in the United States: popular perception v. climate reality. *Natural Hazards*, **29**, 103-112.

Basu, R., F. Dominici, and J.M. Samet, 2005: Temperature and mortality among the elderly in the United States: a comparison of epidemiologic methods. *Epidemiology*, **16(1)**, 58-66.

Bates, D.V., 2005: Ambient ozone and mortality. *Epidemiology*, **16(4)**, 427-429.

Beggs, P.J. and H.J. Bambrick, 2005: Is the global rise of asthma an early impact of anthropogenic climate change? *Environmental Health Perspectives*, **113**, 915-919.

Bell, M. and H. Ellis, 2004: Sensitivity analysis of tropospheric ozone to modified biogenic emissions for the Mid-Atlantic region. *Atmospheric Environment*, **38(1)**, 1879–1889.

Bell, M.L., R. Goldberg, C. Hogrefe, P.L. Kinney, K. Knowlton, B. Lynn, J. Rosenthal, C. Rosenzweig, and J. Patz, 2007: Climate change, ambient ozone, and health in 50 U.S. cities. *Climatic Change*, **82(1-2)**, 61-76.

Bell, M.L., A. McDermott, S.L. Zeger, J.M. Samet, and F. Dominici, 2004: Ozone and mortality in 95 U.S. urban communities, 1987 to 2000. *Journal of the American Medical Association*, **292**, 2372-2378.

Bernard, S.M. and K.L. Ebi, 2001: Comments on the process and product of the health impacts assessment component of the national assessment of the potential consequences of climate variability and change for the United States. *Environmental Health Perspectives*, **109(2)**, 177-184.

Bharti, A.R., J.E. Nally, J.N. Ricladi, M.A. Matthias, M.M. Diaz, M.A. Lovett, P.N. Levett, R.H. Gilman, M.R. Willig, E. Gotuzzo, and J.M. Vinetz, 2003: Leptospirosis: a zoonotic disease of global importance. *The Lancet Infectious Diseases*, **3**, 757-771.

Braga, A.L., A. Zanobetti, and J. Schwartz, 2001: The time course of weather related deaths. *Epidemiology*, **12**, 662-667.

Bronstert, A., 2003: Floods and climate change: interactions and impacts. *Risk Analysis*, **23(3)**, 545-557.

Brown, T. J., B.L. Hall, and A.L. Westerling, 2004: The impact of twenty-first century climate change on wildland fire danger in the western United States: an applications perspective. *Climatic Change*, **62**, 365-388.

Brownstein, J.S., T.R. Holford, and D. Fish, 2003: A climate-based model predicts the spatial distribution of the Lyme disease vector *Ixodes scapularis* in the United States. *Environmental Health Perspectives*, **111(9)**, 1152-1157.

Brownstein, J.S., T.R. Holford, and D. Fish, 2004: Enhancing West Nile virus surveillance, United States. *Emerging Infectious Diseases*, **10**, 1129-1133.

Burkhardt, W. and K.R. Calci, 2000: Selective accumulation may account for shellfish-associated viral illness. *Applied and Environmental Microbiology*, **66(4)**, 1375-1378.

Burr, M.L., J.C. Emberlin, R. Treu, S. Cheng and N.E. Pearce, 2003: Pollen counts in relation to the prevalence of allergic rhinoconjunctivitis, asthma and atopic eczema in the International Study of Asthma and Allergies in Childhood (ISAAC). *Clinical and Experimental Allergy*, **33**, 1675-80.

Cabanes, P.E., F. Wallett, E. Pringuez, and P. Pernin, 2001: Assessing the risk of primary amoebic meningoencephalitis from swimming in the presence of environmental *Naegleria fowleri*. *Applied and Environmental Microbiology*, **67(7)**, 2927-2931.

Callaghan, W.M., S.A. Rasmussen, D.J. Jamieson, S.J. Ventura, S.L. Farr, P.D. Sutton, T.J. Matthews, B.E. Hamilton, K.R. Shealy, D. Brantley, and S.F. Posner, 2007: Health concerns of women and infants in times of natural disasters: lessons learned from Hurricane Katrina. *Maternal and Child Health Journal*, **11(4)**, 307-311.

Carson, C., S. Hajat, B. Armstrong, and P. Wilkinson, 2006: Declining vulnerability to temperature-related mortality in London over the 20th century. *American Journal of Epidemiology*, **164(1)**, 77-84.

Casassa, G., C. Rosenzweig, A. Imeson, D.J. Karoly, C. Liu, A. Menzel, S. Rawlins, T.L. Root, B. Seguin and P. Tryjanowski, 2007: Assessment of observed changes and responses in natural and managed systems. *Climate Change 2007: Impacts, Adaptation and Vulnerability. Contribution of Working Group II to the Fourth Assessment Report of the Intergovernmental Panel on Climate Change.* [Parry, M.L., O.F. Canziani, J.P. Palutikof, P.J. van der Linden, C.E. Hansson (eds.)]. Cambridge University Press, Cambridge, UK.

Casman. E., B. Fischhoff, M. Small, H. Dowlatabadi, J. Rose, and M.G. Morgan, 2001: Climate change and cryptosporidiosis: a qualitative analysis. *Climatic Change*, **50**, 219-249.

CCSP, 2008: *Synthesis and Assessment Product 3.3: Weather and Climate Extremes in a Changing Climate.* A report by the U.S. Climate Change Science Program.

CDC, 2004a: Rapid assessment of the needs and health status of older adults after Hurricane Charley—Charlotte, DeSoto, and Hardee Counties, Florida, August 27-31, 2004. *MMWR—Morbidity & Mortality Weekly Report*, **53(36)**, 837-840.

CDC, 2004b. The impact of malaria, a leading cause of death worldwide. Retrieved November 12, 2007, from *http://www.cdc.gov/malaria/impact/index.htm*.

CDC, 2005a: Norovirus outbreak among evacuees from Hurricane Katrina—Houston, Texas, September 2005. *MMWR—Morbidity & Mortality Weekly Report*, **54(40)**, 1016-1018.

CDC, 2005b: Heat-related mortality—Arizona, 1993-2002, and United States, 1979-2002. *MMWR—Morbidity & Mortality Weekly Report*, **54(25)**, 628-630.

CDC, 2006a: Morbidity surveillance after Hurricane Katrina—Arkansas, Louisiana, Mississippi, and Texas, September 2005. *MMWR—Morbidity & Mortality Weekly Report*, 55(26), 727-731.

CDC, 2006b: Carbon monoxide poisonings after two major hurricanes—Alabama and Texas, August-October 2005. *MMWR—Morbidity & Mortality Weekly Report*, **55(09)**, 236-239.

CDC, 2006c: Mortality associated with Hurricane Katrina—Florida and Alabama, August-October 2005. *MMWR—Morbidity & Mortality Weekly Report*, 55(09), 239-242.

Cefalu, W.T., S.R. Smith, L. Blonde, V. Fonseca, 2006: The Hurricane Katrina aftermath and its impact on diabetes care. *Diabetes Care*, 29(1), 158-160.

Charles, M.D., R.C. Holman, A.T. Curns, U.D. Parashar, R.I. Glass, and J.S. Breeze, 2006: Hospitalizations associated with rotavirus gastroenteritis in the United States, 1993-2002. *The Pediatric Infectious Disease Journal*, **25(6)**, 489-493.

Choi, K.M., G. Christakos, and M.L. Wilson, 2006: El Nino effects on influenza mortality risks in the state of California. *Public Health*, **120(6)**, 505-516.

Civerolo, K., C. Hogrefe, B. Lynn, J. Rosenthal, J.-Y. Ku, W. Solecki, et al., 2006: Estimating the effects of increased urbanization on future surface meteorology and ozone concentrations in the New York City metropolitan region. *Atmospheric Environment*, 41(9), 1803-1818.

CLIMB, 2004: *Infrastructure Systems, Services and Climate Change: Integrated Impacts and Response Strategies for the Boston Metropolitan Area.* National Environmental Trust, Boston, Massachusetts. [Accessed 25 February 2007].

Cook, D.W., 1994: Effect of time and temperature on multiplication of *Vibrio vulnificus* in post-harvest Gulf Coast shellstock oysters. *Applied and Environmental Microbiology*, **60(9)**, 3483-3484.

Cook, J.T. and D.A. Frank, 2007: Food security, poverty, and human development in the United States. *Annals of the New York Academy of Sciences*, 1-16.

Cook, S.M., R.I. Glass, C.W. LeBaron, and M.S. Ho, 1990: Global seasonality of rotavirus infections. *Bulletin of the World Health Organization*, **58(2)**, 171-177.

Curriero, F.C., K.S. Heiner, J.M. Samet, S.L. Zeger, L. Strug, and J.A. Patz, 2002: Temperature and mortality in 11 cities of the eastern United States. *American Journal of Epidemiology*, **155(1)**, 80-87.

Curriero, F.C., J.A. Patz, J.B. Rose, and S. Lele, 2001: The association between extreme precipitation and waterborne disease outbreaks in the United States, 1948-1994. *American Journal of Public Health*, **91(8)**, 1194-1199.

D'Souza, R.M., N.G. Becker, G. Hall, and K.B.A. Moodie, 2004: Does ambient temperature affect food-borne disease? *Epidemiology*, **15(1)**, 86-92.

Davies, C.M., C.M. Ferguson, C. Kaucner, M. Krogh, N. Altavilla, D.A. Deere, and N.J. Ashbolt, 2004: Dispersion and transport of *Cryptosporidium* oocysts from fecal pats under simulated rainfall events. *Applied and Environmental Microbiology*, **70(2)**, 1151-1159.

Davis, R., P. Knappenberger, W. Novicoff, and P. Michaels, 2002: Decadal changes in heat-related human mortality in the eastern United States. *Climate Research*, **22**, 175-184.

Davis, R., P. Knappenberger, P. Michaels, and W. Novicoff, 2004: Seasonality of climate-human mortality relationships in US cities and impacts of climate change. *Climate Research*, **26**, 61-76.

Davis, R.E., P.C. Knappenberger, P.J. Michaels, and W.M. Novicoff, 2003a: Changing heat-related mortality in the United States. *Environmental Health Perspectives*, **111(14)**, 1712-1718.

Davis, R.E., P.C. Knappenberger, W.M. Novicoff, and P.J. Michaels, 2003b: Decadal changes in summer mortality in U.S. cities. *International Journal of Biometeorology*, **47(3)**, 166-175.

Day, J.C., 1996: *Population Projections of the United States by Age, Sex, Race and Hispanic Origin: 1995-2050*. U.S. Bureau of the Census, Current Population Reports, P25-1130, U.S. Government Printing Office, Washington, DC.

DeGaetano, A.T., 2005: Meteorological effects on adult mosquito (*Culex*) populations in metropolitan New Jersey. *International Journal of Biometeorology*, **49(5)**, 345-353.

DeGroot, D.W., G. Havenith, and W.L. Kenney, 2006: Responses to mild cold stress are predicted by different individual characteristics in young and older subjects. *Journal of Applied Physiology*, **101(6)**, 1607-1615.

Dessai, S., 2003: Heat stress and mortality in Lisbon Part II. An assessment of the potential impacts of climate change. *International Journal of Biometeorology*, **48(1)**, 37-44.

Devlin, R.B., W.F. McDonnell, R. Mann, S. Becker, D.E. House, D. Schreinemachers, H.S. Koren, 1991: Exposure of humans to ambient levels of ozone for 6.6 hours causes cellular and biochemical changes in the lung. *American Journal of Respiratory Cell and Molecular Biology*, **4**, 72-81.

Diaz, J., A. Jordan, R. Garcia, C. Lopez, J.C. Alberdi, E. Hernandez, *et al.*, 2002: Heat waves in Madrid 1986-1997: effects on the health of the elderly. *International Archives of Occupational & Environmental Health*, **75(3)**, 163-170.

Dietz, V.J. and J.M. Roberts, 2000: National surveillance for infection with *Cryptosporidium parvum*, 1995-1998: what have we learned? *Public Health Reports*, **115**, 358-363.

Dockery, D.W., C.A. Pope III, X. Xu, J.D. Spengler, J.H. Ware, M.E. Fay, et al., 1993. An association between air pollution and mortality in six U.S. cities. *New England Journal of Medicine*, 329, 1753-1759.

Dominici, F., R.D. Peng, M.L. Bell, L. Pham, A. McDermott, S.L. Zeger, J.M. Samet, 2006: Fine particulate air pollution and hospital admission for cardiovascular and respiratory diseases. *Journal of the American Medical Association*, **295(10)**, 1127-1134.

Donaldson, G.C., H. Rintamaki, and S. Nayha, 2001: Outdoor clothing: its relationship to geography, climate, behaviour and cold-related mortality in Europe. *International Journal of Biometeorology*, **45(1)**, 45-51.

Dushoff, J., J.B. Plotkin, C. Viboud, D.J. Earn, and L. Simonsen, 2005: Mortality due to influenza in the United States—an annualized regression approach using multiple-cause mortality data. *American Journal of Epidemiology*, **163(2)**, 181-187.

Dzuiban, E.J., J.L. Liang, G.F. Craun, V. Hill, P.A. Yu, J. Painter, M.R. Moore, R.L. Calderon, S.L. Roy, and M.J. Beach, 2006: Surveillance for waterborne disease and outbreaks associated with recreational water—United States, 2003—2004. *MMWR—Morbidity & Mortality Weekly Report*, **55(12)**, 1-31.

Ebi, K.L., K.A. Exuzides, E. Lau, M. Kelsh, and A. Barnston, 2001: Association of normal weather periods and El Nino events with hospitalization for viral pneumonia in females: California, 1983–1998. *American Journal of Public Health*, **91(8)**, 1200–1208.

Ebi, K.L., D.M. Mills, J.B. Smith, and A. Grambsch, 2006a: Climate change and human health impacts in the United States: an update on the results of the U.S. national assessment. *Environmental Health Perspectives*, **114(9)**, 1318-1324.

Ebi, K.L. and J.K. Schmier, 2005: A stitch in time: improving public health early warning systems for extreme weather events. *Epidemiologic Reviews*, **27**, 115-121.

Ebi, K.L., T.J. Tiesberg, L.S. Kalkstein, L. Robinson, and R.F. Weiher, 2004: Heat watch/warning systems save lives. *Bulletin of the American Meteorological Society*, **85(8)**, 1067-1073.

Emberlin, J., M. Detandt, R. Gehrig, S. Jaeger, N. Nolard and A. Rantio-Lehtimaki, 2002: Responses in the start of Betula (birch) pollen seasons to recent changes in spring temperatures across Europe. *International Journal of Biometeorology*, **46**, 159-70.

EPA. 2006. Excessive Heat Events Guidebook. Report from the United States Environmental Protection Agency, Washington, DC. EPA 430-B-06-005, 52 pp.

EPRI, 2005: *Interactions of Climate Change and Air Quality: Research Priorities and New Directions.* Electric Power Research Institute, Program on Technology Innovation, Technical Update 1012169.

Fallico, F, K. Nolte, L. Siciliano, and F. Yip, 2005: Hypothermia-related deaths—United States, 2003-2004. *MMWR—Morbidity & Mortality Weekly Report,* **54(07)**, 173-175.

Flahault, A., C. Viboud, K. Pakdaman, P.Y. Boelle, M.L. Wilson, M. Myers, and A.J. Valleron, 2004: Association of influenza epidemics in France and the USA with global climate variability. In: *Proceedings of the International Conference on Options for the Control of Influenza V* [Kawaoka. Y. (ed.)]. Elsevier Inc., San Diego, California. pp. 73–77.

Fleury, M., D.F. Charron, J.D. Holt, O.B. Allen, and A.R. Maarouf, 2006: A time series analysis of the relationship of ambient temperature and common bacterial enteric infections in two Canadian provinces. *International Journal of Biometerology,* **60**, 385-391.

Folinsbee, L.J., W.F. McDonnell, D.H. Horstman, 1988: Pulmonary function and symptom responses after 6.6-hour exposure to 0.12 ppm ozone with moderate exercise. *Journal of the Air Pollution Control Association,* **38**, 28-35.

Fong, T.T., D.W. Griffin, and E.K. Lipp, 2005: Molecular assays for targeting human and bovine enteric viruses in coastal waters and application for library-independent source tracking. *Applied and Environmental Microbiology,* **71(4)**, 2070-2078.

Ford, E.S., A.H. Mokdad, M.W. Link, W.S. Garvin, L.C. McGuire, R.B. Jiles, et al., 2006: Chronic disease in health emergencies: in the eye of the hurricane. *Preventing Chronic Disease,* 3(2), 1-7.

Forkel, R., and R. Knoche, 2006: Regional climate change and its impact on photooxidant concentrations in southern Germany: Simulations with a coupled regional climate-chemistry model. *Journal of Geophysical Research,* **111**, D12302, doi: 10.1029/2005JD006748.

Fried, B.J., M.E. Domino, and J. Shadle, 2005: Use of mental health services after hurricane Floyd in North Carolina. *Psychiatric Services,* **56(11)**, 1367-1373.

Fried, J. S., M. S. Torn, and E. Mills. 2004. The impact of climate change on wildfire severity: a regional forecast for northern California. *Climatic Change,* **64**, 169-191.

Frost, F.J., T.R. Kunde, and G. F. Craun, 2002: Is contaminated groundwater an important cause of viral gastroenteritis in the United States? *Journal of Environmental Quality,* **65(3)**, 9-14.

Furness, B.W., M.J. Beach, and J.M. Roberts, 2000: Giardiasis surveillance—United States, 1992-1997. *MMWR—Morbidity & Mortality Weekly Report,* **49(07)**, 1-13.

Gangarosa, R.E., R.I. Glass, J.F. Lew, and J.R. Boring, 1992: Hospitalizations involving gastroenteritis in the United States, 1985: the special burden of the disease among the elderly. *American Journal of Epidemiology,* **135(3)**, 281-290.

Gatntzer, C., E. Dubois, J.-M. Crance, S. Billaudel, H. Kopecka, L. Schwartzbrod, M. Pommepuy, and F. Le Guyader, 1998: Influence of environmental factors on survival of enteric viruses in seawater. *Oceanologica Acta,* **21(6)**, 883-992.

Gedalof, Z., D.L. Peterson, and N.J. Mantua, 2005: Atmospheric, climatic, and ecological controls on extreme wildfire years in the northwestern United States. *Ecological Applications,* **15(1)**, 154-174.

Gerrity, E.T., and B.W. Flynn, 1997: Mental health consequences of disasters. In: *The Public Health Consequences of Disasters* [Noji. E.K. (ed.)]. Oxford University Press, New York. pp. 101–121.

Goodman, P.G., D.W. Dockery, and L. Clancy, 2004: Cause-specific mortality and the extended effects of particulate pollution and temperature exposure. [erratum appears in *Environmental Health Perspectives,* **112(13)**, A729]. *Environmental Health Perspectives,* **112(2)**, 179-185.

Goodman, R.A., J.W. Buehler, H.B. Greenberg, T.W. McKinley, and J.D. Smith, 1982: Norwalk gastroenteritis associated with a water system in a rural Georgia community. *Archives of Environmental Health,* **37(6)**, 358-360.

Gouveia, N., S. Hajat, and B. Armstrong, 2003: Socio-economic differentials in the temperature-mortality relationship in Sao Paulo, Brazil. *International Journal of Epidemiology,* **32**, 390-397.

Greenberg, J.H., J. Bromberg, C.M. Reed, T.L. Gustafson, R.A. Beauchamp, 1983: The epidemiology of heat-related deaths, Texas—1950, 1970-79, and 1980. *American Journal of Public Health,* **73(7)**, 805-807.

Greene, S.K., E.L. Ionides, and M.L. Wilson, 2006: Patterns of influenza-associated mortality among US elderly by geographic region and virus subtype, 1968-1998. *American Journal of Epidemiology,* **163(4)**, 316-326.

Greenough, G., McGeehin M., S.M. Bernard, J. Trtanj, J. Riad, and D. Engleberg, 2001: The potential impacts of climate variability and change on health impacts of extreme weather events in the United States. *Environmental Health Perspectives*, **109**, 191-198.

Griffin, K., A. Donaldson, J.H. Paul, and J.B. Rose, 2003: Pathogenic human viruses in coastal waters. *Clinical Microbiology Reviews*, **16(1)**, 129-143.

Groisman, P.Y., R.W. Knight, T.R. Karl, D.R. Easterling, B. Sun, and J.H. Lawrimore, 2004: Contemporary changes of the hydrological cycle over the contiguous United States: trends derived from in situ observations. *Journal of Hydrometeorology*, **5**, 64-85.

Gubler, D.J., P. Reiter P., K.L. Ebi, W. Yap, R. Nasci, and J.A. Patz, 2001: Climate variability and change in the United States: potential impacts on vector- and rodent-borne diseases. *Environmental Health Perspectives*, **109(2)**, 223-233.

Guzmán, M.G. and G. Kourí, 2002: Dengue: an update. *Lancet Infectious Diseases*. **2(1)**, 33-42.

Haines, A., R.S. Kovats, D. Campbell-Lendrum, and C. Corvalan, 2006a: Climate change and human health: impacts, vulnerability, and public health. *Lancet*, **367(9528)**, 2101-2109.

Haines, A., R.S. Kovats, D. Campbell-Lendrum, and C. Corvalan, 2006b: Climate change and human health: impacts, vulnerability and public health. *Public Health*, **120(7)**, 585-596.

Hajat, S., K.L. Ebi, R.S. Kovats, B. Menne, S. Edwards, A. Haines, 2003: The health consequences of flooding in Europe and the implications for public health: a review of the evidence. *Applied Environmental Science and Public Health*, 1(1), 13-21.

Hajat, S., R. Kovats, and K. Lachowycz, 2007: Heat-related and cold-related deaths in England and Wales: who is at risk? *Occupational Environmental Medicine*, **64**, 93-100.

Haley, B.J., 2006: *Ecology of Salmonella in a Southeastern Watershed*. University of Georgia, M.S. Thesis. Athens, Georgia.

Halverson, J.B., 2006: A climate conundrum: the 2005 hurricane season has been touted as proof of global warming and an indication of worse calamities to come. Where is the line between fact and speculation? *Weatherwise*, **59(2)**,18-23.

Havenith, G., 2001: Individualized model of human thermoregulation for the simulation of heat stress response. *Journal of Applied Physiology*, **90(5)**, 1943-1954.

Havenith, G., J.M. Coenen, L. Kistemaker, and W.L. Kenney, 1998: Relevance of individual characteristics for human heat stress response is dependent on exercise intensity and climate type. *European Journal of Applied Physiology & Occupational Physiology*, **77(3)**, 231-241.

Havenith, G., Y. Inoue, V. Luttikholt, and W.L. Kenney, 1995: Age predicts cardiovascular, but not thermoregulatory, responses to humid heat stress. *European Journal of Applied Physiology & Occupational Physiology*, **70(1)**, 88-96.

Hayhoe, K., D. Cayan, C.B. Field, P.C. Frumhoff, E.P. Maurer, N.L. Miller, *et al.*, 2004: Emissions pathways, climate change, and impacts on California. *Proceedings of the National Academy of Sciences of the United States of America*, **101(34)**, 12,422-12,427.

Healy, J.D., 2003: Excess winter mortality in Europe: a cross country analysis identifying key risk factors. *Journal of Epidemiology and Community Health*, **57(10)**, 784-789.

Herikstad, H., S. Yang, T.J. Van Gilder, D. Vugia, J. Hadler, P. Blake, V. Deneen, B. Shiferaw, F.J. Angulo and the Foodnet Working Group, 2002: A population-based estimate of the burden of diarrhoeal illness in the United States: FoodNet, 1996-7. *Epidemiology and Infection*, **19**, 9-17.

Hogrefe, C., J. Biswas, B. Lynn, K. Civerolo, J-Y. Ku, J. Rosenthal, C. Rosenzweig, R. Goldberg, and P.L. Kinney, 2004a: Simulating regional-scale ozone climatology over the Eastern United States: model evaluation results, *Atmospheric Environment*, **38**, 2627-2638.

Hogrefe, C., K. Civerolo, J-Y. Ku, B. Lynn, J. Rosenthal, K. Knowlton, B. Solecki, C. Small, S. Gaffin, R. Goldberg, C. Rosenzweig, and P.L. Kinney, 2004b: *Modeling the Air Quality Impacts of Climate and Land Use Change in the New York City Metropolitan Area*. Models-3 Users' Workshop, October 18-20, Research Triangle Park, North Carolina.

Hogrefe, C., K. Civerolo, J-Y. Ku, B. Lynn, J. Rosenthal, B. Solecki, C. Small, S. Gaffin, K. Knowlton, R. Goldberg, C. Rosenzweig, and P.L. Kinney, 2006: Air quality in future decades—determining the relative impacts of changes in climate, anthropogenic and biogenic emissions, global atmospheric composition, and regional land use. In: *Air Pollution Modeling and its Application XVII* [Borrego, C. and A.L. Norman (eds.)]. Proceedings of the 27th NATO/CCMS International Technical Meeting on Air Pollution Modeling and its Application, Springer, October 25-29, 2004, Banff, Canada, 772 pp.

Hogrefe, C., R. Leung, L. Mickley, S. Hunt, and D. Winner, 2005a: Considering climate change in air quality management. *Environmental Manager*, 19-23.

Hogrefe, C., B. Lynn, K. Civerolo, J-Y. Ku, J. Rosenthal, C. Rosenzweig, *et al.*, 2004c: Simulating changes in regional air pollution over the eastern United States due to changes in global and regional climate and emissions. *Journal of Geophysical Research*, **109**, D22301.

Hogrefe, C., B. Lynn, C. Rosenzweig, R. Goldberg, K. Civerolo, J-Y. Ku, J. Rosenthal, K. Knowlton, and P.L. Kinney, 2005b: *Utilizing CMAQ Process Analysis to Understand the Impacts of Climate Change on Ozone and Particulate Matter.* Models-3 Users' Workshop, September 26-28, Chapel Hill, North Carolina.

Holley, K., M. Arrus, K.H. Ominiski, M. Tenuta, and G. Blank, 2006: *Salmonella* survival in manure-treated soils during simulated seasonal temperature exposure. *Journal of Environmental Quality*, **35**, 1170-1180.

Howell, D. and D. Cole, 2006: Leptospirosis: a waterborne zoonotic disease of global importance. *Georgia Epidemiology Report*, **22(8)**, 1-2.

Hrudey, S.E., P. Payment, P.M. Houck, R.W. Gillham, and E.J. Hrudry, 2003: A fatal waterborne disease epidemic in Walkerton, Ontario: comparison with other waterborne outbreaks in the developed world. *Water Science and Technology*, **47(3)**, 7-14.

Huynen, M., and B. Menne, 2003: *Phenology and human health: allergic disorders.* Report of a WHO meeting, 16-17 January 2003, Rome, Italy,. Health and Global Environmental Series. EUR/03/5036791, World Health Organization, Copenhagen, 64 pp.

IPCC, 1996: *Climate Change 1995: Economic and Social Dimensions of Climate Change. Contribution of Working Group III to the Second Assessment Report of the Intergovernmental Panel on Climate Change* [Bruce, J., H. Lee, and E. Haites (eds.)]. Cambridge University Press, Cambridge, United Kingdom and New York, USA.

IPCC, 2007a: *Climate Change 2007: The Physical Science Basis.* Contribution of Working Group I to the Fourth Assessment Report of the Intergovernmental Panel on Climate Change [Solomon, S., D. Qin, M. Manning, Z. Chen, M. Marquis, K.B. Averyt, M. Tignor and H.L. Miller (eds.)]. Cambridge University Press, Cambridge, United Kingdom and New York, New York, USA, 996 pp.

IPCC, 2007b: *Climate Change 2007: Impacts, Adaptation and Vulnerability.* Contribution of Working Group II to the Fourth Assessment Report of the Intergovernmental Panel on Climate Change [M.L. Parry, O.F. Canziani, J.P. Palutikof, P.J. van der Linden, and C.E. Hanson (eds.)]. Cambridge University Press, Cambridge, UK, 976 pp.

Jacobson M.Z., 2008: On the causal link between carbon dioxide and air pollution mortality. *Geophysical Research Letters*, 35, L03809, doi: 10.1029/2007GL031101

Jamieson, D.J., R.N. Theiler, S.A. Rasmussen, 2006: Emerging infections and pregnancy. Emerging *Infectious Diseases*, 12(11), 1638-1643.

Janda, J.M., C. Powers, R.G. Bryant, and S.L. Abbott, 1988: Clinical perspectives on the epidemiology and pathogenesis of clinically significant *Vibrio* spp. *Clinical Microbiology Reviews*, **1(3)**, 245-267.

Jones, T.S., A.P. Liang, E.M. Kilbourne, M.R. Griffin, P.A. Patriarca, S.G. Wassilak, *et al.*, 1982: Morbidity and mortality associated with the July 1980 heat wave in St. Louis and Kansas City, MO. *Journal of the American Medical Association*, **247(24)**, 3327-3331.

Kalkstein, L.S. 1993: Health and Climate Change: Direct Impacts in Cities. *Lancet*, **342**, 1397-1399.

Kalkstein, L.S., 2000: Saving lives during extreme weather in summer. *British Medical Journal*, **321(7262)**, 650-651.

Katz, A.R., V.E. Ansdell, P.V. Effler, C.R. Middleton, and D.M. Sasaki, 2002: Leptospirosis in Hawaii, 1974-1988: epidemiologic analysis of 353 laboratory-confirmed cases. *American Journal of Tropical Medicine and Hygiene*, **66(1)**, 61-70.

Keatinge, W.R. and G.C. Donaldson, 2001: Mortality related to cold and air pollution in London after allowance for effects of associated weather patterns. *Environmental Research*, **86(3)**, 209-216.

Khetsuriani, N., A. LaMonte-Fowlkes, M.S. Oberste, and M.A. Pallansch, 2006: Enterovirus surveillance—United States, 1970—2005. *MMWR—Morbidity & Mortality Weekly Reports*, **55(08)**, 1-20.

Khosla, R. and K.K. Guntupalli, 1999: Heat-related illnesses. *Critical Care Clinics*, **15(2)**, 251-263.

Kim, J., 2003: Effects of climate change on extreme precipitation events in the Western US. In: *AMS Symposium on Global Change and Climate Variations, V. 14.* American Meteorological Society, Boston, Massachusetts.

King, B.J. and P.T. Monis, 2006: Critical processes affecting Cryptosporidium oocyst survival in the environment. *Parasitology*, 1-15.

Kinney, P.L., and H. Ozkaynak, 1991: Associations of daily mortality and air pollution in Los Angeles County. *Environmental Research*, **54**, 99-120.

Kinney, P.L., J.E. Rosenzweig, C. Hogrefe, W. Solecki, K. Knowlton, C. Small, *et al.*, 2006: Chapter 6. Assessing the potential public health impacts of changing climate and land use: NY Climate & Health Project. In: *Climate Change and Variability: Impacts and Responses* [Ruth M., K. Donaghy K, and P. Kirshen (eds.)]. New Horizons in Regional Science, Edward Elgar, Cheltenham, UK.

Kistemann, T., T. Classen, C. Koch, F. Dangendorf, R. Fischeder, J. Gebel, V. Vacata, and M. Exner, 2002: Microbial load of drinking water reservoir tributaries during extreme rainfall and runoff. *Applied Environmental Microbiology*, **68**, 2188-97.

Klinenberg, E., 2002: *Heat wave: A Social Autopsy of Disaster in Chicago.* The University of Chicago Press, Chicago.

Knowlton, K., B. Lynn, R. Goldberg, C. Rosenzweig, C. Hogrefe, J. Rosenthal, *et al.*, 2007: Projecting heat-related mortality impacts under a changing climate in the New York City region. *American Journal of Public Health*, **97(11)**, 2028-2034.

Knowlton, K., J. Rosenthal, C. Hogrefe, B. Lynn, S. Gaffin, R. Goldberg, C. Rosenzweig, K. Civerolo, J-Y. Ku, and P.L. Kinney, 2004: Assessing ozone-related health impacts under a changing climate. *Environmental Health Perspectives*, **112**, 1557–1563.

Kolivras, K.N. and A.C. Comrie, 2003: Modeling valley fever (coccidioidomycosis) incidence on the basis of climate conditions. *International Journal of Biometeorology*, **47**, 87-101.

Kosatsky, T., M. Baccini, A. Biggeri, G. Accetta, B. Armstrong, B. Menne, *et al.*, 2006: Years of life lost due to summertime heat in 16 European cities. *Epidemiology*, **17(6)**, 85.

Kovats, R.S. and K.L. Ebi, 2006: Heat waves and public health in Europe. *European Journal of Public Health*, **16(6)**, 592-599.

Kovats, R.S., S.J. Edwards, D. Charron, J. Cowden, R.M. D'Souza, K.L. Ebi, C. Gauci, P.G Smidt, S. Hajat, S. Hales, G.H. Pezzi, B. Kriz, K. Kutsar, P. McKeown, K. Mellou, B. Menne, S. O'Brien, W. van Pelt, and H. Schmidt, 2005: Climate variability and campylobacter infection: an international study. *International Journal of Biometerology*, **49**, 207-214.

Kovats, R.S., S.J. Edwards, S. Hajat, B. Armstrong, K.L. Ebi, B. Menne, and The Collaborating Group, 2004a: The effect of temperature on food poisoning: a time-series analysis of salmonellosis in ten European countries. *Epidemiology and Infection*, **132**, 443-453.

Kovats, R.S., S. Hajat S, and P. Wilkinson, 2004b: Contrasting patterns of mortality and hospital admissions during hot weather and heat waves in Greater London, UK. *Occupational & Environmental Medicine*, **61(11)**, 893-898.

Kunkel, K.E., 2003: North American trends in extreme precipitation. *Natural Hazards*, **29**, 291-305.

Kunkel, K.E., R.J. Novak, R.L. Lampman, and W. Gu, 2006: Modeling the impact of variable climatic factors on the crossover of *Culex restauns* and *Culex pipiens* (Diptera: Culicidae), vectors of West Nile virus in Illinois. *American Journal of Tropical Medicine & Hygiene*, **74**, 168-173.

Künzli, N., M. Jerrett, W.J. Mack, B. Beckerman, L. LaBree, F. Gilliland, D. Thomas, J. Peters, and H.N. Hodis, 2005. Ambient air pollution and atherosclerosis in Los Angeles. *Environmental Health Perspectives*, **113**, 201-206.

Lachowsky, K. and R. Kovats, 2006: Estimating the burden of disease due to heat and cold under current and future climates. *Epidemiology*, **17(6)**, S50.

Lacy, R.W., 1993: Food-borne bacterial infections. *Parasitology*, **107**, S75-S93.

Laden, F., Schwartz, J., Speizer, F.E., Dockery, D.W. (2006). Reduction in fine particulate air pollution and mortality: extended. *American Journal of Respiratory and Critical Care Medicine*, **173**, 667-672.

Landsea, **C. W.**, 2005: Hurricanes and global warming. *Nature*, **438**, 11-12.

Lee, S.H., D.A. Levy, G.F. Craun, M.J. Beach, and R.L. Calderon, 2002: Surveillance for waterborne disease outbreaks—United States, 1999—2000. *MMWR—Morbidity & Mortality Weekly Report*, **51(08)**, 1-28.

Leung, R.L. and W.I. Gustafson Jr., 2005: Potential regional climate change and implications to U.S. air quality. *Geophysical Research Letters*, **32(16)**, L16711, doi: 10.1029/2005GL022911 .

Liang, J.L., E.J. Dziuban, G.F. Craun, V. Hill, M.R. Moore, R.J. Gelting, R.L. Calderon, M.J. Beach, and S.L. Roy, 2006: Surveillance for waterborne disease and outbreaks associated with drinking water and water not intended for drinking— United States, 2003-2004. *MMWR—Morbidity & Mortality Weekly Report*, **55(12)**, 32-65.

Lindgren, E., L. Talleklint, and T. Polfeldt, 2000: Impact of climatic change on the northern latitude limit and population density of the disease-transmitting European tick *Ixodes ricinus*. *Environmental Health Perspectives*, **108(2)**, 119-123.

Lipp, E.K., A. Huq, and R.R. Colwell, 2002: Effects of global climate in infectious disease: the cholera model. *Clinical Microbiology Reviews*, **15(4)**, 757-770.

Lipp, E.K., R. Kurz, R. Vincent, C. Rodriguez-Palacios, S.R. Farrah, and J.B. Rose, 2001a: The effects of seasonal variability and weather on microbial fecal pollution and enteric pathogens in a subtropical estuary. *Estuaries*, **24(2)**, 266-276.

Lipp, E.K., C. Rodriguez-Palacios, and J.B. Rose, 2001b: Occurrence and distribution of the human pathogen *Vibrio vulnificus* in a subtropical Gulf of Mexico estuary. *Hydrobiologia*, **460**, 165-173.

Lipp, E.K. and J.B. Rose, 1997: The role of seafood in food-borne diseases in the United States of America. *Revue Scientifique et Technique (Office International des Epizooties)*, **16(2)**, 620-640.

Lobitz, B., L. Beck, A. Huq, B. Wood, G. Fuchs, A.S.G. Faruque and R. Colwell, 2000: Climate and infectious disease: use of remote sensing for detection of *Vibrio cholerae* by indirect measurement. *Proceedings of the National Academy of Sciences*, **97**, 1438-1443.

Louis, V.R., I.A. Gillespie, S.J. O'Brien, E. Russek-Cohen, A.D. Pearson, and R.R. Colwell, 2005: Temperature-driven campylobacter seasonality in England and Wales. *Applied and Environmental Microbiology*, **71(1)**, 85-92.

Louis, V.R., E. Russek-Choen, N. Choopun, I.N. Rivera, B. Gangle, S.C. Jiang, A. Rubin, J.A. Patz, A. Hug, R.R. Colwell, 2003: Predictability of Vibrio cholerae in Chesapeake Bay. *Applied Environmental Microbiology*, **69**, 2773-2785.

Louisiana Department of Health and Hospitals (LDHH), 2006: *Vital Statistics of All Bodies at St. Gabriel Morgue*. 23 February 2006.

Lynch, M., J. Painter, R. Woodruff, and C. Braden, 2006: Surveillance for food-borne-disease outbreaks—United States, 1998-2002. *MMWR—Morbidity & Mortality Weekly Reports*, **55(10)**, 1-42.

Marciano-Cabral, F., R. MacLean, A. Mensah, and L. LaPat-Polasko, 2003: Identification of *Naegleria fowleri* in domestic water sources by nested PCR. *Applied and Environmental Microbiology*, **69(10)**, 5864-5869.

McCabe, G.J. and J.E. Bunnell, 2004: Precipitation and the occurrence of Lyme disease in the northeastern United States. *Vector Borne & Zoonotic Diseases*, **4(2)**, 143-148.

McConnell, R., K. Berhane, F. Gilliland, S.J. London, T. Islam, W. Gauderman, *et al.*, 2002: Asthma in exercising children exposed to ozone: a cohort study. *Lancet*, **359**, 386-391.

McGeehin, M.A. and M. Mirabelli, 2001: The potential impacts of climate variability and change on temperature-related morbidity and mortality in the United States. *Environmental Health Perspectives*, **109(2)**, 185-189.

McLaughlin, J.B., A. DePaola, C.A. Bopp, K.A. Martinek, N.P. Napolilli, C.G.Allison, S.L. Murray, E.C. Thompson, M.M. Bird, and J.P. Middaugh, 2005: Outbreaks of *Vibrio parahaemolyticus* gastroenteritis associated with Alaskan oysters. *New England Journal of Medicine*, **353(14)**, 1463-1470.

Mead, P.S., L. Slutsker, V. Dietz, L.F. McCaig, J.S. Bresee, C. Shapiro, P.M. Griffin, and R.V. Tauxe, 1999: Food-related illness and death in the United States. *Emerging Infectious Diseases*, **5(5)**, 607-625.

Medina-Ramon, M., A. Zanobetti, D.P. Cavanagh, and J. Schwartz, 2006: Extreme temperatures and mortality: assessing effect modification by personal characteristics and specific cause of death in a multi-city case-only analysis. *Environmental Health Perspectives*, **114(9)**, 1331-1336.

Meehl, G.A. and C. Tebaldi, 2004: More intense, more frequent, and longer lasting heat waves in the 21st century. *Science*, **305(5686)**, 994-997.

Meites, E., M.T. Jay, S. Deresinski, W.J. Shieh, S.R. Zaki, L. Tomkins, and D.S. Smith, 2004: Reemerging leptospirosis, California. *Emerging Infectious Diseases*, **10(3)**, 406-412.

Mickley, L.J., D.J. Jacob, B.D. Field, and D. Rind, 2004: Effects of future climate change on regional air pollution episodes in the United States. *Geophysical Research Letters*, **31**, L24103.

Middleton, K.L., J. Willner, and K. M. Simmons, 2002: Natural disasters and posttraumatic stress disoder symptom complex: evidence from the Oklahoma tornado outbreak. *International Journal of Stress Management*, **9(3)**, 229-236.

Miossec, L., F. Le Guyader, L. Haugarreau, and M. Pommepuy, 2000: Magnitude of rainfall on viral contamination of the marine environment during gastroenteritis epidemics in human coastal population. *Revue Epidemiologie Sante Publique*, **38(2)**, 62-71.

Mississippi Department of Health (MSDH), 2005: *Mississippi Vital Statistics 2005*. 14 February 2007.

Mohan, J.E., L.H. Ziska, W.H. Schlesinger, R.B. Thomas, R.C. Sicher, K. George and J.S. Clark, 2006: Biomass and toxicity responses of poison ivy (Toxicodendron radicans) to elevated atmospheric CO_2. *Proceedings of the National Academy of Sciences*, **103**, 9086-9089.

Morris, J.G., 2003: Cholera and other types of vibriosis: a story of human pandemics and oysters on the half shell. *Clinical Infectious Diseases*, **37**, 272-280.

Mounts, A.W., T. Ando, M. Koopmans, J.S. Breese, J. Noel and R.I. Glass, 2000: Cold weather seasonality of gastroenteritis associated with Norwalk-like Viruses. *Journal of Infectious Diseases*, **181(2)**, S284-S287.

Mouslin C., F. Hilber, H. Huang, F.A. Groisman FA. 2002. Conflicting needs for a Salmonella hypervirulence gene in host and non-host environments. *Molecular Microbiology*, **45**, 1019-27.

Murazaki, K., and P. Hess, 2006: How does climate change contribute to surface ozone change over the United States? *Journal of Geophysical Research*, **111**.

NAS Committee on Climate Ecosystems Infectious Disease and Human Health Board on Atmospheric Sciences and Climate and National Research Council (NRC), 2001: *Under the Weather: Climate, Ecosystems, and Infectious Disease.* National Academics Press, Washington, DC.

NationalAtlas.gov™, *Geographic Vulnerability of US Residents to Selected Climate Related Health Impacts* [maps]. 2008. Generated using NationalAtlas.gov™ Map Maker. http://nationalatlas.gov/natlas/Natlasstart.asp.

Naumova, E.N., J.S. Jjagai, B. Matyas, A. DeMaria, I.B. MacNeill, and J.K. Griffiths, 2006: Seasonality in six enterically transmitted diseases and ambient temperature. *Epidemiology and Infection*, 1-12.

New England Governors and Eastern Canadian Premiers (NEG/ECP), 2001: *Report to New England Governors and Eastern Canadian Premiers Climate Change Action Plan.* New England Governor's Conference Inc., Boston, Massachusetts.

Newel, D.G., 2002: The ecology of *Campylobacter jejuni* in avian and human hosts and in the environment. *International Journal of Infectious Diseases*, **6**, 16-21.

NOAA, 2005a. 65-year list of severe weather fatalities. Retrieved February 23, 2007, from http://www.weather.gov/os/severe_weather/65yrstats.pdf.

NOAA, 2005b. *NOAA Heat/Health Watch Warning System Improving Forecasts and Warnings for Excessive Heat.* NOAA Air Resources Laboratory. Retrieved March 4, 2005, from http://www.arl.noaa.gov/ss/transport/archives.html.

NOAA, 2006. *Galveston Storm of 1900.* Retrieved February 23, 2007, from http://www.noaa.gov/galveston1900

NOAA, 2007a. *67-Year List of Severe Weather Fatalities.* Retrieved October 18, 2007, from http://www.weather.gov/om/hazstats.shtml.

NOAA, 2007b. *Billion Dollar Climate and Weather Disasters 1980-2006.* Retrieved January 31, 2007, from www.ncdc.noaa.gov/oa/reports/billionz.html.

NOAA, 2007c. Climate models suggest warming-induced wind shear changes could impact hurricane development, intensity. Retrieved May 28, 2008, from www.noaanews.noaa.gov/stroies2007/s2840.htm.

North, C.S., A. Kawasaki, E.L. Spitznagel, and B.A. Hong, 2004: The course of PTSD, major depression, substance abuse, and somatization after a natural disaster. *The Journal of Nervous and Mental Disease*, **192(12)**, 823-829.

Ogden, N.H., A. Maarouf, I.K. Barker, M. Bigras-Poulin, L.R. Lindsay, M.G. Morshed, C.J. O'Callaghan, F. Ramay, D. Waltner-Toews, and D.F. Charron, 2006: Climate change and the potential for range expansion of the Lyme disease vector *Ixodes scapularis* in Canada. *International Journal for Parasitology*, **36(1)**, 63-70.

Lipp, E.K. and J.B. Rose, 1997: The role of seafood in food-borne diseases in the United States of America. *Revue Scientifique et Technique (Office International des Epizooties)*, **16(2)**, 620-640.

O'Neill, M.S., M. Jerrett, I. Kawachi, J.I. Levy, A.J. Cohen, N. Gouveia, *et al.*, 2003a: Health, wealth, and air pollution: advancing theory and methods. *Environmental Health Perspectives*, **111(16)**, 1861-1870.

O'Neill, M.S. 2003: Air conditioning and heat-related health effects. *Applied Environmental Science and Public Health*, **1(1)**, 9-12.

O'Neill, M.S., S. Hajat, A. Zanobetti, M. Ramirez-Aguilar, and J. Schwartz, 2005: Impact of control for air pollution and respiratory epidemics on the estimated associations of temperature and daily mortality. *International Journal of Biometeorology*.

O'Neill, M.S., A. Zanobetti, and J. Schwartz, 2003: Modifiers of the temperature and mortality association in seven US cities. *American Journal of Epidemiology*, **157(12)**, 1074-1082.

O'Neill, M.S., Zanobetti A, Schwartz J. 2005b: Disparities by race in heat-related mortality in four U.S. cities: the role of air conditioning prevalence. *Journal of Urban Health*, **82(2)**, 191-197.

Ostfeld, R.S., C.D Canham, K. Oggenfuss, R.J. Winchcombe, and F. Keesing, 2006: Climate, deer, rodents, and acorns as determinants of variation in Lyme disease risk. *PLoS Biology*, **4(6)**, e145.

Parkinson, A.J., J.C. Butler, 2005: Potential impacts of climate change on infectious diseases in the Arctic. *International Journal of Circumpolar Health*, **64**, 478-486.

Parry, M., C. Rosenzweig, and M. Livermore, 2005: Climate change, global food supply and risk of hunger. *Philosophical Transactions of the Royal Society B*, **360**, 2125-2138.

Patz, J.A., M.A. McGeehin, S.M. Bernard, K.L. Ebi, P.R. Epstein, A. Grambsch, D.J. Gubler, P. Reiter, I. Romieu, J.B. Rose, J.M. Samet, and J. Trtanj, 2000: The potential health impacts of climate variability and change for the United States: executive summary of the report of the health sector of the U.S. National Assessment. *Environmental Health Perspectives*, **108**, 367-376.

Pfeffer, C.S., M.F. Hite, and J.D. Oliver, 2003: Ecology of *Vibrio vulnificus* in estuarine waters of eastern North Carolina. *Applied and Environmental Microbiology*, **69(6)**, 3526-3531.

Piacentino, J.D. and B.S. Schwartz, 2002: Occupational risk of lyme disease: an epidemiological review. *Occupational and Environmental Medicine*, **59**, 75-84.

Pielke, Jr., R.A., C. Landsea, M. Mayfield, J. Laver, and R. Pasch, 2005: Hurricanes and global warming. *Bulletin of the American Meteorological Society*, 1571-1575.

Pinho, O.S. and M.D. Orgaz, 2000: The urban heat island in a small city in coastal Portugal. *International Journal of Biometeorology*, **44(4)**, 198-203.

Pope, C.A. III, M.J. Thun, M.M. Namboodiri, D.W. Dockery, J.S. Evans, F.E. Speizer, *et al.*, 1995: Particulate air pollution as a predictor of mortality in a prospective study of U.S. adults. *American Journal of Respiratory and Critical Care Medicine*, 151, 669-674.

Pope, C.A., III, R.T. Burnett, M.J. Thun, E.E. Calle, D. Krewski, K. Ito, and G.D. Thurston, 2002: Lung Cancer, Cardiopulmonary Mortality, and Long-Term Exposure to Fine Particulate Air Pollution. *Journal of the American Medical Association*, **287**, 1132-1141.

Pope, C.A. III, R.T. Burnett, G.D. Thurston, M.J. Thun, E.E. Calle, D. Krewski, J.J. Godleski, 2004: Cardiovascular mortality and long-term exposure to particulate air pollution: epidemiological evidence of general pathophysiological pathways of disease. *Circulation*, **109(1)**, 71-77.

Pope, C.A., D.W. Dockery, 2006: Health effects of fine particulate air pollution: lines that connect. *Journal of Air and Waste Management Association*,**54**, 709-742.

Powell, D. and B. Chapman, 2007: Fresh threat: what's lurking in your salad bowl? *Journal of the Science of Food and Agriculture*, **87**, 1799-1801.

Purse, B.V., P.S. Mellor, D.J. Rogers, A.R. Samuel, P.P. Mertens, and M. Baylis, 2005: Climate change and the recent emergence of bluetongue in Europe. *Nature Reviews Microbiology*, **3(2)**, 171-181.

Randa, M.A., M.F. Polz, and E. Lim, 2004: Effects of temperature and salinity on *Vibrio vulnificus* population dynamics as assessed by quantitative PCR. *Applied and Environmental Microbiology*, **70(9)**, 5469-5476.

Randolph, S.E., 2004a: Evidence that climate change has caused 'emergence' of tick-borne diseases in Europe? *International Journal of Medical Microbiology*, **293(37)**, 5-15.

Randolph, S.E. and D.J. Rogers, 2000: Fragile transmission cycles of tick-borne encephalitis virus may be disrupted by predicted climate change. *Proceedings of the Royal Society/ Biological Sciences*, **267(1454)**, 1741-1744.

Reisen, W.K., Y. Fang, and V.M. Martinez, 2006: Effects of temperature on the transmission of West Nile virus by Culex tarsalis (Diptera: Culicidae). *Journal of Medical Entomology*, **43(2)**, 309-17.

Reiter, P., 1996: Global warming and mosquito-borne disease in USA. *Lancet*, **348(9027)**, 622.

Reiter, P., S. Lathrop, M. Bunning, B. Biggerstaff, D. Singer, T. Tiwari, L. Baber, M. Amador, J. Thirion, J. Hayes, C. Seca, J. Mendez, B. Ramirez, J. Robinson, J. Rawlings, V. Vorndam, S. Waterman, D. Gubler, G. Clark, and E. Hayes, 2003: Texas lifestyle limits transmission of dengue virus. *Emerging Infectious Diseases*. **9(1)**, 86-89.

Ren, C., G.M. Williams, and S. Tong, 2006: Does particulate matter modify the association between temperature and cardiorespiratory diseases? *Environmental Health Perspectives*, **114(11)**, 1690-1696.

Rogers, C., P. Wayne, E. Macklin, M. Muilenberg, C. Wagner, P. Epstein and F. Bazzaz, 2006: Interaction of the onset of spring and elevated atmospheric CO2 on ragweed (Ambrosia artemisiifolia L.) pollen production. *Environmental Health Perspectives*, **114(6)**, 865-869. doi:10.1289/ehp.8549.

Rose, J.B., S. Daeschner, D.R. Easterling, F.C. Curriero, S. Lele, and J.A. Patz, 2000: Climate and waterborne disease outbreaks. *Journal of the American Water Works Association*, **92(9)**, 77-87.

Running, S.W., 2006: Is global warming causing more, larger wildfires? *Science*, **313**, 927-928.

Russoniello, C.V., T.K. Skalko, K. O'Brien, S.A. McGhee, D. Bingham-Alexander, and J. Beatley, 2002: Childhood posttraumatic stress disorder and efforts to cope after Hurricane Floyd. *Behavioral Medicine*, **28**, 61-71.

Rzezutka, A., and N. Cook, 2004: Survival of human enteric viruses in the environment and food. *FEMS Microbiology Reviews*, **28**, 441-453.

Samet, J.M., F. Domenici, F. Curriero, I. Coursac, and S.L. Zeger, 2000: Fine Particulate Air Pollution and Mortality in 20 U.S. Cities, 1987–1994. *New England Journal of Medicine*, **343**, 1742-1749.

Schwartz, B.S. and M.D. Goldstein, 1990: Lyme disease in outdoor workers: risk factors, preventive measures, and tick removal methods. *American Journal of Epidemiology*, **131(5)**, 877-885.

Schwartz, J., 1995: Short term fluctuations in air pollution and hospital admissions of the elderly for respiratory disease. *Thorax*, **50**, 531–538.

Schwartz, J., 2005: Who is sensitive to extremes of temperature? A case-only analysis. *Epidemiology*, **16(1)**, 67-72.

Schwartz, J., J.M. Samet, and J.A. Patz, 2004: Hospital admissions for heart disease: The effects of temperature and humidity. *Epidemiology*, **15(6)**, 755-761.

Seidell, J.C., 2000: Obesity, insulin resistance and diabetes—a worldwide epidemic. *British Journal of Nutrition*, **83(1)**, S5-8.

Semenza, J.C., J.E. McCullough, W.D. Flanders, M.A. McGeehin, and J.R. Lumpkin, 1999: Excess hospital admissions during the July 1995 heat wave in Chicago. *American Journal of Preventive Medicine*, **16(4)**, 269-277.

Semenza, J.C., C.H. Rubin, K.H. Falter, J.D. Selanikio, W.D. Flanders, H.L. Howe, *et al.*, 1996: Heat-related deaths during the July 1995 heat wave in Chicago. *New England Journal of Medicine*, **335(2)**, 84-90.

Senior, C.A., R.G. Jones, J.A. Lowe, C.F. Durman, and D. Hudson, 2002: Predictions of extreme precipitation and sea level rise under climate change. *Philosophical Transactions of the Royal Society of London*, **360(A)**, 1301-1311.

Setzer, C. and M.E. Domino, 2004: Medicaid outpatient utilization for waterborne pathogenic illness following Hurricane Floyd. *Public Health Reports*, **119**, 472-478.

Sheridan, S. and T. Dolney, 2003: Heat, mortality, and level of urbanization: measuring vulnerability across Ohio, USA. *Climate Research*, **24**, 255-266.

Sheridan, S.C., 2006: A survey of public perception and response to heat warnings across four North American cities: an evaluation of municipal effectiveness. *International Journal of Biometeorology*, **52**, 3-15.

Shone, S.M., F.C. Curriero, C.R. Lesser, and G.E. Glass, 2006: Characterizing population dynamics of *Aedes sollicitans* (Diptera: Culicidae) using meteorological data. *Journal of Medical Entomology*, **43(2)**, 393-402.

Sibold, J.S. and T.T. Veblen, 2006: Relationships of subalpine forest fires in the Colorado Front Range with interannual and multidecadal-scale climatic variation. *Journal of Biogeography*, **33**, 833-842.

Skelly, C. and P. Weinstein, 2003: Pathogen survival trajectories: an eco-environmental approach to the modeling of human campylobacteriosis ecology. *Environmental Health Perspectives*, **111(1)**, 19-28.

Southern, J.P., R.M. Smith and S.R. Palmer, 1990: Bird attack on milk bottles: possible mode of transmission of *Campylobacter jejuni* to man. *Lancet*, **336**, 1425-1427.

Srikantiah, P., J.C. Lay, S. Hand, J.A. Crump, J. Campbell, M.S. Van Duyne, R. Bishop, R. Middendor, M. Currier, and P.S. Mead, 2004: *Salmonella enterica* serotype Javiana infections associated with amphibian contact, Mississippi, 2001. *Epidemiology and Infection*, **132**, 273-281.

Stanley, K.N., J.S. Wallace, J.E. Currie, P.J. Diggle and K. Jones, 1998: The seasonal variation of thermophilic campylobacters in beef cattle, dairy cattle and calves. *Journal of Applied Microbiology*, **85(3)**, 472-480.

Steiner, A.L., S. Tonse, R.C. Cohen, A.H. Goldstein, and R.A. Harley, 2006: Influence of future climate and emissions on regional air quality in California. *Journal of Geophysical Research*, **111**.

Subak, S., 2003: Effects of climate on variability in Lyme disease incidence in the northeastern United States. *American Journal of Epidemiology*, **157(6)**, 531–538.

Tapsell, S.M., E.C. Penning-Rowsell, S.M. Tunstall, and T.L. Wilson, 2002: Vulnerability to flooding: health and social dimensions. *Philosophical Transactions of the Royal Society of London A*, 360, 1511-1525.

Thomas, M.K., D.F. Charron, D.Waltner-Toews, C. Schuster, A.R. Maarouf, and J.D. Holt, 2006: A role of high impact weather events in waterborne disease outbreaks in Canada, 1975—2001. *International Journal of Environmental Health Research*, **16(3)**, 167-180.

Thompson, J.R., M.A. Randa, L.A. Marcelino, A.Tomita-Mitchell, E. Lim, and M.F. Polz, 2004: Diversity and dynamics of a North Atlantic coastal *Vibrio* community. *Applied and Environmental Microbiology*, **70(7)**, 4103-4110.

Trenberth, K., 2005: Uncertainty in hurricanes and global warming. *Science*, **308**, 1753-1754.

U.S. Census Bureau, 2004: *U.S. Interim Projections by Age, Sex, Race, and Hispanic Origin: 2000-2050*. Retrieved September 12, 2007, from http://www.census.gov/ipc/www/usinterimproj/.

U.S. EPA, 2005: *Heat Island Effect*. U.S. Environmental Protection Agency.

U.S. EPA, 2006: *Associated project details for RFA: The impact of climate change & variability on human health (2005)*. U.S. Environmental Protection Agency.

U.S. Senate Committee on Homeland Security and Governmental Affairs (CHSGA), 2006: *Hurricane Katrina: A Nation Still Unprepared*. 109th Congress, 2nd Session, S. Rept. 109-322, Washington, DC.

Vereen, E., R.R. Lowrance, D.J. Cole, and E.K. Lipp, 2007: Distribution and ecology of campylobacters in coastal plain streams (Georgia, United States of America). *Applied and Environmental Microbiology*, **73(5)**, 1395-1403.

Verger, P., M. Rotily, C. Hunault, J. Brenot, E. Baruffol, and D. Bard, 2003: Assessment of exposure to a flood disaster in a mental-health study. *Journal of Exposure Analysis and Environmental Epidemiology*, **13**, 436-442.

Viboud, C., K. Pakdaman, P-Y. Boelle, M.L. Wilson, M.F. Myers, A.J. Valleron, and A. Flahault, 2004: Association of influenza epidemics with global climate variability. *European Journal of Epidemiology*, **19(11)**, 1055-1059.

Visscher, T.L. and J.C. Seidell, 2001: The public health impact of obesity. *Annual Review of Public Health*, **22**, 355-375.

Vose, R., T. Karl, D. Easterling, C. Williams, and M. Menne, 2004: Climate (communication arising): Impact of land-use change on climate. *Nature*, **427(6971)**, 213-214.

Vugia, D., A. Cronquist, J. Hadler, M. Tobin-D'Angelo, D. Blythe, K. Smith, et al., 2006: Preliminary FoodNet data on the incidence of infection with pathogens transmitted commonly through food—10 states, United States, 2005. *MMWR—Morbidity & Mortality Weekly Reports*, **55(14)**, 392-395.

Wade, T.J., S.K. Sandu, D. Levy, S. Lee, M.W. LeChevallier, L. Katz, and J.M. Colford, Jr., 2004: Did a severe flood in the Midwest cause an increase in the incidence of gastrointestinal symptoms? *American Journal of Epidemiology*, **159(4)**, 398-405.

Wan, S.Q., T. Yuan, S. Bowdish, L. Wallace, S.D. Russell and Y.Q. Luo, 2002: Response of an allergenic species Ambrosia psilostachya (Asteraceae), to experimental warming and clipping: implications for public health. *American Journal of Botany*, **89**, 1843-1846.

Watkins, S.J., D. Byrne, and M. McDevitt, 2001: Winter excess morbidity: is it a summer phenomenon? *Journal of Public Health Medicine*, **23(3)**, 237-241.

Wayne, P., S. Foster, J. Connolly, F. Bazzaz and P. Epstein, 2002: Production of allergenic pollen by ragweed (Ambrosia artemisiifolia L.) is increased in CO_2-enriched atmospheres. *Annuals of Allergy Asthma and Immunology*, **88**, 279-82.

Wegbreit, J. and W.K. Reisen, 2000: Relationships among weather, mosquito abundance, and encephalitis virus activity in California: Kern County 1990-98. *Journal of the American Mosquito Control Association*, **16(1)**, 22-27.

Weisler, R.H., J.G.I. Barbee, and M.H. Townsend, 2006: Mental health and recovery in the Gulf coast after hurricanes Katrina and Rita. *The Journal of the American Medical Association*, **296(5)**, 585-588.

Weisskopf, M.G., H.A. Anderson, S. Foldy, L.P. Hanrahan, K. Blair, T.J. Torok, *et al.*, 2002: Heat wave morbidity and mortality, Milwaukee, Wis, 1999 vs 1995: An improved response? *American Journal of Public Health*, **92(5)**, 830-833.

Wellings, F.M., P.T. Amuso, S.L. Chang, and A.L. Lewis, 1977: Isolation and identification of pathogenic *Naegleria* from Florida lakes. *Applied and Environmental Microbiology*, **34(6)**, 661-667.

Westerling, A.L., A. Gershunov, T.J. Brown, D.R. Cayan, and M.D. Dettinger, 2003: Climate and wildfire in the western United States. *Bulletin of the American Meteorological Society*, 595-604.

Westerling, A.L., H.G. Hidalgo, D. R. Cayan, and T. W. Swetnam, 2006: Warming and earlier spring increase western U.S. forest wildfire activity. *Science*, **313**, 940-943.

Wetz, J.J., E.K. Lipp, D.W. Griffin, J. Lukasik, D. Wait, M.D. Sobsey, T.M. Scott, and J.B. Rose, 2004: Presence, infectivity and stability of enteric viruses in seawater: relationship to marine water quality in the Florida Keys. *Marine Pollution Bulletin*, **48**, 700-706.

Whitman, S., G. Good, E.R. Donoghue, N. Benbow, W. Shou, and S. Mou, 1997: Mortality in Chicago attributed to the July 1995 heat wave. *American Journal of Public Health*, **87(9)**, 1515-1518.

Wilkinson, P., S. Pattenden, B. Armstrong, A. Fletcher, R.S. Kovats, P. Mangtani, *et al.*, 2004: Vulnerability to winter mortality in elderly people in Britain: population based study. *British Medical Journal*, **329(7467)**, 647.

Woodruff, R.E., S. Hales, C.D. Butler, and A.J. McMichael, 2005: *Climate change health impacts in Australia: Effects of dramatic CO2 emissions reductions*. Report for the Australian Conservation Foundation and the Australian Medical Association, 45 pp.

World Health Organization, 1999: Leptospirosis worldwide, 1999. *Weekly Epidemiological Record*, **74**, 237-244.

Xu, H.Q. and B.Q. Chen, 2004: Remote sensing of the urban heat island and its changes in Xiamen City of SE China. *Journal of Environmental Sciences*, **16(2)**, 276-281.

Zender, C.S. and J. Talamantes, 2006: Climate controls on valley fever incidence in Kern County, California. *International Journal of Biometeorology*, **50**, 174-82.

Zhuang, R-Y., L.R. Beuchat, and F.J. Angulo, 1995: Fate of *Salmonella* Montevideo on and in raw tomatoes as affected by temperature and treatment with chlorine. *Applied and Environmental Microbiology*, **61(6)**, 2127-2131.

Ziska, L.H., S.D. Emche, E.L. Johnson, K. George, D.R. Reed, and R.C. Sicher, 2005: Alterations in the production and concentration of selected alkaloids as a function of rising atmospheric carbon dioxide and air temperature: implications for ethno-pharmacology. *Global Change Biology*, **11**, 1798-1807.

Effects of Global Change on Human Settlements

Lead Author: Thomas J. Wilbanks, Oak Ridge National Laboratory

Contributing Authors: Paul Kirshen, Tufts University; Dale Quattrochi, NASA/Marshall Space Flight Center; Patricia Romero-Lankao, NCAR; Cynthia Rosenzweig, NASA/Goddard; Matthias Ruth, University of Maryland; William Solecki, Hunter College; Joel Tarr, Carnegie Mellon University

Contributors: Peter Larsen, University of Alaska-Anchorage; Brian Stone, Georgia Tech

3.1 INTRODUCTION

3.1.1 Purpose

Human settlements are where people live and work, including all population centers ranging from small rural communities to densely developed metropolitan areas. This chapter addresses climate change impacts, both positive and negative, on human settlements in the United States. First, the chapter summarizes current knowledge about the vulnerability of human settlements to climate change, in a context of concurrent changes in other non-climate factors. Next, the chapter summarizes opportunities within settlements for adaptation to climate change. Finally, the chapter provides an overview of recommendations for expanding the current knowledge base with respect to climate change and human settlements.

3.1.2 Background

Events such as Hurricane Katrina in 2005 and electric power outages during the hot summer of 2006 have demonstrated how climate-related events can dramatically impact U.S. settlements. Climate affects the costs of assuring comfort at home and work. Climate affects inputs for a good life: water; products and services from agriculture and forestry; pleasures and tourist potentials from nature, biodiversity, and outdoor recreation. Climate also affects the presence and spread of diseases and other health problems, and it is associated with threats from natural disasters, including floods, fires, droughts, wind, hail, ice, and heat and cold waves.

Some U.S. settlements may find opportunities in climate change. Warmer winters are not necessarily undesirable. Periods of change tend to reward forward-looking, effectively governed communities. Considering climate change effects may help to focus attention on other important issues for the long-term sustainable development of settlements and communities. Furthermore, planning for the future is an essential part of public policy decision-making in urban areas.

Since infrastructure investments in urban areas are often both large and difficult to reverse, climate considerations are increasingly perceived as one of a number of relevant issues to consider when planning for the future (Ruth, 2006a). If U.S. settlements, especially larger cities, respond effectively to climate change concerns, their actions could have far-reaching implications for human well-being, because these areas are where most of the U.S. population lives, large financial decisions are made, political influence is often centered, and technological and social innovations take place.

Meanwhile, the pattern of human settlements in the United States is changing. In addition to shifts of population from frost-belt to sun-belt settlements, patterns are changing in other ways as well. For instance, the trend of households moving from urban centers to peripheries is reversing as many city centers renew and metropolitan areas continue to expand across multiple jurisdictions (Solecki and Leichenko, 2006). Modern information technologies are enabling people to perform what were historically urban functions from relatively remote locations (Riebsame, 1997).

3.1.3 Current State of Knowledge

The current knowledge base provides limited grounds for developing conclusions and recommendations related to climate impacts on human settlements. In many cases, the best that can be done is to sketch out the issue "landscape" that should be considered by both policy-makers and the research community as a basis for further discussions and offer illustrations from the relatively limited research literature that is now available.

The fact is that little research has been done to date specifically on the effects of climate change in U.S. cities and towns. Reasons appear to include (i) limitations in capacities to project climate change impacts at the geographic scale of a metropolitan area (or smaller) and (ii) the fact that none of the federal agencies currently active in climate science research has a clear responsibility for settlement impact issues. Improvements are required in our understanding of the impacts of and adaptation to climate change across different sectors and geographic regions, differential vulnerabilities, and interventions to build resilience. (NRC, 2007).

To some degree, gaps can be filled by referring to several comprehensive analyses that do exist, including literature on effects of climate variation on settlements and their responses, research on climate change impacts on cities in other parts of the world, and historical analogs of responses of urban areas to significant environmental changes. Box 3.1 presents a historical perspective of U.S. urban responses to environmental change. This perspective examines how American cities have been affected by environmental change over the past two centuries. But this is little more than a place to start.

At the current state of knowledge, vulnerabilities to possible impacts are easier to project than actual impacts because they estimate risks or opportunities associated with possible consequences rather than estimating the consequences themselves, which requires far more detailed information about future conditions. Vulnerabilities are shaped not only by existing exposures, sensitivities, and adaptive capacities but also by the ability of settlements to develop responses to risks.

3.2 CLIMATE CHANGE IMPACTS AND THE VULNERABILITIES OF HUMAN SETTLEMENTS

This section examines possible impacts of climate change on settlements in the United States including the determinants of vulnerability to such impacts and how those impacts could affect settlement patterns and various systems related to those patterns.

3.2.1 Determinants of Vulnerability

It has been difficult to project impacts of climate change on human settlements in the United States, in part because climate change forecasts are not specific enough for the scale of decision-making, but more so because climate change is not the only change being confronted by settlements. More often, attention is paid to vulnerabilities to climate change, if those changes should occur.

Vulnerabilities to or opportunities from climate change are related to three factors, both in absolute terms and in comparison to other elements (Clark *et al.*, 2000):

1. *Exposure to climate change.* To what climate changes are settlements likely to be exposed: Changes in temperature or precipitation? Changes in storm exposures and/or intensities? Changes in sea level?

BOX 3.1. U.S. Urban Responses to Environmental Change: An Historical Perspective

Over time, American cities have been affected by environmental change. City founders often showed an important disregard with respect to siting of settlements, focusing on aspects of location such as commercial or recreational opportunities rather than on risks such as flood potential, limited water, food or fuel supplies, or the presence of health threats. Oftentimes settlers severely exploited their environments, polluting ground water and adjacent water bodies, building in unsafe and fragile locations, changing landforms, and filling in wetlands. Construction of the urban built environment involved vast alterations in the landscape, as forests and vegetation and wildlife species were eliminated and replaced by highways, suburbs, and commercial buildings. The building of wastewater and water supply systems had the effect of altering regional hydrology and creating large vulnerabilities. In other cases settlers concluded that the weather was changing for the good, that technology would solve problems, or that new resources could be discovered.

Technological fixes were pursued to seek ways to modify or control environmental change. Cities exposed to flooding built levees and seawalls and channelized rivers. When urbanites depleted and polluted local water supplies, cities went outside their boundaries to seek new supplies: building reservoirs, aqueducts, and creating protected watersheds. When urban consumption exhausted local fuel sources, cities adapted to new fuels, embraced new technologies, or searched far beyond city boundaries for new supplies. Many of these actions resulted in the extension of the urban ecological footprint, so that urban growth and development affected not only the urban site but also increasingly the urban hinterland and beyond.

There are few examples of environmental disasters or climate change actually resulting in the abandonment of an urban site. One case appears to be that of the Hohokam Indians of the Southwest, who built extensive irrigation systems, farmed land, and built large and dense settlements over a period of approximately 1,500 years (Krech, 1999: 45-72). Yet, they abandoned their settlements and disappeared into history. The most prominent explanation for their disappearance is an ecological one—that the Hohokam irrigation systems suffered from salinization and water logging, eventually making them unusable. Other factors besides ecological ones may have also entered into the demise of their civilization and abandonment of their cities, but the ecological explanation appears to have the most supporters.

In the case of America in the 19th and 20th centuries, however, no city has been abandoned because of environmental or climatic factors. Galveston, Texas suffered from a catastrophic tidal wave but still exists as a human settlement, now protected by an extensive sea wall. Johnstown, Pennsylvania has undergone major and destructive flooding since the late 19th century, but continues to survive as a small city. Los Angeles and San Francisco are extremely vulnerable to earthquakes, but still continue to increase in population. And, in coming years New Orleans almost certainly will experience a hurricane as or more severe than Katrina, and yet rebuilding goes on, encouraged by the belief that technology will protect it in the future. Whether or not ecological disaster or extreme risk will eventually convince Americans to abandon some of their settlements, as the Hohokam did, has yet to be determined (Colten, 2005; Steinberg, 2006; Vale and Campanella, 2005).

2. *Sensitivity to climate change.* If primary climate changes occur, how sensitive are the activities and populations of a settlement to those changes? For instance, a city dependent substantially on a regional agricultural or forestry economy, or the availability of abundant water resources, might be considered more sensitive than a city whose economy is based mainly on an industrial sector less sensitive to climate variation.

3. *Adaptive capacity.* Finally, if effects are experienced due to a combination of exposure and sensitivity, how able is a settlement to handle those impacts without disabling damages, perhaps even while realizing new opportunities?

3.2.2 Impacts of Climate Change on Human Settlements

Impacts of climate change on human settlements vary regionally (see Boxes 3.2 and 3.3), and generally relate to some of the following issues:

1. *Effects on health.* It is well-established that higher temperatures in urban areas are related to higher levels of ozone, which cause respiratory and cardiovascular problems. There is also some evidence that combined effects of heat stress and air pollution may be greater than simple additive effects (Patz and Balbus, 2001). Moreover, historical data show relationships between mortality and temperature extremes (Rozenzweig and Solecki, 2001a). Other health concerns include changes in exposure to water and food-borne diseases, vector-borne diseases, concentrations of plant species associated with allergies, and exposures to extreme weather events such as storms, floods, and fires (see Chapter 2).

2. *Effects on water and other urban infrastructures.* Changes in precipitation patterns may lead to reductions in meltwater, river flows, groundwater levels, and in coastal areas may lead to saline intrusion in rivers and groundwater, affecting water supply. Meanwhile, warming may increase water demands (Gleick *et al.*, 2000; Kirshen, 2002; Ruth *et al.*, 2007). Moreover, storms, floods, and other severe weather events may affect other infrastructure, including sanitation systems, transportation, supply lines for food and energy, and communication. Exposed structures such as bridges and electricity transmission networks are especially vulnerable. In many cases, infrastructures are interconnected; an impact on one can also affect others (Kirshen, *et al.*, 2007). An example is an interruption in energy supply, which increases heat stress for vulnerable populations (Ruth *et al.*, 2006a). Many of the infrastructures in older cities are aging and are already under stress from increasing demands.

3. *Effects on energy requirements.* Warming is virtually certain to increase energy demand in U.S. cities for cooling in buildings while reducing demand for heating in buildings

(see SAP 4.5). Demands for cooling during warm periods could jeopardize the reliability of service in some regions by exceeding the supply capacity, especially during periods of unusually high temperatures (see Vignettes in Boxes 3.2 and 3.3). Higher temperatures also affect costs of living and business operation by increasing costs of climate control in buildings (Amato *et al.*, 2005; Ruth and Lin, 2006c; Kirshen *et al.*, 2007).

4. *Effects on the urban metabolism.* An urban area is a living complex mega-organism, associated with a host of inputs, transformations, and outputs: heat, energy, materials, and others (Decker *et al.*, 2000). An example is the Urban Heat Index, which measures the degree to which built/paved areas are associated with higher temperatures relative to surrounding areas (see Box 3.4: Climate Change Impacts on the Urban Heat Island Effect (UHI)). Imbalances in the urban metabolism can aggravate climate change impacts, such as roles of UHI in the formation of smog in cities. The maps in this box demonstrate how the built environment creates and retains heat in metropolitan settings.

5. *Effects on economic competitiveness, opportunities, and risks.* Climate change has the potential not only to affect settlements directly but also to affect them through impacts on other areas linked to their economies at regional, national, and international scales (Rosenzweig and Solecki, 2006). In addition, it can affect a settlement's economic base if it is sensitive to climate, as in areas where settlements are based on agriculture, forestry, water resources, or tourism (IPCC, 2001a).

6. *Effects on social and political structures.* Climate change can add to stress on social and political structures by increasing management and budget requirements for public services such as public health care, disaster risk reduction, and even public security. As sources of stress grow and combine, the resilience of social and political structures that are already somewhat unstable is likely to suffer, especially in areas with relatively limited resources (Sherbinin *et al.*, 2006).

BOX 3.2. Vignettes of Vulnerability—I

Alaskan Settlements

No other region in the United States is likely to be as profoundly changed by climate change as Alaska, our nation's part of the polar region of Earth (ACIA, 2004). Because warming is more pronounced closer to the poles, and because settlement and economic activities in Alaska have been shaped and often constrained by Arctic conditions, in this region warming is especially likely to reshape patterns of human settlement.

Human settlements in Alaska are already being exposed to impacts from global warming (ACIA, 2004), and these impacts are expected to increase. Many coastal communities see increasing exposure to storms, with significant coastal erosion, and in some cases facilities are being forced either to relocate or to face increasing risks and costs. Thawing ground is beginning to destabilize transportation, buildings, and other facilities, posing needs for rebuilding, with ongoing warming adding to construction and maintenance costs. And indigenous communities are facing major economic and cultural impacts. One recent estimate of the value of Alaska's public infrastructure at risk from climate change set the value at tens of billions of today's dollars by 2080, with the replacement of buildings, bridges, and other structures with long lifetimes having the largest public costs (Larsen *et al.*, 2007).

Besides impacts on built infrastructures designed for permafrost foundations and effects on indigenous societies, many observers expect warming in Alaska to stimulate more active oil and gas development (and perhaps other natural resource exploitation), and if thawing of Arctic ice permits the opening of a year-round Northwest sea passage it is virtually certain that Alaska's coast will see a boom in settlements and port facilities (ACIA, 2004).

Coastal Southeast Settlements

While there is currently no evidence for a long-term increase in North American mainland land-falling hurricanes, concerns remain that certain aspects of hurricanes, such as wind speed and rainfall rates may increase (CCSP, 2008). In addition, sea level rise is expected to increase storm surge levels (CCSP, 2008). Recent hurricanes striking the coast of the U.S. Southeast cannot be attributed clearly to climate change, but they suggest a range of possible impacts. As an extreme case, consider the example of Hurricane Katrina. In 2005, the city of New Orleans had a population of about half a million, located on the delta of the Mississippi River along the U.S. Gulf Coast. Urban development throughout the 20th Century has significantly increased land use and settlement in areas vulnerable to flooding, and a number of studies had indicated growing vulnerabilities to storms and flooding. In late August 2005, Hurricane Katrina moved onto the Louisiana and Mississippi coast with a storm surge, supplemented by waves, reaching up to 8.5 m above sea level. In New Orleans, the surge reached around 5 m, overtopping and breaching sections of the city's 4.5 m defenses, flooding 70 to 80 percent of New Orleans, with 55 percent of the city's properties inundated by more than 1.2 m and maximum flood depths up to 6 m. Approximately 1,101 people died in Louisiana, nearly all related to flooding, concentrated among the poor and elderly.

Across the whole region, there were 1.75 million private insurance claims, costing in excess of $40 billion (Hartwig, 2006), while total economic costs are projected to be significantly in excess of $100 billion. Katrina also exhausted the federally backed National Flood Insurance Program (Hunter, 2006), which had to borrow $20.8 billion from the Government to fund the Katrina residential flood claims. In New Orleans alone, while flooding of residential structures caused $8-$10 billion in losses, $3-6 billion was uninsured. 34,000-35,000 of the flooded homes carried no flood insurance, including many that were not in a designated flood risk zone (Hartwig, 2006). Six months after Katrina, it was estimated that the population of New Orleans was 155,000, with the number projected to rise to 272,000 by September 2008 – 56 percent of its pre-Katrina level *(McCarthy et al.*, 2006).

BOX 3.3. Vignettes of Vulnerability—II

Arid Western Settlements

Human settlements in the arid West are affected by climate in a variety of ways, but perhaps most of all by water scarcity and risks of fire. Clearly, access to water for urban populations is sensitive to climate, although the region has developed a vast system of engineered water storage and transport facilities, associated with a very complex set of water rights laws (NACC, 2001). It is very likely that climate change will reduce winter snowfall in the West, reducing total runoff – increasing spring runoff while decreasing summer water flows. Meanwhile, water demands for urban populations, agriculture, and power supply are expected to increase, and conflicts over water rights are likely to increase. If total precipitation decreases or becomes more variable, extending the kinds of drought that have affected much of the interior West in recent years, water scarcity will be exacerbated, and increased water withdrawals from wells could affect aquifer levels and pumping costs. Moreover, drying increases risks of fire, which has threatened urban areas in California and other Western areas in recent years. The five-year average of acres burned in the West is more than 5 million, and urban expansion is increasing the length of the urban-wild lands interface (Morehouse et al., 2006). Drying would lengthen the fire season, and pest outbreaks such as the pine beetle could affect the scale of fires.

Summer 2006 Heat Wave

In July and August 2006, a severe heat wave spread across the United States, with most parts of the country recording temperatures well above the average for that time of the year. For example, temperatures in California were extraordinarily high, setting records as high as 130°. As many as 225 deaths were reported by press sources, many of them in major cities such as New York and Chicago. Electric power transformers failed in several areas, such as St. Louis and Queens, New York, causing interruptions of electric power supply, and some cities reported heat-related damages to water lines and roads. In many cities, citizens without home air-conditioning sought shelter in public and office buildings, and city/county health departments expressed particular concern for the elderly, the young, pregnant women, and individuals in poor health. Although this heat wave cannot be attributed directly to climate change, it suggests a number of issues for human settlements in the United States as they contemplate a prospect of temperature extremes in the future that are higher and/or longer-lasting than historical experience.

7. *Effects on vulnerable populations* (see Chapter 1). Where climate change stresses settlements, it is likely to be especially problematic for vulnerable parts of the population: the poor, the elderly, those already in poor health, the disabled, those living alone, those with limited rights and power (*e.g.*, recent in-migrants with limited English skills), and/or indigenous populations dependent on one or a few resources. As one example, warmer temperatures in urban summers have a more direct impact on populations who live and work without air-conditioning. Implications for environmental justice are clear; see, for instance, Congressional Black Caucus Foundation, 2004.

8. *Effects on vulnerable regions.* Approximately half of the U.S. population, 160 million people, will live in one of 673 coastal counties by 2008 (Crossett *et al.*, 2004). Obviously, settlements in coastal areas—particularly on gently sloping coasts—should be concerned about sea level rise in the longer term, especially if they are subject to severe storms and storm surges and/or if their regions are showing gradual land subsidence (Neumann *et al.*, 2000; Kirshen *et al.*, 2004). Settlements in risk-prone regions have reason to be concerned about severe weather events, ranging from severe storms combined with sea level rise in coastal areas to increased risks of fire in drier arid areas. Vulnerabilities may be especially great for rapidly growing and/or larger metropolitan areas, where the potential magnitude of both impacts and coping requirements could be very large (IPCC, 2001a; Wilbanks *et al.*, 2007b).

BOX 3.4. Climate Change Impacts on the Urban Heat Island Effect (UHI)
(Lo and Quattrochi, 2003; Brazel and Quattrochi, 2006; Ridd, 2006; Stone, 2006)

Climate change impacts on the UHI will primarily depend upon the geographic location of a specific city, its urban morphology (*i.e.*, landscape and built-up characteristics), and areal extent (*i.e.*, overall spatial "footprint"). These factors will mitigate or exacerbate how the UHI phenomenon (Figure 3.1) is affected by climate change, but overall, climate change is likely to impact the UHI in the following ways:

- Exacerbation of the intensity and areal extent of the UHI as a result of warmer surface and air temperatures along with the overall growth of urban areas around the world. Additionally, as urban areas grow and expand, there is a propensity for lower albedos, which forces a more intense UHI effect. (There is also some indication that sustained or prolonged higher nighttime air temperatures over cities that may result from warmer global temperatures will have a more significant impact on humans than higher daytime temperatures.)

- As the UHI intensifies and increases, there could be a subsequent impact on deterioration of air quality, particularly on ground level ozone caused by higher overall air temperatures and an increased background effect produced by the UHI as an additive air temperature factor that helps to elevate ground level ozone production. Additionally, particulate matter ($PM_{2.5}$) could increase due to a number of human induced and natural factors (e.g., more energy production to support higher usage of air conditioning).

- The UHI has an impact on local meteorological conditions by forcing rainfall production either over, or downwind, of cities. As the UHI intensifies, there will be a higher probability for urban-induced rainfall production (dependent upon geographic location) with a subsequent increase in urban runoff and flash flooding.

- Exacerbation and intensification of the UHI would have the following impacts on human health:

  - increased incidence of heat stress

  - impact on respiratory illnesses such as asthma due to increases in particulate matter caused by deterioration in air quality as well as increased pollination production because of earlier pollen production from vegetation in response to warmer overall temperatures

The image on the left illustrates daytime surface heating for urban surfaces across the Georgia Central Business District (CBD). White and red colors indicate very warm surfaces (~40-50°C).

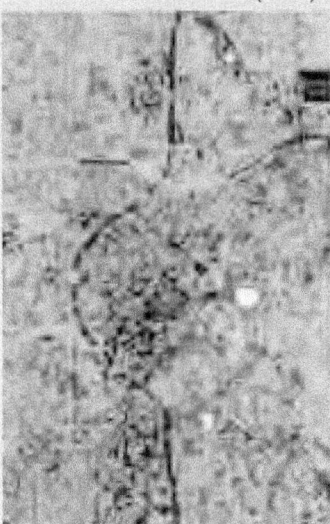 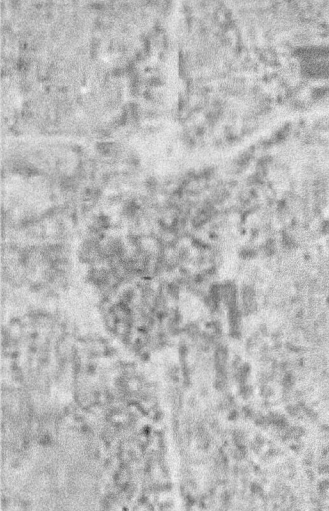

Green relates to surfaces of moderately warm temperatures (~25-30°C). Blue indicates cool surfaces (e.g., vegetation, shadows) (~15-20°C). Surface temperatures are reflected in the albedo image on the right where warm surfaces are dark (*i.e.*, low reflectivity) and cooler surfaces are in red and green (*i.e.*, higher reflectivity). The images exemplify how urban surface characteristics influence temperature and albedo as drivers of the UHI (Quattrochi *et al.*, 2000).

Figure 3.1. Example of urban surface temperatures and albedo for the Atlanta, Georgia Central Business District area derived from high spatial resolution (10m) aircraft thermal remote sensing data.

Different combinations of circumstances are likely to cause particular concerns for cities and towns in the United States as they consider possible implications of climate change.

3.2.3 The Interaction of Climate Impacts with Non-Climate Factors

In general, climate change effects on human settlements in the United States are imbedded in a variety of complexities that make projections of quantitative impacts over long periods of time very difficult. For instance, looking out over a period of many decades, it seems likely that other kinds of change—such as technological, economic, and institutional—will have more impact on the sustainability of most settlements rather than climate change per se (Wilbanks, *et al.*, 2007b). Climate change will interact with other processes, driving forces, and stresses; and its significance, positive or negative, will largely be determined by these interactions. It is therefore difficult to assess effects of climate change without a reasonably clear picture of future scenarios for these other processes.

In many cases, these interactions involve not only direct impacts such as warming or more or less precipitation but, sometimes more important, second, third, or higher order impacts, as direct impacts cascade through urban systems and other settlement-determined processes (*e.g.*, warming which affects urban air pollution which affects health which affects public service requirements which affect social harmony: Kirshen *et al.*, 2007). Some of

these higher order impacts, in turn, may feed back to create ripple effects of their own. For example, a heat wave may trigger increased energy demands for cooling, which may cause more air conditioners and power generators to be operated, which could lead to higher UHI effects, inducing even higher cooling needs.

Besides this "multi-stress" perspective, it is highly likely that effects of climate change on settlements are shaped by certain "thresholds," below which effects are incidental but beyond where effects quickly become major when a limiting or inflection point is reached. An example might be a city's capacity to cope with sustained heat stress combined with a natural disaster. In general, these climate-related thresholds for human settlements in the United States are not well-understood. For multi-stress assessments of thresholds, changes in climate **extremes** are very often of more concern than changes in climate **averages**. Besides extreme weather events, such as hurricanes or tornadoes, ice storms, winds, heat waves, drought, or fire, settlements may be affected by changes in daily or seasonal high or low levels of temperature or precipitation, which have not always been projected by climate change models.

Finally, human settlements may be affected by climate change mitigation initiatives as well as by climate change itself. Examples include effects on policies related to energy sources and uses, environmental emissions, and land use. The most direct and short-term effects would likely be on settlements in regions whose economies are closely related to the production and consumption of large quantities of fossil fuels. Indirect and longer term effects are less predictable.

As climate change affects settlements in the United States, impacts are realized at the intersection of climate change with underlying forces. Most of the possible effects are linked with changes in regional comparative advantage, with consequent migration of population and economic activities (Ruth and Coelho, in press). Examples of these complex interactions and issues include:

1. *Regional risks and availability of insurance.* It is possible that regions exposed to risks from climate change will see movement of population and economic activity to other locations. One reason is public perceptions of risk, but a more powerful driving force may be the availability of insurance. The insurance sector is one of the most adaptable of all economic sectors, and its exposure to costs from severe storms and other extreme weather events is likely to lead it to withdraw (or to make much more expensive) private insurance coverage from areas vulnerable to climate change impacts (Wilbanks, *et al.*, 2007b), which would encourage both businesses and individual citizens to consider other locations over a period of several decades.

2. *Areas whose economies are linked with climate-sensitive resources or assets.* Settlements whose economic bases are related to such sectors as agriculture, forestry, tourism, water availability, or other climate-related activities could be affected either positively or negatively by climate change, depending partly on the adaptability of those sectors (*i.e.*, their ability to adapt to changes without shifting to different locations).

3. *Shifts in comparative living costs, risks, and amenities.* Related to a range of possible climate change effects—higher costs for space cooling in warmer areas, higher costs of water availability in drier areas, more or less exposure to storm impacts in some areas, and sea level rise—regions of the United States and their associated settlements are likely to see gradual changes over the long term in their relative attractiveness for a variety of human activities. One example, although its likelihood is highly uncertain, would be a gradual migration of the "Sun Belt" northward, as retirees and businesses attracted by environmental amenities find that regions less exposed to very high temperatures and seasonal major storms are more attractive as places to locate.

4. *Changes in regional comparative advantage related to shifts in energy resource use.* If climate mitigation policies result in shifts from coal and other fossil resources toward non-fossil energy sources, or if climate changes affect the prospects of renewable energy sources (especially hydropower), regional economies related to the production and/or use of energy from these sources could be affected, along with regional economies more closely linked with alternatives (Wilbanks, 2007c)

5. *Urban "footprints" on other areas.* Resource requirements for urban areas involve larger areas than their own bounded territories alone. Ecologists have sought to estimate the land area required to supply the consumption of resources and compensate for emissions and other wastes from urban areas (*e.g.*, Folke *et al.*, 1997). By possibly affecting settlements, along with their resource capacities for their inputs and destinations of their outputs, climate change could affect the nature, size, and geographic distribution of these footprints.

Human settlements are foci for many economic, social, and governmental processes, and historical experience has shown that catastrophes in cities can have significant economic, financial, and political effects much more broadly. The case that has received the most attention to date is insurance and finance (Wilbanks, *et al.*, 2007b).

3.2.4 Realizing Opportunities from Climate Change in the United States

Climate change can have positive as well as negative implications for settlements. Examples of potential positive effects include:

1. *Reduced winter weather costs and stresses.* Warmer temperatures in periods of the year that are normally cold are not necessarily undesirable. They reduce cold-related stresses and costs (*e.g.*, costs of warming buildings and costs of clearing ice and snow from roads and streets), particularly for cold-vulnerable populations. They expand opportunities for warmer-weather recreational opportunities over larger parts of the year, and they expand growing seasons for crops, parks, and gardens.

2. *Increased attention to long-term sustainability.* One of the most positive aspects of climate change can be its capacity to stimulate a broader discussion of what sustainability means for settlements (Wilbanks, 2003; Ruth, 2006). Even if climate change itself may not be the most serious threat to sustainability, considering climate change impacts in a multi-change, multi-stage context can encourage and facilitate processes that lead to progress in dealing with other sources of stress.

3. *Improved competitiveness compared with settlements subject to more serious adverse impacts.* While some settlements may turn out to be "losers" due to climate change impacts, others may be "winners," as changes in temperature or precipitation result in added economic opportunities (see the following section), at least if climate change is not severe. In addition, for many settlements climate change can be an opportunity not only to compare their net impacts with others, seeking advantages as a result, but to present a progressive image by taking climate change (and related sustainability issues) seriously.

3.2.5 Examples of Impacts on Metropolitan Areas in the United States

Possible impacts of climate change on settlements in the United States are usually assessed by projecting climate changes at a regional scale: temperature, precipitation, severe weather events, and sea level rise (see Table 3.2 and Boxes 3.2 and 3.3). Ideally, these regional projections are at a relatively detailed scale, and ideally they consider seasonal as well as annual changes and changes in extremes as well as in averages; but these conditions cannot always be met.

The most comprehensive assessments of possible climate change impacts on settlements in the United States have been two studies of major metropolitan areas:

1. New York: This assessment concluded that impacts of climate change on this metropolitan area are likely to be primarily negative over the long term, with potentially significant costs increasing as the magnitude of climate change increases, although there are substantial uncertainties (Rosenzweig and Solecki, 2001a; Rosenzweig and Solecki, 2001b; Solecki and Rosenzweig, 2006).

2. Boston: This assessment concluded that long-term impacts of climate change are likely to depend at least as much on behavioral and policy changes over this period as on temperature and other climate changes (Kirshen *et al.*, 2004; Kirshen *et al.*, 2006; Kirshen *et al.*, 2007).

Other U.S. studies include Seattle (Hoo and Sumitani, 2005) and Los Angeles (Koteen *et al.*, 2001) (Table 3.1). Internationally, studies have included several major metropolitan areas, such as London (London Climate Change Partnership, 2004) and Mexico City (Molina *et al.*, 2005) as well as possible impacts on smaller settlements (*e.g.*, AIACC: see www.aiaccproject.org). A relevant historical study of effects of an urban heat wave in the United States is reported by Klinenberg (2003).

Table 3.1. Overview of Integrated Assessments of Climate Impacts and Adaptation in U.S. Cities. "X" Indicates that the Reference Addresses a Category of Interest.

	Bloomfield et al., 1999	Kooten et al., 2001	Rosenzweig et al., 2000	Kirshen et al., 2004	Hoo and Sumitani, 2005
Location:	Greater Los Angeles	New York	Metropolitan New York	Metropolitan Boston	Metropolitan Seattle
Coverage:					
Water Supply	X	X	X	X	
Water Quality				X	
Water Demand				X	
Sea level rise	X		X	X	X
Transportation				X	X
Communication					
Energy			X	X	
Public Health					
Vector-borne Diseases					
Food-borne Diseases		X			
Temperature-related Mortality				X	
Temperature-related Morbidity	X	X			
Air-quality Related Mortality					
Air-quality Related Morbidity			X		
Other Health Issues	X	X	X		
Ecosystems					
Wetlands					
Other Ecol. (Wildfires)	X		X		
Urban Forests (Trees and Vegetation)		X			
Air Quality		X			X
Extent of:					
Quantitative Analysis	Low	Medium	Medium	High	Low
Computer-based Modeling	None	Low	Low	High	None
Scenario Analysis	None	None	Medium	High	Medium
Explicit Risk Analysis	None	None	None	Medium	None
Involvement of:					
Local Planning Agencies	None	None	High	High	High
Local Government Agencies	None	None	High	High	High
Private Industry	None	None	None	Low	None
Non-profits	None	None	Low	High	None
Citizens	None	None	None	Medium	None
Identification of:					
Adaptation Options	X	X	X	X	X
Adaptation Cost			X	X	
Extent of Integration Across Systems	None	None	Low	Medium	Low
Attention to Differential Impacts (e.g., on individual types of businesses, populations)	None	None	Low	Low	Low

Table 3.2. Regional Vulnerabilities of Settlements to Impacts of Climate Change in the United States

Region	Vulnerabilities	Major Uncertainties
Metro NE	Flooding, infrastructures, health, water supply, sea level rise	Storm behavior, precipitation
Larger NE	Changes in local landscapes, tourism, water, energy needs	Ecosystem impacts
Mid-Atlantic	Multiple stresses; e.g., interactions between climate change and aging infrastructures	Ecosystem impacts
Coastal SE	More intense storms, sea level rise, flooding, heat stress	Storm behavior, coastal land use, sea level rise
Inland SE	Water shortages, heat stress, UHI, economic impacts	Precipitation change, development paths
Upper Midwest	Lake and river levels, extreme weather events, health	Precipitation change, storm behavior
Inner Midwest	Extreme weather events, health	Storm behavior
Appalachians	Ecological change, reduced demand for coal	Ecosystem impacts, energy policy impacts
Great Plains	Water supply, extreme events, stresses on communities	Precipitation changes, weather extremes
Mountain West	Reduced snow, water shortages, fire, tourism	Precipitation changes, effects on winter snowpack
Arid Southwest	Water shortages, fire	Development paths, precipitation changes
California	Water shortages, heat stress, sea level rise	Temperature and precipitation changes, infrastructure impacts
Northwest	Water shortages, ecosystem stresses, coastal effects	Precipitation changes, sea level rise
Alaska	Effects of warming, vulnerable populations	Warming, sea level rise
Hawaii	Storms and other weather extremes, freshwater supplies, health, sea level rise	Storm characteristics, precipitation change

3.3 OPPORTUNITIES FOR ADAPTATION OF HUMAN SETTLEMENTS TO CLIMATE CHANGE

Settlements are important in considering prospects for adaptation to climate change, both because they represent concentrations of people and because buildings and other infrastructures offer ways to manage risk and monitor/control threats associated with climate extremes and other non-climate stressors.

Where climate change presents risks of adverse impacts for U.S. settlements and their populations, there are two basic options to respond to such concerns (a third is combining the two). One response is to contribute to climate change mitigation strategies, *i.e.*, by taking actions to reduce greenhouse gas emissions and by showing leadership in encouraging others to support such actions

(see Box 3.5: Roles of Settlements in Climate Change Mitigation). The second response is to consider strategies for adaptation, *i.e.*, finding ways either to reduce sensitivity to projected changes or to increase the settlement's coping capacities. Adaptation can rely mainly on anticipatory actions to avoid damages and costs, such as "hardening" coastal structures to sea level rise; or adaptation can rely mainly on response potentials, such as emergency preparedness; or it can include a mix of the two approaches. Research to date suggests that anticipatory adaptation may be more cost-effective than reactive adaptation (Kirshen *et al.*, 2004).

Adaptation strategies will be important to the well-being of U.S. settlements as climate change evolves over the next century. As just one example, the New York climate impact assessment (Rosenzweig and Solecki, 2001a) projects significant increases in heat-related deaths based on historical relationships between

---

BOX 3.5. Roles of Settlements in Climate Change Mitigation

Although U.S. government commitments to climate change mitigation policies at the national level have emerged only recently, an increasing number of state and local authorities are involved in strategies to mitigate greenhouse gas emissions (GHG) (Selin and Vandeveer, 2005; Rabe, 2006; Selin, 2006). U.S. states and cities are joining such initiatives as the International Council for Local Environmental Initiatives (ICLEI) (ICLEI, 2006), the U.S. Mayor Climate Protection Agreement, the Climate Change Action Plan, the Regional Greenhouse Gas Initiative (RGGI) (Selin, 2006), and the Large Cities Climate Leadership Group.[a] These initiatives focus on emissions inventories; on such actions aimed at reducing GHG emissions as switching to more energy efficient vehicles, using more efficient furnaces and conditioning systems, and introducing renewable portfolio standards. These strategies, which mandate an increase in the amount of electricity generated from renewable resources also adapt to negative social, economic, and environmental impacts; and on actions to promote public awareness (see references in footnote[a]).

Different drivers lie behind these mitigation efforts. Public and private entities have begun to "perceive" such possible impacts of climate change as rising sea level, extreme shifts in weather, and losses of key resources. They have realized that a reduction of GHG emissions opens opportunities for longer economic development (e.g., investment in renewable energy: Rabe, 2006). In addition, climate change can become a political priority if it is reframed in terms of local issues (i.e., air quality, energy conservation) already on the policy agenda (Betsill, 2001; Bulkeley and Betsill, 2003; Romero Lankao, 2007)

The promoters of these initiatives face challenges related partly to inertia (e.g., the time it takes to replace energy facilities and equipment with a relatively long life of 5 to 50 years: Haites et al., 2007). They can also face opposition from organizations who do not favor actions to reduce GHG emissions, some of whom are prepared to bring legal challenges against state and local initiatives (Rabe, 2006:17). But the number of bottom-up grassroots activities currently under way in the United States is considerable, and that number appears to be growing.

---

a   Local governments participating in ICLEI's Cities for Climate Protection Campaign commit to a) conduct an energy- and emissions-inventory and fore-cast, b) establish an emissions target, c) develop and obtain approval for the Local Action Plan, d) implement policies and measures, and e) monitor and verify results (ICLEI, 2006: April 20 2006 www.iclei.org). The Large Cities Climate Leadership Group is a group of cities committed to the reduction of urban carbon emissions and adapting to climate change. It was founded following the World Cities Leadership Climate Change Summit organized by the Mayor of London in October 2005. For more information on the US Mayor Climate Protection Agreement see http://www.seattle.gov/mayor/climate/.

heat stress and mortality, unchanged by adaptation. The Climate's Long Term Impacts on Metro Boston (CLIMB) assessment (Kirshen et al., 2004) projects that, despite similar projections of warming, heat-related deaths will decline over the coming century because of adaptation. Whether or not adaptation to climate change occurs in U.S. cities is therefore a potentially serious issue. The CLIMB assessment includes analyses showing that in many cases adaptation actions taken now are better than adaptation actions delayed until a later time (Kirshen et al., 2006).

3.3.1 Perspectives on Adaptation by Settlements

For decision-makers in U.S. settlements, climate change is yet one more source of possible risks that need to be addressed. Climate change is different as an issue because it is relatively long-term in its implications, future impacts are uncertain, and public awareness is growing from a relatively low level to a higher level of concern. Because climate change is different in these ways, it is seldom attractive to consider allocating massive amounts of funding or management attention to current climate change actions. What generally makes more sense is to consider actions that reduce vulnerabilities to climate change impacts (or increase prospects for realizing benefits from climate change impacts) and have other desirable aspects, often referred to as "co-benefits." Examples include actions that reduce vulnerabilities to current climate variability regardless of long-term climate change, actions that add resilience to water supply and other urban infrastructures that are already stressed, and actions that make metropolitan areas more attractive for their citizens in terms of their overall quality of life.

Cities and towns have used both "hard" approaches such as developing infrastructure and "soft" approaches such as regulations to address impacts of climate variability. Examples include water supply and waste water systems, drainage networks, buildings, transportation systems, land use and zoning controls, water quality standards, and emission caps and tax incentives. All of these are designed in part with climate and environmental conditions in mind. The setting of regulations has always been in a context of benefit-cost analysis and political realities; and infrastructure is also designed in a benefit-cost framework, subject to local design codes. The fact that both regulations and infrastructures vary considerably across the United States reflects cultural, economic, and environmental factors. This suggests that mechanisms exist to respond to concerns about climate change. Urban designers and managers deal routinely with uncertainties because they must consider uncertain demographic and other socioeconomic changes. Thus if climate change is properly institutionalized into the

urban planning process, it can be handled as yet another uncertainty.

3.3.2 Major Categories of Adaptation Strategies

Adaptation strategies for human settlements, large and small, include a wide range of possibilities such as:

1. *Changing the location of people or activities (within or between settlements)*—especially addressing the costs of sustaining built environments in vulnerable areas: *e.g.*, siting and land-use policies and practices to shift from more vulnerable areas to less, adding resilience to new construction in vulnerable areas, increased awareness of changing hazards and associated risks, and assistance for the less-advantaged (including actions by the private insurance sector as a likely driving force).

2. *Changing the spatial form of a settlement*—managing growth and change over decades without excluding critical functions (*e.g.*, architectural innovations improving the sustainability of structures, reducing transportation emissions by reducing the length of journeys to work, seeking efficiencies in resource use through integration of functions, and moving from brown spaces to green spaces). Among the alternatives receiving the most attention are encouraging "green buildings" (*e.g.*, green roofs: Parris, 2007; see Rosenzweig *et al.*, 2006a; Rosenzweig *et al.*, 2006b) and increasing "green spaces" within urban areas (*e.g.*, Bonsignore, 2003).

3. *Technological change to reduce sensitivity of physical and linkage infrastructures*—*e.g.*, more efficient and affordable interior climate control, surface materials that reduce heat island effects (Quattrochi *et al.*, 2000), waste reduction and advanced waste treatment, and better warning systems and controls. Physical design changes for long-lived infrastructure may also be appropriate, such as building water-treatment or storm-water runoff outflow structures based on projected sea level rather than the historical level.

4. *Institutional change to improve adaptive capacity,* including assuring effective governance, providing financial mechanisms for increasing resiliency, improving structures for coordinating among multiple jurisdictions, targeting assistance programs for especially impacted segments of the population, adopting sustainable community development practices, and monitoring changes in physical infrastructures at an early stage (Wilbanks *et al.*, 2007a). Policy instruments include zoning, building and design codes, terms for financing, and early warning systems (Kirshen *et al.*, 2005).

5. *"No regrets" or low net cost policy initiatives* that add resilience to the settlement and its physical capital—*e.g.*, in coastal areas changing building codes for new construction to require coping with projected amounts of sea level rise over the expected lifetimes of the structures.

The choice of strategies from among the options is likely to depend on co-benefits in terms of other social, economic, and ecological driving forces; the availability of fiscal and human resources; and political aspects of "who wins" and "who loses."

3.3.3 Examples of Current Adaptation Strategies

In most cases in the United States, settlements have been more active in climate change mitigation than climate change adaptation (see Box 3.5), but there are some indications that adaptation is growing as a subject of interest (Solecki and Rosenzweig, 2005; Ruth, 2006). Bottom-up grassroots activities currently under way in the United States are considerable, and that number appears to be growing. For example, Boston has built a new wastewater treatment plant at least one-half meter higher than currently necessary to cope with sea level rise, and in a coastal flood protection plan for a site north of Boston the U.S. Corps of Engineers incorporated sea level rise into their analysis (Easterling *et al.*, 2004). California is considering climate change adaptation strategies as a part of its more comprehensive attention to climate change policies (Franco, 2005), and Alaska is already pursuing ways to adapt to permafrost melting and other climate change effects.

Meanwhile, in some cases, settlements are taking actions for other reasons that add resilience to climate change effects. An example is the promotion of water conservation, which is reducing per capita water consumption in cities that could be subject to increased water scarcity (City of New York, 2005).

It seems very likely that local governments will play an important role in climate change responses in the United States. Many adaptation options must be evaluated at a relatively local scale in terms of their relative costs and benefits and their relationships with other urban sustainability issues, and local governments are important as guardians of public services, able to mobilize a wide range of stakeholders to contribute to broad community-based initiatives (as in the case of the London Climate Change Partnership, 2004). Because climate change impact concerns and adaptation potentials tend to cross jurisdictional boundaries in highly fragmented metropolitan areas, local actions might encourage cross-boundary interactions that would have value for other reasons as well.

While no U.S. communities have developed comprehensive programs to ameliorate the effects of heat islands, some localities are recognizing the need to address these effects. In Chicago, for example, several municipal buildings have been designed to accommodate "green" rooftops. Atlanta has had a Cool Communities "grass roots" effort to educate local and state officials and developers on strategies that can be used to mitigate the UHI.

This Cool Communities effort was instrumental in getting the State of Georgia to adopt the first commercial building code in the country emphasizing the benefits of cool roofing technology (Young, 2002; Estes, Jr. *et al.*, 2003). The "Excessive Heat Events Guidebook," developed by the Environmental Protection Agency in collaboration with National Oceanic and Atmospheric Administration, the Centers for Disease Control and Prevention, and the Department of Human Services provides information for municipal officials in the event of an excessive heat event.[1]

3.3.4 Strategies to Enhance Adaptive Capacity

In most cases, the likelihood of effective adaptation is related to the capacity to adapt, which in turn is related to such variables as knowledge and awareness, access to fiscal and human resources, and good governance (IPCC, 2001a). Strategies for enhancing such capacities in U.S. settlements are likely to include the development and use of local expertise on climate change issues (AAG, 2003); attention to the emerging experience with climate change effects and response strategies globally and in other U.S. settlements; information sharing about adaptation potentials and constraints among settlements and their components (likely aided by modern information technology); and an emphasis on participatory decision-making where local industries, institutions, and community groups are drawn into discussions of possible responses.

3.4 CONCLUSIONS

Even from a current knowledge base that is very limited, it is possible to conclude several things about effects of climate change on human settlements in the United States:

1. Climate change takes place in the context of a variety of factors driving an area's development: it is likely to be a secondary factor in most places, with its importance determined mainly by its interactions with other factors, except in the case of major abrupt climate change (very likely).

2. Effects of climate change will vary considerably according to location-specific vulnerabilities, and the most vulnerable areas are likely to be Alaska, coastal and river basins susceptible to flooding, arid areas where water scarcity is a pressing issue, and areas whose economic bases are climate-sensitive (very likely).

3. The main impact concerns, in areas other than Alaska, have to do with changes in the intensity, frequency, and/or location of extreme weather events and, in some cases, water availability rather than changes in temperature (very likely).

4. Over the time period covered by current climate change projections, the potential for adaptation through technological and institutional development as well as behavioral changes are considerable, especially where such developments meet other sustainable development needs and especially considering the initiatives already being shown at the local level across the United States (extremely likely).

5. While uncertainties are very large about specific impacts in specific time periods, it is possible to talk with a higher level of confidence about vulnerabilities to impacts for most settlements in most parts of the United States (virtually certain).

---

1 For more information please see: http://www.epa.gov/hiri/about/heatguidebook.html.

3.5 EXPANDING THE KNOWLEDGE BASE

A number of sources, including NACC, 1998; Parson *et al.*, 2003; Ruth, 2006; and Ruth *et al.*, 2004, have considered research pathways for improving the understanding of effects of climate change on human settlements in the United States.

The following list suggests a number of research topics that would help expand the knowledge base about the linkages between climate change and human settlements.

- Advance understanding of settlement vulnerabilities, impacts, and adaptive responses in a variety of different local contexts around the country through case studies. In addition to identifying vulnerable settlements, these studies should also identify vulnerable populations (such as the urban poor and native populations on rural and/or tribal lands) that have limited capacities for response to climate change within those settlements. Better understanding of climate change at the community scale would provide a basis for adaptation research that addresses social justice and environmental equity concerns.

- Develop better projections of climate change at the scale of U.S. metropolitan areas or smaller, including scenarios projecting extremes and scenarios involving abrupt changes.

- Improve abilities to associate projections of climate change in U.S. settlements with changes in other driving forces related to impacts, such as changes in metropolitan/urban patterns and technological change.

- Design practically implementable, socially acceptable strategies for shifting human populations and activities away from vulnerable locations.

- Improve the understanding of vulnerabilities of urban inflows and outflows to climate change impacts, as well as second and third-order impacts of climate change in urban environments, including interaction effects among different aspects of the urban system.

- Improve the understanding of the relationships between settlement patterns

(both regional and intra-urban) and resilience/adaptive capacity.

- Improve understanding of how urban decision-making is changing as populations become more heterogeneous and decisions become more decentralized, especially as this affects adaptive responses.

- Review current policies and practices related to climate change responses to help inform community decision-makers and other stakeholders about potentials for relatively small changes to make a large difference.

- Evaluate and document experiences with urban/settlement climate change responses while involving decision-making, research and stakeholder communities more actively in discussions of climate change impacts and response issues. Focus attention on the costs, benefits, and possible limits and potentials of adaptation to climate change vulnerabilities in U.S. cities and smaller settlements.

- Improve tools and approaches for infrastructure planning and design to reduce exposure and sensitivity to climate change effects while increasing adaptive capacity.

- Enhance coordination within federal government agencies to improve understanding about impacts, vulnerabilities, and responses to climate change for the nation's cities and smaller settlements. Connections with U.S. urban decision-makers can enable integration of climate change considerations into what they do with building codes, zoning, lending practices, etc. as mainstreamed urban decision processes.

3.6 REFERENCES

ACIA, 2004: *Impacts of a Warming Arctic: Arctic Climate Impact Assessment.* Arctic Climate Impact Assessment, Cambridge University Press, Cambridge.

AIACC, *Assessments of Impacts and Adaptations to Climate Change in Multiple Regions and Sectors.* Retrieved January 25, 2007, from http://www.aiaccproject.org

Amato, A., M. Ruth, P. Kirshen, and J. Horwitz, 2005: Regional energy demand responses to climate change: methodology and application to the Commonwealth of Massachusetts. *Climatic Change,* **71(1)**, 175-201.

Association of American Geographers (AAG), 2003: *Global Change In Local Places: Estimating, Understanding, And Reducing Greenhouse Gases* [GCLP Research Team: Kates, R., T. Wilbanks, and R. Abler (eds.)]. Association of American Geographers, Cambridge University Press, Cambridge, Massachusetts.

Betsill, M.M., 2001: Mitigating climate change in US cities: Opportunities and obstacles. *Local Environment,* **4**, 393-406.

Bloomfield, J., M. Smith, and N. Thompson, 1999: *Hot Nights in the City: Global Warming, Sea level rise and the New York Metropolitan Region.* Environmental Defense Fund, Washington, DC.

Bonsignure, R., 2003: *Urban Green Space: Effects on Water and Climate,* Center for American Urban Landscape, Design Brief Number 3, University of Minnesota.

Brazel, A.J. and D.A. Quattrochi, 2005: Urban climates. In: *Encyclopedia of World Climatology.* [J.E. Oliver, (ed.)]. Springer, Dordrecht, The Netherlands, pp. 766-779.

Bulkeley, H. and M. Betsill, 2003: *Cities and Climate Change; Urban Sustainability and Global Environmental Governance. Routledge,* London.

City of New York, 2005: *New York City's Water Supply System.* The City of New York Department of Environmental Protection, New York.

Clark, W.C., J. Jaeger, R. Corell, R. Kasperson, J.J. McCarthy, D. Cash, et al., 2000: *Assessing Vulnerability to Global Environmental Risks.* Discussion Paper 200-12, Environment and Natural Resources Program, Kennedy School of Government, Harvard University, Cambridge, Massachusetts.

Colten, C.E., 2005. *An Unnatural Metropolis: Wrestling New Orleans from Nature.* Louisiana State University, Baton Rouge, Louisiana.

Congressional Black Caucus Foundation, 2004: *African Americans and Climate Change: An Unequal Burden.* Redefining Progress, Washington, DC.

Crossett, K. M., T. J. Culliton, P.C. Wiley, and T.R. Goodspeed, 2004: *Population trends along the coastal United States:* 1980-2008. NOAA National Ocean Service, Management and Budget Office, 54 pp.

Decker, E., S. Elliot, F. Smith, D. Blake, and F. Rowland, 2000: Energy and material flow through the urban ecosystem. *Annual Review of Energy and the Environment,* **25**, 685-740.

Easterling, W.E., B.H. Hurd, and J.B. Smith, 2004: *Coping with Global Climate Change: The Role of Adaptation in the United States,* prepared for the Pew Center on Global Climate Change.

Emanuel, K.A., 2005: Increasing destructiveness of tropical cyclones over the past 30 years. *Nature,* **436**, 686-688.

Estes, Jr., M.D., D. Quattrochi, and E. Stasiak, 2003: The urban heat island phenomena: how its effect can influence environmental decision making in your community. *Public Management,* **85**, 8-12.

Folke, C., A. Janssen, J. Larsson, and R. Costanza, 1997: Ecosystem appropriation by cities, *Ambio,* **26**, 167-172.

Franco, G., 2005: *Climate Change Impacts and Adaptation in California,* prepared for the California Energy Commission.

Gleick, P.H., D.B. Adams, 2000: Water: *The Potential Consequences of Climate Variability and Change.* A Report of the National Water Assessment Group, U.S. Global Change Research Program, U.S. Geological Survey, U.S. Department of the Interior, and the Pacific Institute, Oakland, California.

Haites, E., K. Caldeira, P. Romero Lankao, A. Rose, and T. Wilbanks, 2007: What are the options that could significantly affect the carbon cycle? In: *State of the Carbon Cycle Report (SOCCR).* A Report by the U.S. Climate Change Science Program and the Subcommittee on Global Change Research [Karl, T.R., S. Hassol, C.D. Miller, and W.L. Murray (eds.)]. National Oceanic and Atmospheric Administration. National Climatic Data Center.

Hartwig, R., 2006: *Hurricane Season Of 2005, Impacts On U.S. P/C Markets, 2006 And Beyond.* Presentation to the Insurance Information Institute, March 2006, New York.

Hoo, W. and M. Sumitani, 2005: *Climate Change Will Impact the Seattle Department of Transportation.* Office of the City Auditor, Seattle, Washington, USA

Hunter, J.R., 2006: Testimony before the Committee on Banking, Housing and Urban Affairs of the United States Senate Regarding Proposals to Reform the National Flood Insurance Program.

ICLEI, 2006: *ICLEI Local Governments for Sustainability.* Retrieved May 28, 2008, from http://www.iclei.org.

IPCC, 2001a: *Climate Change 2001: Impacts, Adaptation, and Vulnerability. Contribution of Working Group II to the Third Assessment Report of the Intergovernmental Panel on Climate Change* [McCarthy, J.J., O.F. Canziani, N.A. Leary, D.J. Dokken, and K.S. White (eds.)]. Cambridge University Press, Cambridge, 967 pp.

IPCC, 2001b: Human settlements, energy, and industry. In: IPCC, 2001a.

IPCC, 2001c: Insurance and other financial services. In: IPCC, 2001a.

Kinney, P., J. Rosenthal, C. Rosenzweig, C. Hogrefe, W. Solecki, K. Knowlton, *et al.*, 2006: Assessing potential public health impacts of changing climate and land uses: The New York climate and health project. In: *Regional Climate Change and Variability* [Ruth, M., K. Donaghy, and P. Kirshen, (eds.)]. New Horizons in Regional Science, Edward Elgar, Cheltenham, UK.

Kirshen, P., 2002: Potential impacts of global warming in eastern Massachusetts. *Journal of Water Resources Planning and Management,* **128(3)**, 216-226.

Kirshen, P., M. Ruth and W. Anderson. 2005: Climate change in metropolitan Boston, *New England Journal of Public Policy*, **20(2)**, 89-103.

Kirshen, P., M. Ruth, and W. Anderson, 2006: Climate's long-term impacts on urban infrastructures and services: The case of metro Boston. In: *Climate Change and Variability: Impacts and Responses* [M. Ruth, P. Kirshen, and Donaghy, (eds.)]. Edward Elgar, Cheltenham, UK, pp. 190-252.

Kirshen, P., M. Ruth, and W. Anderson, 2007: Interdependencies of urban climate change impacts and adaptation strategies: a case study of metropolitan Boston. *Climatic Change,* **66**, 105-122.

Kirshen, P., M. Ruth, W. Anderson, T. R. Lakshmanan, S. Chapra, W. Chudyk, L. Edgers, D. Gute, M. Sanayei, and R. Vogel, 2004: *Climate's Long-term Impacts on Metro Boston.* Final Report to the U.S. Environmental Protection Agency, Office of Research and Development, Washington, DC.

Klinenberg, E., 2003: *Heat wave: A Social Autopsy of Disaster in Chicago.* University of Chicago, Chicago, Illinois.

Koteen, L., J. Bloomfield, T. Eichler, C. Tonne, R. Young, H. Poulshock, and A. Sosler, 2001: *Hot Prospects: The Potential Impacts of Global Warming on Los Angeles and the Southland.* Environmental Defense, Washington, DC.

Kretch, S. III, 1999. *Myth and History: The Ecological Indian.* W.W. Norton, New York.

Larsen, Peter, O.S. Goldsmith, O. Smith, and M. Wilson, 2007: *A Probabilistic Model to Estimate the Value of Alaska Public Infrastructure at Risk to Climate Change.* ISER Working Paper, Institute of Social and Economic Research, University of Alaska, Anchorage, Alaska.

Lo, C.P. and D.A. Quattrochi, 2003: Land-use and land-cover change, urban heat island phenomenon, and health implications: a remote sensing approach. *Photogrammetric Engineering and Remote Sensing.*

London Climate Change Partnership, 2004: *London's Warming, A Climate Change Impacts in London Evaluation Study.* London, 293 pp.

McCarthy, K., D. Peterson, N. Sastry, and M. Pollard, 2006: *The Repopulation of New Orleans after Hurricane Katrina.* Rand, Santa Monica, California.

Molina, L.T., M.J. Molina, 2005: *Air Quality in the Mexico Megacity: An Integrated Assessment.* Kluwer, Dordrecht, Netherlands.

Morehouse, B., G. Christopherson, M. Crimmins, B. Orr, J. Overpeck, T. Swetnam, and S. Yool, 2006: Modeling interactions among wildland fire, climate and society in the context of climatic variability and change in the southwest US. In: *Regional Climate Change and Variability* [M. Ruth, K. Donaghy, and P. Kirshen (eds.)]. Edward Elgar Publishing, pp. 58-78.

NACC, 1998: *Climate Change and a Global City: An Assessment of the Metropolitan East Coast Region.* Columbia Earth Institute/NASA Goddard Institute for Space Studies.

NACC, 2000: *Climate Change Impacts on the United States: The Potential Consequences of Climate Variability and Change.* U.S. Global Change Research Program, Washington, DC.

NACC, 2001: *Climate Change Impacts on the United States: The Potential Consequences of Climate Variability and Change,* U.S. Global Change Research Program, Washington, DC.

Neumann, J.E., G. Yohe, R. Nichols, and M. Manion, 2000: *Sea Level Rise and Global Climate Change: A Review of Impacts to U.S. Coasts.* Pew Center on Global Climate Change, Washington, DC.

NRC, 2007: *Evaluating Progress of the U.S. Climate Change Science Program: Methods and Preliminary Results.* Committee on Strategic Advice on the U.S. Climate Change Science Program, National Academies Press, Washington, DC.

Parris, T., 2007: Green buildings. *Environment,* **49(1)**, 3.

Parson, E.A., R.W. Corell, E.J. Barron, V. Burkett, A. Janetos, L. Joyce, T.R. Karl, M.C. MacCracken, J. Melillo, M.G. Morgan, D.S. Schimel, T. Wilbanks, 2003: Understanding climatic impacts, vulnerabilities, and adaptation in the United States: building a capacity for assessment, *Climatic Change,* **57(1-2)**, 9–42.

Patz, J.A. and J.M. Balbus, 2001: Global climate change and air pollution. In: *Ecosystem Change and Public Health.* A Global Perspective [Aron J.L. and J.A. Patz (eds.)]. The Johns Hopkins University Press, Baltimore, pp. 379–408.

Quattrochi, D.A., J.C. Luvall, D.L. Rickman, M.G. Estes, Jr., C.A. Laymon, and B.F. Howell, 2000: A decision support system for urban landscape management using thermal infrared data. *Photogrammetric Engineering and Remote Sensing,* 66, 1195-1207.

Rabe, B., 2006: *Second Generation Climate Policies in the States.* Paper presented at the Conference Climate Change Politics in North America, Washington, DC, 18-19.

Ridd, M.K., 2006: 10.3 Environmental dynamics of human settlements. In: *Remote Sensing of Human Settlements, Manual of Remote Sensing, Third Edition, Volume 5* [Ridd, M.K. and J.D. Ripple, (eds.)]. American Society for Photogrammetry and Remote Sensing, Bethesda, Maryland, pp. 564-642.

Riebsame, W., 1997: *Atlas of the New West: Portrait of a Changing Region.* W.W. Morton, New York.

Romero Lankao, P., 2007: Are we missing the point? Particularities of urbanization, sustainability and carbon emissions in Latin American cities. *Urbanization and Environment,* **19(1)**.

Rosenzweig, C., S. Gaffin, and I. Parshall (eds.), 2006a: *Green Roofs In The New York Metropolitan Region.* Research Report, Columbia University Center for Climate Systems Research and NASA Goddard Institute for Space Studies, New York.

Rosenzweig, C., and W. Solecki (eds.), 2001a: *Climate Change and a Global City: The Potential Consequences of Climate Variability and Change – Metro East Coast.* Columbia Earth Institute, New York.

Rosenzweig, C. and W.D. Solecki, 2001b: Global environmental change and a global city: lessons for New York. *Environment,* **43(3)**, 8-18.

Rosenzweig, C., W. Solecki, C. Paine, V. Gornitz, E. Hartig, K. Jacob, D. Major, P. Kinney, D. Hill, and R. Zimmerman, 2000: *Climate Change and a Global City: An Assessment of the Metropolitan East Coast Region.* U.S. Global Change Research Program.

Rosenzweig, C., W. Solecki, L. Parshall, M. Chopping, G. Pope, and R. Goldberg, 2005: Characterizing the urban heat island in current and future climates in New Jersey, *Global Environmental Change Part B: Environmental Hazards,* **(6)**, 51-62.

Rosenzweig, C., W. Solecki, and R. Slosberg, 2006b: *Mitigating New York City's Heat Island with Urban Forestry, Living Roofs, and Light Surfaces.* Final Report, New York City Regional Heat Island Initiative, Prepared for the New York State Energy Research and Development Authority, NYSERDA Contract #6681, New York.

Ruth, M. (ed.), 2006: *Smart Growth and Climate Change.* Edward Elgar Publishers, Cheltenham, England, 403 pp.

Ruth, M., A. Amato, and P. Kirshen, 2006a: Impacts of changing temperatures on heat-related mortality in urban areas: the issues and a case study from metropolitan Boston. In: *Smart Growth and Climate Change* [Ruth, M. (ed.)]. Edward Elgar Publishers, Cheltenham, England, pp. 364 – 392.

Ruth, M., C. Bernier, N. Jollands, and N. Golubiewski, 2007: Adaptation to urban water supply infrastructure to impacts from climate and socioeconomic changes: the case of Hamilton, New Zealand. *Water Resources Management,* **21**, 1031-1045.

Ruth, M. and D. Coelho. In press: Managing the interrelations among urban infrastructure, population, and institutions. *Global Environmental Change.*

Ruth, M., K. Donaghy, and P.H. Kirshen (eds.), 2006b: *Regional Climate Change and Variability: Impacts and Responses.* Edward Elgar Publishers, Cheltenham, England, 260 pp.

Ruth, M. and A-C Lin, 2006c: Regional energy and adaptations to climate change: methodology and application to the state of Maryland, *Energy Policy,* **34**, 2820-2833.

Selin, H. and S.D. VanDeveer, 2006: Political science and prediction: What's next for U.S. climate change policy? *Review of Policy Research* (forthcoming).

Sherbinin, A., A. Schiller, and A. Pulsipher, 2006: The vulnerability of global cities to climate hazards. *Environment and Urbanization*, **12(2)**, 93-102.

Solecki, W. and R. Leichenko, 2006: Urbanization and the metropolitan environment. *Environment*, **48(4)**, 8-23.

Solecki, W.D. and C. Rosenzweig, 2006: Climate change and the city: observations from metropolitan New York. In: *Cities and Environmental Change*, [Bai, X. (ed.)]. Yale University Press, New York.

Steinberg, T., 2006. *Acts of God: The Unnatural History of Natural Disaster in America.* Oxford, New York.

Stone, B., 2006: Physical planning and urban heat island formation, 2006. In: *Smart Growth and Climate Change* [Ruth, M. (ed.)]. Edward Elgar Publishers, Cheltenham, England, pp. 318-341.

Vale, L.J., and T J. Campanella, 2005. *The Resilient City: How Modern Cities Recover from Disaster*, Oxford, New York.

Wilbanks, T.J., 2003: Integrating climate change and sustainable development in a place-based context. *Climate Policy*. Supplement on Climate Change and Sustainable Development, **3(1)**, 147-154.

Wilbanks, T.J., P. Leiby, R. Perlack, J.T. Ensminger, and S.B. Wright, 2007a: Toward an integrated analysis of mitigation and adaptation: some preliminary findings. *Mitigation and Adaptation Strategies for Global Change*, **12(5)**, 713-725.

Wilbanks, T.J., P. Romero Lankao, M. Bao, F. Berkhout, S. Cairncross, J.-P. Ceron, M. Kapshe, R. Muir-Wood and R. Zapata-Marti, 2007b: Industry, settlement and society. *Climate Change 2007: Impacts, Adaptation and Vulnerability. Contribution of Working Group II to the Fourth Assessment Report of the Intergovernmental Panel on Climate Change* [Parry, M.L., O.F. Canziani, J.P. Palutikof, P.J. van der Linden and C.E. Hanson (eds.)]. Cambridge University Press, Cambridge, UK, pp. 357-390.

Wilbanks, T.J., M.J., Sale, V. Bhatt, W.C. Horak, D. Billello, S.R. Bull, J. Ekmann, Y. Huang, D. Levine, S. Schmalzer, and M.J. Scott, 2007c: *Effects of Climate Change on Energy Production and Use in the United States.* Synthesis and Assessment Product 4.5. U.S. Climate Change Science Program, Washington, D.C.

Young, B., 2002: Thinking clean and green. *Georgia Trend*, **18**, 93-100.

Effects of Global Change on Human Welfare

Lead Author: Frances G. Sussman, Environmental Economics Consulting

Contributing Authors: Maureen L. Cropper, University of Maryland at College Park; Hector Galbraith, Galbraith Environmental Sciences LLC; David Godschalk, University of North Carolina at Chapel Hill; John Loomis, Colorado State University; George Luber, Centers for Disease Control and Prevention; Michael McGeehin, Centers for Disease Control and Prevention; James E. Neumann, Industrial Economics, Incorporated; W. Douglass Shaw, Texas A&M University; Arnold Vedlitz, Texas A&M University; Sammy Zahran, Colorado State University

4.1 INTRODUCTION

Human welfare is an elusive concept. There is no single, commonly accepted definition or approach to thinking about welfare. Clearly there is a shared understanding that human welfare, well-being, and quality of life (terms that are often used interchangeably) refer to aspects of individual and group life that improve living conditions and reduce chances of injury, stress, and loss. The physical environment is one factor, among many, that may improve or reduce human well-being. Climate is one aspect of the physical environment, and can affect human well-being via economic, physical, psychological, and social pathways and influence individual perceptions of quality of life.

Climate change may result in lifestyle changes and adaptive behavior with both positive and negative implications for well-being. For example, warmer temperatures may change the amount of time that individuals are comfortable spending outdoors in work, recreation, or other activities, and temperature combined with other climatic changes may alter (or induce) changes in intra- and inter-country human migration patterns. More generally, studies of climate change and the United States identify an assortment of impacts on human health, the productivity of human and natural systems, and human settlements. Many of these impacts—ranging from changes in livelihoods to changes in water quality and supply—are linked to some aspect of human well-being.

Communities are an integral determinant of human well-being. Climate change that affects public goods—such as damaged infrastructure or interruptions in public services—or disrupts the production of goods and services, will affect economic performance including overall health, poverty, employment, and other measures. These changes may have consequences, such as a lost job or a more difficult commute, that affect individual well-being directly. In other cases, individual well-being may be indirectly affected due to concern for the well-being of other individuals, or for a lack of cohesion within the community. The sustainability or resilience of a community (*i.e.,* its ability to cope with climate change and other stressors over the long term) may be reduced by climate change weakening the physical and social environment. In the extreme, such changes may undermine the individual's sense of security or faith in government's capacity to accommodate change.

Completely cataloging the effects of global change on human well-being or welfare would be an immense undertaking. Despite its importance, no well-accepted structure for doing so has been developed and applied. Moreover, little (if any) research focuses explicitly on the impact of global change on human well-being, per se. The chapter seeks to make a review of this topic manageable by focusing on several discrete issues:

- Alternative approaches to defining and studying human well-being;

- Identifying human well-being and quality of life measures and indicators (qualitative and quantitative);

- Describing economic welfare and monetary methods of assigning value to climate change's potential impacts; and,

- Providing examples of climate change impacts on selected categories of well-being and reporting indicators of economic welfare for these categories.

Section 4.2 focuses on valuation and non-monetary metrics and draws on the literature to provide insights into a possible foundation for future research into the effects of climate change on human well-being. This section first discusses the literature defining human well-being. Next, it presents an illustrative place-based indicators approach (the typical approach of planners and policy makers to evaluating quality of life in communities, cities, and countries). Approaches of this type represent a commonly accepted way of thinking about well-being that is linked to objective (and sometimes subjective) measures. While a place-based indicators approach has not been applied to climate change, it has the potential to provide a framework for identifying categories of human well-being that might be affected by climate change, and for making the identification of measures or metrics of well-being a more concrete enterprise in the future. To illustrate

that potential, the section draws links between community welfare and some of the negative impacts of climate change.

Economics has been at the forefront of efforts to quantify the welfare impacts of climate change. Economists employ, however, a very specific definition of well-being—*economic welfare*—for valuing goods and services or, in this case, climate impacts. This approach is commonly used to support environmental policy decision making in many areas. Section 4.3 very briefly describes the basis of this approach, and the techniques that economists use (focusing on those that have been applied to estimate impacts of climate change). This section next summarizes the existing economic estimates of the *non-market* impacts of climate change.[1] An accompanying appendix provides more information on the economic approach to valuing changes in welfare, and highlights some of the challenges in applying valuation techniques to climate impacts.

The fourth section of the chapter summarizes some of the key points of the chapter, and concludes with a brief discussion of research gaps.

4.2 HUMAN WELFARE, WELL-BEING, AND QUALITY OF LIFE

No single, widely accepted definition exists for the term human welfare, or for related terms such as well-being and quality of life. They are all often used interchangeably (Veenhoven, 1988, 1996, 2000; Ng, 2003; Rahman, 2007). Economists, epidemiologists, health scientists, psychologists, sociologists, geographers, political scientists, and urban planners have all rendered their own definitions and statistical indicators of life quality at both individual and

---

1 Because more concrete aspects of welfare, such as impacts on prices or income, may be covered by other synthesis and assessment products (see, for example, discussions of dollar values in SAP 4.3, *The Effects of Climate Change on Agriculture, Land Resources, Water Resources, and Biodiversity*), this report focuses exclusively on the types of intangible amenities that directly impact quality of life, but are not traded in markets, including health, recreation, ecosystems, and climate amenities.

community levels.[2] For purposes of clarity in this chapter, we adopt the convention of the Millenium Assessment (MA, 2005) and the Intergovernmental Panel on Climate Change (IPCC, 2007a), which use "well-being" as an umbrella term—referring broadly to the extent to which human conditions satisfy the range of constituents of well-being, including health, social relations, material needs, security, and freedom of choice. "Quality of life" is here used synonymously with well-being, to reflect usage in a wide range of disciplines, including medical, sociological, psychological, and urban planning literatures. The term "welfare" is generally used to refer narrowly to economic measures of individual well-being, although it is also used in the context of communities in a broader sense.

Despite differences in definitions, human well-being—in its broadest sense—is typically a multi-dimensional concept, addressing the availability, distribution, and possession of economic assets, and non-economic goods such as life expectancy, morbidity and mortality, literacy and educational attainment, natural resources and ecosystem services, and participatory democracy. These conceptualizations often also include social and community resources (sometimes referred to as social capital in social scientific literature), such as the presence of voluntary associations, arts, entertainment, and shared recreational amenities (see Putnam, 1993, 2000). The quantity of community resources shared by a population is often called social capital.[3] These components of life quality are interrelated and correlate with subjective valuations of life satisfaction, happiness, pleasure, and the operation of successful democratic political systems (Putnam, 2000).

The concepts of well-being, economic welfare, and quality of life play important roles not only in academic research, but also in practical analysis and policy making. Quality of life measures may be used, for example, to gauge progress in meeting policy or normative goals in particular cities by planners. Municipalities in New Zealand, England, Canada, and the United States have constructed their own metrics of quality of life to estimate the overall well-being and life chances available to citizens. Similarly, health-related quality of life measures can indicate progress in meeting goals. For example, the U.S. Medicare program uses metrics to track quality of life for beneficiaries and to monitor and improve health care quality (HCFR, 2004). Moreover, international agencies from the United Nations Human Development Programme (UNDP), to the Millenium Ecosystem Assessment on Ecosystems and Human Well-Being, and highly regarded periodicals like *The Economist*, have built composite measures of human and societal well-being to compare and rank nations of the world.[4]

Life quality and human well-being are increasingly important objects of theoretical and empirical research in diverse disciplines. Two analytic approaches characterize the research literature: (1) studies that emphasize well-being as an individual attribute or possession;

2 For example, in sociological literature, the terms well-being and welfare are used interchangeably to refer to objectively measurable life chances and experiences, and the term quality of life is used to describe subjective assessments and experiences of individuals.

3 The concept of social capital has been defined, in different ways, by Putnam (1993, 1995, 2000) and by Coleman (1988, 1990, 1993). For Coleman, social capital is a store of community value that is embodied in social structures and the relations between social actors, from which individuals can draw in the pursuit of private interest. Putnam's definition is similar, but places a stronger emphasis on altruism and community resources.

4 See, for example, the discussion of the sources of Table 1 subsequently in this chapter, which include a number of country-level quality of life assessments. The UNDP Human Development Index, a country by country ranking of quality of life indicators, can be accessed at http://hdr.undp.org/en/statistics/.

and (2) studies that treat well-being as a social or economic phenomenon associated with a geographic place.

4.2.1 Individual Measures of Well-being

Approaches focusing on individuals are generally found in medical, health, cognitive, and economic sciences. We turn to these first, and then next to place-focused indicators.

4.2.1.1 Health-focused Approaches

In medical science, quality of life is used as an outcome variable to evaluate the effectiveness of medical, therapeutic, and/or policy interventions to promote population health. Quality of life is an individual's physiological state constituted by body structure, function, and capability that enable pursuit of stated and revealed preferences. In medical science, the concept of life quality is synonymous with good health–a life free of disease, illness, physical, and/or cognitive impairment (Raphael et al., 1996, 1999, 2001).

In addition to objective measures of physical and occupational function, disease absence, or somatic sensation, life quality scientists measure an individual's perception of life satisfaction. The scientific basis of such research is that pain and/or discomfort associated with a physiological impairment are registered and experienced variably. Based on patient reports or subjective valuations, psychologists and occupational therapists have developed

valid and reliable instruments to assess how mental, developmental, and physical disabilities interfere with the performance and enjoyment of life activities (Bowling, 1997; Guyatt et al., 1993).

4.2.1.2 Economic and Psychological Approaches

Individual valuations of life quality also anchor economic and psychological investigations of happiness and utility. In the new science of happiness, scholars use the tools of neuroscience, experimental research, and modern statistics to discover and quantify the underlying psychological and physiological sources of happiness (for reviews see Kahneman et al., 1999; Frey and Stutzer, 2002; Kahneman and Krueger, 2006). Empirical studies show, for example, that life satisfaction and happiness correlate predictably with marital status (married persons are generally happier than single people), religiosity (persons that practice religion report lower levels of stress and higher levels of life satisfaction), and individual willingness to donate time, money and effort to charitable causes. Similarly, the scholarly literature notes interesting statistical associations between features of climate (such as variations in sunlight, temperature, and extreme weather events) and self-reported levels of happiness, utility, or life satisfaction.

Individual valuations of health, psychological, and emotional well-being are sometimes summed across representative samples of a population or country to estimate correspondences between life satisfaction and "hard" indicators of living standards such as income, life expectancy, educational attainment, and environmental quality. Cross-national analyses generally find that population happiness or life satisfaction increases with income levels and material standards of living (Ng, 2003) and greater personal autonomy (Diener et al., 1995; Diener and Diener, 1995).[5] In such studies, subjective valuations of life satisfaction are embedded in

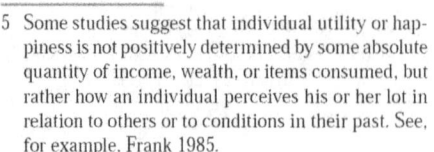

---

5 Some studies suggest that individual utility or happiness is not positively determined by some absolute quantity of income, wealth, or items consumed, but rather how an individual perceives his or her lot in relation to others or to conditions in their past. See, for example, Frank 1985.

broader conceptions of quality of life associated with the conditions of a geographic place, community, region, or country—the social indicators approach.

4.2.2 The Social Indicators Approach

In a second strand of research, what some refer to as the social indicators approach, scholars assemble location-specific measures of social, economic, and environmental conditions, such as employment rates, consumption flows, the availability of affordable housing, rates of crime victimization and public safety, public monies invested in education and transportation infrastructure, and local access to environmental, cultural, and recreational amenities. These place-specific variables are seen as exogenous sources of individual life quality. Scholars reason that life quality is a bundle of conditions, amenities, and lifestyle options that shape stated and revealed preferences. In technical terms, the social indicators approach treats quality of life as a latent variable, jointly determined by several causal variables that can be measured with reasonable accuracy.

The indicators approach has several advantages in the context of understanding the impacts of climate change on human well-being. First, social indicators have considerable intuitive appeal, and their widespread use has not only made it familiar to both researchers and the general public, but has subjected them to considerable debate and discussion. Second, they offer considerable breadth and flexibility in terms of categories of human well-being that can be included. Third, for many of the indicators or dimensions of well-being, objective metrics exist for measurement.

In addition, while its strength is in providing indicators of progress on individual dimensions of quality of life, the indicators approach has also been used to support aggregate or composite measures, at least for purposes of ranking or measuring progress. Various techniques are also available, or being developed, that aggregate or combine measures of well-being. These

range from pure data reduction procedures to stakeholder input models where variables are evaluated on their level of social and economic importance. For example, Richard Florida (2002a) has constructed a statistical index of technology, talent, and social tolerance variables to estimate the human capital of cities in the United States. Given the analytical strengths of the social indicators approach, it may be a good starting point for understanding the relationships between human well-being and climate change.

4.2.2.1 A Taxonomy of Categories of Well-being

Taxonomies of place-specific well-being or quality of life typically converge on six categories or dimensions: (1) economic conditions; (2) natural resources, environment, and amenities; (3) human health; (4) public and private infrastructure; (5) government and public safety; and (6) social and cultural resources. These categories represent broad aspects of personal and family circumstances, social structures, government, environment, and the economy that influence well-being. Table 4.1 illustrates these categories, which are listed in Column 1. The third column, "components/indicators of well-being," provides examples of the way in which these categories are often interpreted. These components represent what, in an ideal world, researchers would wish to measure in order to determine how a specific society fares from the perspective of well-being. The fourth column provides illustrative metrics, *i.e.,* objective or quantifiable measures that are often used by researchers as indicators of

well-being for each category.[6] Finally, the last column provides some examples of climate impacts that may be linked to that category. This column should not be viewed as an attempt to create a comprehensive list of impacts, or even to list impacts with equal weights, in terms of importance or likelihood of occurrence. Further, while Table 4.1 focuses on negative impacts (as potentially more troubling for quality of life), in some categories there are also opportunities or potential positive impacts.

These categories of well-being or life quality are interrelated. For example, as economic or social conditions in a society improve (*e.g.*, as measured by Gross Domestic Product (GDP), GDP per capita, and rates of adult literacy), improvements occur in human health outcomes such as infant mortality, rates of morbidity, and life expectancy at birth. Thus, while the categories and corresponding metrics of well-being presented in Table 4.1 are analytically separable, in reality they are highly interconnected.[7]

Economics as a source of quality of life refers to a mix of production, consumption, and exchange activities that constitute the material well-being of a geographic place, community, region, or country. Standard components of economic well-being include income, wealth, poverty, employment opportunities, and costs of living. Localities characterized by efficient and equitable allocation of economic

rewards and opportunities enable material security and subjective happiness of residents (Florida, 2002a).

Natural resources, environment, and amenities as a source of well-being refers to natural features, such as ecosystem services, species diversity, air and water quality, natural hazards and risks, parks and recreational amenities, and resource supplies and reserves. Natural resources and amenities directly and indirectly affect economic productivity, aesthetic and spiritual values, and human health (Blomquist *et al.*, 1988; Glaeser *et al.*, 2001; Cheshire and Magrini, 2006).

Human health as a source of well-being includes features of a community, locality, region, or country that influence risks of mortality, morbidity, and the availability of health care services. Good health is desirable in itself as a driver of life expectancy (and the quality of life during those years), and is also critical to economic well-being by enabling labor force participation (Raphael *et al.*, 1996, 1999, 2001).

Public and private infrastructure sources of well-being include transportation, energy and communication technologies that enable commerce, mobility, and social connectivity. These technologies provide basic conditions for individual pursuits of well-being (Lambiri *et al.*, 2007).

Government and public safety as a source of well-being are activities by elected representatives and bureaucratic officials that secure and maximize the public services, rights, liberties, and safety of citizens. Individuals derive happiness and utility from the employment, educational, civil rights, public service, and security efforts of their governments (Suffian, 1993).

Finally, *social and cultural resources* as a source of well-being are conditions of life that promote social harmony, family and friendship, and the availability of arts, entertainment, and leisure activities that facilitate human happiness. The terms social and creative capital have become associated with these factors. Communities with greater levels of social and

---

6 Sources that contributed to the development of Table 4.1 include: MA (2005); Sufian, 1993; Rahman, *2007, and* Lambiri, *et al.*, 2007. Insights were also derived from quality of life studies of individual cities and countries, including: http://www.bigcities.govt.nz/indicators.htm *Quality of Life in New Zealand's Large Urban Areas*; http://www.asu.edu/copp/morrison/public/qof199.htm *What Matters in Greater Phoenix 1999 Edition: Indicators of Our Quality of Life*; and http://www.jcci.org/statistics/qualityoflife.aspx *Tracking the Quality of Life in Jacksonville.*

7 More recently, scholars (Costanza *et al.* 2007) and government agencies (like NOAA's Coastal Service Center) have moved toward the global concept of *capital* to integrate indicators and assess community quality of life. The term capital is divided into four types: economic; physical; ecological or natural; and socio-cultural. Various metrics constitute these types of capital, and are understood to foster community resilience and human needs of subsistence, reproduction, security, affection, understanding, participation, leisure, spirituality, creativity, identity, and freedom. See also Rothman, Amelung, and Poleme (2003).

Table 4.1 Categorization of Well-being

Category of Well-being	Description and Rationale	Components/Indicators of Well-being	Illustrative Metrics/Measures of Well-being	Examples of Negative Climate Linkages*
Economic conditions	The economy supports a mix of activities: opportunities for employment, a strong consumer market, funding for needed public services, and a high standard of living shared by citizens.	• Income and production • Economic standard of living, e.g., wealth and income, cost of living, poverty • Economic development, e.g., business and enterprise, employment • Availability of affordable housing • Equity in the distribution of income	• GDP • Wage rates (e.g., persons at minimum wage) • Employment rates • Business startups and job creation • Housing prices • Dependence on public assistance • Families/children living in poverty • Utility costs, gasoline prices, and other prices	Reduced job opportunities and wage rates in areas dependent on natural resources, such as agricultural production in a given region that faces increased drought. Higher electricity prices resulting from increased demand for air conditioning as average temperatures and frequency of heat waves rise.
Natural resources, environment, and amenities	Resources enhance the quality of life of citizens; pollution and other negative environmental effects are kept below levels harmful to ecosystems, human health, and other quality of life considerations; and natural beauty and aesthetics are enhanced.	• Air, water, and land pollution • Recreational opportunities • Water supply and quality • Natural hazards and risks • Ecosystem condition and services • Biodiversity • Direct climate amenity effects	• Air and water quality indices • Waste recycling rates • Acreage, visitation, funding of recreational and protected/preserved areas • Water consumption and levels • Deaths, injuries, and property loss due to natural hazards • Endangered and threatened species	Sea level rise could both inundate coastal wetland habitats (with negative effects on marsh and estuarine environments necessary to purify water cycle systems and support marine hatcheries) and erode recreational beaches.
Human health	Health care institutions provide medical and preventive health-care services with excellence; citizens have access to services regardless of financial means, and physical and mental health is generally high.	• Mortality risks • Morbidity and risk of illness • Quality and accessibility of health care • Health status of vulnerable populations • Prenatal and childhood health • Psychological and emotional health	• Deaths from various causes (suicide, cancer, accidents, heart disease) • Life expectancy at birth • Health insurance coverage • Hospital services and costs • Infant mortality and care of elderly • Subjective measure of health status	Increased frequency of heat waves in a larger geographical area will directly affect health, resulting in higher incidence of heat-related mortality and illness. Climate can also affect human health indirectly via effects on ecosystems and water supplies.

Category of Well-being	Description and Rationale	Components/Indicators of Well-being	Illustrative Metrics/ Measures of Well-being	Examples of Negative Climate Linkages*
Public and private infrastructure	Transportation and communication infrastructure enable citizens to move around efficiently and communicate reliably.	• Affordable, and accessible public transit • Adequate road, air, and rail infrastructure • Reliable communication systems • Waste management and sewerage • Maintained and available public and private facilities • Power generation	• Mass transit use and commute times • Rail lines, and airport use and capacity • Telephones, newspapers, and internet • Waste tonnage and sewerage safety • Congestion and commute to work • Transportation accident rates • Noise pollution	Melting permafrost due to warming in the arctic damages road transport, pipeline, and utility infrastructure, which in turn leads to disrupted product and personal movements, increased repair costs, and shorter time periods for capital replacement.
Government and public safety	Governments are led by competent and responsive officials, who provide public services effectively and equitably, such as order and public safety; citizens are well-informed and participate in civic activities.	• Electoral participation • Civic engagement • Equity and opportunity • Municipal budgets and finance • Public safety • Emergency services	• Voter registration, turnout, approval • Civic organizations membership rates • Availability of public assistance programs • Debt, deficits, taxation, and spending • Crime rates and victimization • Emergency first-responders per capita	Dislocations and pressures created by climate change stressors can place significant new burdens on police, fire and emergency services.
Social and cultural resources	Social institutions provide services to those in need, support philanthropy, volunteerism, patronage of arts and leisure activities, and social interactions characterized by equality of opportunity and social harmony.	• Volunteerism • Culture, arts, entertainment, and leisure activities • Education and human capital services • Social harmony • Family and friendship networks	• Donations of time, money, and effort • Sports participation, library circulation, and support for the arts • Graduation rates and school quality • Hate, prejudice, and homelessness • Divorce rates, social supports	Disruptions in economic and political life caused by climate change stressors or extreme weather events associated with climate change could create new conflicts and place greater pressure on social differences within communities.

* The focus is on negative impacts as potentially more troubling for quality of life; there are also positive impacts and opportunities in some categories.

creative capital are expected to have greater individual and community quality of life (Putnam, 2000; Florida, 2002b).

In thinking about these indicators, it is important to keep two important contextual realities about climate change and well-being at the forefront. First, while discussions of climate change usually have a global context to them, the fact is that the effects of any specific changes in temperature, rainfall, storm frequency/intensity, and sea level rise will be felt at the local and regional level by citizens and communities living and working in those vulnerable areas. Therefore, not all populations will be placed under equal amounts of climate change-generated stress. Some will experience greater impacts, will suffer greater damage, and will need more remediation and better plans and resource allocations for adaptation and recovery efforts to protect and restore quality of life (see, for example, Zahran *et al.*, 2008; Liu, Vedlitz and Alston, 2008; Vedlitz *et al.*, 2007).

Second, not all citizens in areas more vulnerable to climate change effects are equally at risk. Some population groupings, within the same community, will be more vulnerable and at risk than others. Those who are poorer, minorities, aged or infirmed, and children are at greater risk than others to the stresses of climate change events (Lindell and Perry, 2004; Peacock, 2003). Recognizing that not all citizens of a particular vulnerable area share the same level of risk is something that planners and decision makers must take into account in projecting the likely impacts of climate change events on their populations, and in dealing with recovery of those populations (Murphy and Gardoni, 2008).

Finally, the situation is further complicated as climate stressors negatively affect disease conditions in other nations with particularly vulnerable and mobile populations. Increased communicable disease incidence in developing nations has the potential, through legal and illegal tourism and immigration, to affect community welfare and individual well-being in the United States.

4.2.2.2 Climate Change and Quality of Life Indicators

Social indicators are generally used to evaluate progress towards a goal: How is society doing? Who is being affected? Tracking performance for these indicators—using the types of metrics or measures indicated in Table 4.1—could provide information to the public on how communities and other entities are reacting to, and successfully adapting to (or failing to adapt to), climate change. The indicators and metrics included in Table 4.1 are intended to be illustrative of the types of indicators that might be used, rather than a comprehensive or recommended set. In any category, multiple indicators could be used; and any one of the indicators could have several measures. For example, exposure to natural hazards and risks could be measured by the percentage of a locality's tax base located in a high hazard zone, the number of people exposed to a natural hazard, the funding devoted to hazard mitigation, or the costs of hazard insurance, among others. Similarly, some indicators are more amenable to objective measurement; others are more difficult to measure, such as measures of social cohesion. The point to be taken from Table 4.1 is that social indicators provide a diverse and potentially rich perspective on human well-being.

The taxonomy presented in Table 4.1—or a similar taxonomy—might also provide a basis for analyses of the impacts of climate change on human well-being, providing a list of important categories for research (the components or indicators of life quality), as well as appropriate metrics (*e.g.,* employment, mortality or morbidity, etc.). The social indicators approach, and the specific taxonomy presented here, are only one of many that could be developed.[8] At the least, different conditions and stakeholder

8 In addition to variants on the social indicators approach, other types of taxonomies are possible—for example a taxonomy based on broad systems (atmospheric, aquatic, geologic, biological, and built environment), or on forms of capital that make up the productive base of society (natural, manufactured, human, and social). Well-being can also be viewed in terms of its endpoints: necessary material for a good life, health and bodily well-being, good social relations, security, freedom and choice, and peace of mind and spiritual existence (Rothman, Amelung, and Poleme, 2003).

mixes may demand different emphases. All taxonomies, however, face a common problem: how to interpret and use the diverse indicators, in order to compare and contrast alternative adaptive or mitigating responses to climate change. For some purposes, metrics have been developed that aggregate across individuals or individual categories of well-being and present a composite measure of well-being; or otherwise operationalize related concepts, such as vulnerability (see, for example the discussion of Figure 4.1).

4.2.3 A Closer Look at Communities

Looking beyond well-being of individuals to the welfare (broadly speaking) of communities— networks of households, businesses, physical structures, and institutions—provides a broader perspective on the impacts of climate change. The categories and metrics in Table 4.1 are appealing from an analytical perspective in part because they represent dimensions of well-being that are clearly important to individuals, but that also have counterparts and can generally be measured objectively at the community level. Thus, for example, the counterparts of individual income or health status are, at the social level, per capita income or mortality/ illness rates. The concept of community welfare is linked to human communities, but is not confined to communities in urban areas, or even in industrialized cultures. Human communities in remote areas, or subsistence economies, face the same range of quality of life issues—from health to spiritual values—although they may place different weights on different values; thus, the weights placed on different components of welfare are not determined *a priori*, but depend on community values and decision making.

Viewing social indicators and metrics through the lens of the community can be instructive in several ways. First, communities are dynamic entities, with multiple pathways of interactions among people, places, institutions, policies, structures, and enterprises. Thus, while the social indicators described in Table 4.1 have metrics that can be measured independently of each other, they are not determined independently within the complex reality of interdependent

human systems. Second, in part because of this interdependence, the aggregate welfare of a community is more than a composite of its quality of life metrics; sustainability provides one means of approaching a concept of aggregate welfare. Third, vulnerability and adaptation are typically analyzed at the sectoral level: "what should agriculture, or the public health system, do to plan for or adapt to climate change." The issue can also, however, be addressed at the level of the community. Each of these issues is touched on below.

4.2.3.1 Community Welfare and Individual Well-being

Rapid onset extreme weather events, such as hurricanes or tornadoes, can do serious damage to community infrastructure, public facilities and services, the tax base, and overall community reputation and quality of life, from which recovery may take years and never be complete (see additional discussion in Chapter 3). More gradual changes in temperature and precipitation will have both negative and positive effects. For example, as discussed elsewhere in this chapter, warmer average temperatures increase risks from heat-related mortality in the summer, but decrease risks from cold-related mortality in the winter, for susceptible populations. Effects such as these will not, however, be confined to a few individual sectors, nor are the effects across all sectors independent.

To illustrate the interdependence of impacts and, by extension, the analogous social indicators and metrics, consider a natural resource that faces additional stresses from climate change: fish populations in estuaries, such as the Chesapeake Bay, that are already stressed by air and water pollution from industry, agriculture, and cities. In this case, while the direct effects of climate will occur to the resource itself, indirect effects can alter welfare as measured by economic, social, and human health indicators. Table 4.2 presents some of the pathways by which resource changes could affect diverse categories of quality of life; the purpose of Table 4.2 is not to assert that all these effects will occur or that they will be significant if they do occur as a result of climate change, but

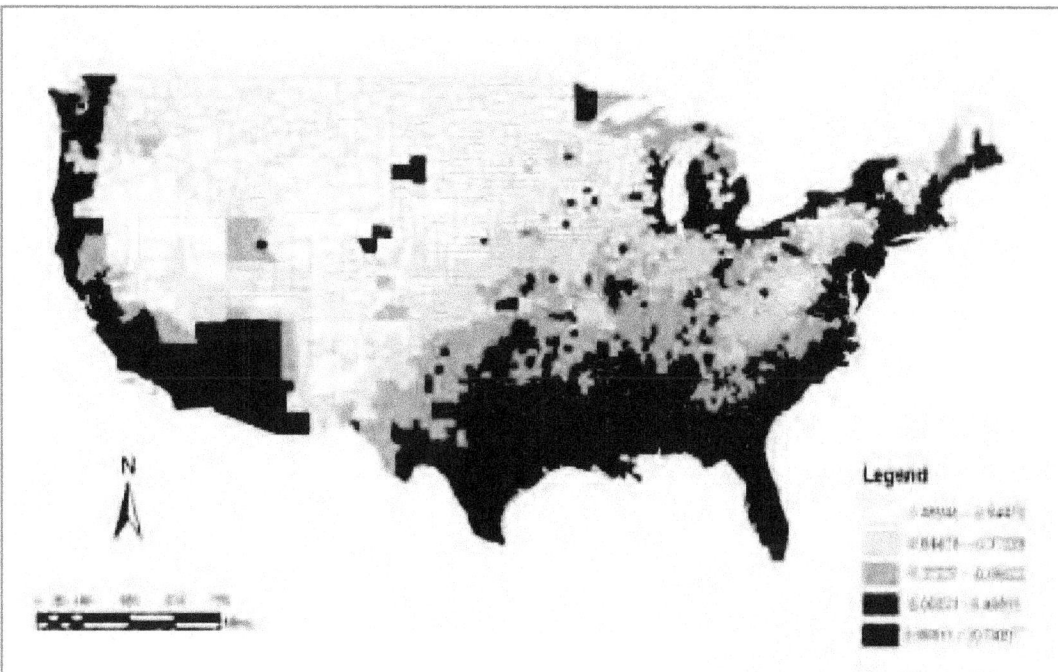

Three measures of climate change risk are used to create the vulnerability index: expected temperature change, extreme weather event history, and coastal proximity. Risk measures are geo-referenced at the county scale. The expected temperature change variable is measured as the expected unit change in average minimum temperature (in degrees Celsius) for a county from 2004 to 2099. Temperature data are from the Hadley Center. Hadley Center monthly time series data on average minimum temperature for the United States are plotted at the 0.5 x 0.5 degree of spatial resolution. In cases where climate cells intersect county boundaries, temperature data are averaged across intersecting climate cells. To estimate extreme weather event history, we summarize the number of reported injuries and fatalities from hydo-meterological hazard events at the county level Jan 01, 1960 to Jul 31, 2004. Higher values on our natural hazard casualty variable reflect more pronounced histories of injury and death from extreme weather events. Casualty data were collected from the Spatial Hazard Events and Losses Database for the United States (SHELDUS). The coastal proximity variable is measured dichotomously. A country receives a score of 1 if it is designated by the National Oceanic and Atmospheric Administration (NOAA) as an "at-risk-coastal" county, and a score of 0 if it is not. NOAA defines a county as at-risk-coastal if at least 15 percent of its total area is located in a coastal watershed. The vulnerability index was created by standardizing then summing each measure of climate change risk (z-score). The distribution of vulnerability is divided into equal quintles, with darker colors reflecting higher vulnerability to climate change.

Figure 4.1 Geography of Climate Change Vulnerability at the County Scale

Table 4.2 An illustration of Possible Effects of Climate Change on Fishery Resources

Linkages/Pathways	Category of Welfare Effect	Possible Metrics
Fishery resource declines as climate changes	Natural resources, environment, and amenities	Fish populations
Recreational opportunities decline	Natural resources, environment, and amenities	Fish catch, visitation days
Related species and habitats are affected	Natural resources, environment, and amenities	Species number and diversity
Employment and wages in resource-based jobs (including recreation) fall as resources decline	Economic conditions	Number of jobs, unemployment rate, wages
Incomes fall as jobs are lost	Economic conditions	Per capita income
More children live in poverty as jobs are lost and incomes fall	Economic conditions	Families, children below poverty level
Access to health care that is tied to jobs and income falls	Human health	Households without health insurance increase
Increased mortality and morbidity as a result of reduced health care	Human health	Disease and death rates increase
Lack of jobs results in out-migration	Economic conditions	Working age population decreases
Fewer new residents attracted, because of reduced jobs and amenities (recreation)	Social and cultural resources	Population growth rate slows
Less incentive/drive to participate in community activities	Social and cultural resources	Drop in volunteerism, civic participation, completion of high school

rather to illustrate the linkages. These linkages underscore the importance of understanding interdependencies within the community or, from another perspective, across welfare indicators. Table 4.2 illustrates the general principle of complex linkages in which a general equilibrium approach can be used to model climate change impacts.

4.2.3.2 Sustainability of Communities

Understanding how climate change and extreme events affect community welfare requires a different conceptual framework than that for understanding individual level impacts, such as quality of life.[9] Communities

are more than the sum of their parts; they have unique aggregate identities shaped by dynamic social, economic, and environmental components. They also have life cycles, waxing and waning in response to societal and environmental changes (Diamond, 2005; Fagan, 2001; Ponting, 1991; Tainter, 1988). Sustainability is a paramount community goal, typically expressed in terms of sustainable development in order to express the ongoing process of adaptation into the long-term future. "Climate change involves complex interactions between climatic, environmental, economic, political, institutional, social, and technological processes. It cannot be addressed or comprehended in isolation from broader societal goals (such as sustainable development)…" (Banuri and Weyant, 2001). Even for a country as developed as the United States, continuing growth and development creates both pressures on the natural and built environments and opportunities for moving in sustainable directions.

---

9 Measures of quality of life provide a database of relevant individual characteristics at various points in time, including economic conditions, natural resources and amenities, human health, public and private infrastructure, government and public safety, and social and cultural resources. Sustainable development measures are similar, but reflect more emphasis on long-term and reciprocal effects, as well as a concern for community-wide and equitable outcomes.

While the term sustainability does not have a single, widely accepted definition, a central guideline is to balance economic, environmental, and social needs and values (Campbell, 1996; Berke *et al.*, 2006). It is distinguished from quality of life by its dynamic linking of economic, environmental, and social components, and by its future orientation (Campbell, 1996; Porter, 2000). Sustainability is seen as living on nature's "interest," while protecting natural capital. Sustainability is a comprehensive social goal that transcends individual sector or impact measurements, although it can include narrower community welfare concepts such as the healthy city. Thinking about the impacts of climate change on communities through the lens of sustainable development allows us to envision cross-sector economic, environmental, and social dynamics.

4.2.4 Vulnerable Populations, Communities, and Adaptation

Responding to climate change at the community level requires understanding both vulnerability and adaptive responses that the community can take. Vulnerability of a community depends on its exposure to climate risk, how sensitive systems within that community are to climate variability and change, and the adaptive capacity of the community (*i.e.*, how it is able to respond and protect its citizens from climate change). Different groups within the community will be differentially vulnerable to climate changes (such as extreme events), and infrastructure and community coping capacity will be more or less effective in invoking a resilient response to climate change.

4.2.4.1 Vulnerable Populations

Categories of persons susceptible to environmental risks and hazards include racial and ethnic groups (Bolin, 1986; Fothergill *et al.*, 1999; Lindell and Perry, 2004; Cutter, 2006), and groups defined by economic variables of wealth, income, and poverty (Peacock, 2003; Dash *et al.*, 1997; Fothergill and Peek, 2004). Overall, research indicates that minorities and the poor are differentially harmed by disaster events. Economic disadvantage, lower human capital, limited access to social and political resources, and residential choices are social and economic reasons that contribute to observed differences in disaster vulnerability by race and ethnicity, and by economic status. While the literature on climate change and vulnerable populations is relatively underdeveloped, Chapter 2 on Human Health and Chapter 3 on Human Settlements each address population vulnerabilities.

Economic, social, and health effects are not neatly bounded by geographic or political regions, and so the damage and stresses that occur in a specific locality are not limited in their effects to only that community. As Hurricane Katrina made clear, impacts felt in one community ripple throughout the region and nation. Many of the persons made homeless in New Orleans resettled in Baton Rouge, Lafayette, and Houston, creating stresses on those communities. Vulnerable groups migrate from stricken areas to more hospitable ones, taking their health, economic, and educational needs and problems with them across both national and state lines.

4.2.4.2 Vulnerable Communities

While most analyses of vulnerability tend to be conducted at the regional scale, Zahran *et al.* (2008) have brought the analysis closer to the community level by mapping the geography of climate change vulnerability at the county scale. Their study uses measures of both physical vulnerability (expected temperature change, extreme weather events, and coastal proximity) and adaptive capacity (as represented by economic, demographic, and civic participation variables that constitute a locality's socioeconomic capacity to commit to costly climate change policy initiatives). Their map identifies the concentrations of highly vulnerable counties as lying along the east and west coasts and Great Lakes, with medium vulnerability counties mostly inland in the southeast, southwest, and northeast. (See Figure 4.1, in which darker areas represent higher vulnerability.)

Many possible dimensions can be used to identify and measure vulnerabilities to climate change impacts and stressors. The one presented in Figure 4.1 illustrates that the concept

of vulnerability is a viable one and can be measured and applied to communities in a Geographic Information System context. It is not the purpose of this chapter to focus in great detail on vulnerability measurement issues (for those interested in other formulations of the vulnerability concept, see Dietz *et al.*, In Press).

4.2.4.3 Adaptation

From the perspective of the community, the goal of successful adaptation to climate impacts—particularly potentially adverse impacts—is to maintain the long-term sustainability and survival of the community. Thus, a resilient community is capable of absorbing climate changes and the shocks of extreme events without breakdowns in its economy, natural resource base, or social systems (Godschalk, 2003). Given their control over shared resources, communities have the capacity to adapt to climate change in larger and more coordinated ways than individuals, by creating plans and strategies to increase resilience in the face of future shocks, while at the same time ensuring that the negative impacts of climate change do not fall disproportionately on their most vulnerable populations and demographic groups (Smit and Pilifosova, 2001).

Public policies and programs are in place in the United States to enhance the capacity of communities to mitigate[10] damage and loss from natural hazards and extreme events (Burby, 1998; Mileti, 1999; Godschalk, 2007). A considerable body of research looks at responses to natural hazards, and recent research has shown that the benefits of natural hazard mitigation at the national level outweigh its costs by a factor of four to one on average (Multihazard Mitigation Council, 2005; Rose *et al.*, 2007). Research also has been done on the social vulnerability of communities to natural hazards (Cutter *et al.*, 2003) and the economic resilience of businesses to natural hazards (Tierney, 1997; Rose, 2004). However, there

is scant research on U.S. policies dealing with community adaptation to the broader impacts of climate change.

4.3 AN ECONOMIC APPROACH TO HUMAN WELFARE

Welfare, well-being, and quality of life are often viewed as multi-faceted concepts. In subjective assessments of happiness or quality of life (see the discussion in Section 4.2), the individual makes a net evaluation of his or her current state, taking into account (at least implicitly) and balancing all the relevant facets or dimensions of that state of being. Constructing an overall statement regarding welfare from a set of objective measures, however, requires a means of weighting or ranking, or otherwise aggregating, these measures. The economic approach supplies one—although not the only possible—approach to aggregation.[11]

Quantitative measures of welfare that use a common metric have two potential advantages. First, the ability to compare welfare impacts across different welfare categories makes it possible to identify and rank categories with regard to the magnitude or importance of effects. Welfare impacts can then provide a signal about the relative importance of different impacts, and so help to set priorities with regard to adaptation or research. Second, if the concept of welfare is (ideally) a net measure, then it should be possible to aggregate the effects of climate across disparate indicators. Quantitative measures that use the same metric can, potentially, be summed to generate net measures of welfare, and gauge progress over time, or under different policy or adaptation scenarios.

---

10 In the natural hazards and disasters field, a single term—mitigation—refers both to adaptation to hazards and mitigation of their stresses (see the Disaster Mitigation Act of 2000, Public Law 106-390).

11 In part because of the difficulty in compiling the information needed for aggregation of economic measures, Jacoby (2004) proposes a portfolio approach to benefits estimation, focusing on a limited set of indicators of global climate change, of regional impact, and one global monetary measure. The set of measures would not be the only information generated and made available, but it would represent a set of variables continuously maintained and used to describe policy choices.

Given the value of welfare both as a multi-dimensional concept, and as one that facilitates comparisons, the economic approach to welfare analysis—which monetizes or puts dollar values on impacts—is one means of comparing disparate impacts. Further—and this is the second advantage of the economic approach—dollar values of impacts can be aggregated, and so provide net measures of changes in impacts that can be useful to policy makers. This section of the chapter discusses the foundation of economic valuation, the distinction between market and non-market effects (only the latter are covered in this paper), and describes some of the valuation tools that economists use for non-market effects. An appendix covers these issues in additional detail, and also describes the challenges that economic valuation faces when used as a tool for policy analysis in the long term context of climate change.

Fundamental to the economic approach is a notion that a key element of support for decision-making is an understanding of the magnitude of costs and benefits, so that the tradeoffs implicit in any decision can be balanced and compared. However, the economic approach, when interpreted as requiring a strict cost-benefit test, is not appropriate in all circumstances, and is viewed by some as controversial in the context of climate change.[12] Cost-benefit analysis is one tool available to decision makers. In the context of climate change, other decision rules and tools, or other definitions of welfare, may be equally, or more relevant. For example, the recent Synthesis Report of the IPCC Fourth Assessment (IPCC, 2007b) presents an average social cost (*i.e.*, damages) of carbon in 2005 of $12 per ton of CO_2, but also notes that the range of the roughly 100 peer-reviewed estimates of this value is -$3 to $95/t$CO_2$.[13] The IPCC attributes this very broad

range to differences in assumptions on climate sensitivity, response lags, the treatment of risk and equity, economic and non-economic impacts, the inclusion of potentially catastrophic losses, and discount rates. The IPCC therefore suggests consideration of an "iterative risk management process" to support decision-making.[14] Estimated benefits and costs therefore can provide information relevant to decision makers, but some of the methodologies and data necessary to provide a relatively complete assessment may be unavailable, as discussed subsequently in this section.[15]

---

12 See Arrow *et al.*, 1996 - at page 7, "There may be factors other than economic benefits and costs that agencies will want to weigh in decisions, such as equity within and across generations. In addition, a decision maker may want to place greater weight on particular characteristics of a decision, such as potential irreversible consequences."

13 See IPCC 2007b, page 23.

14 The IPCC further notes that existing analyses suggest costs and benefits of mitigation are roughly comparable in magnitude, "but do not as yet permit an unambiguous determination of an emissions pathway or stabilization level where benefits exceed costs." (IPCC 2007b page 23).

15 Other factors that might be considered, in addition to economic estimates, include emotions, perceptions, cultural values, and other subjective factors, all of which can play a role in creating preferences and reaching decisions. Those factors are beyond what we can evaluate in this chapter.

4.3.1 Economic Valuation

The framework that economists employ reflects a specific view of human welfare and how to measure it. Economists define the value of something—be it a good, service, or state of the world—by focusing on the well-being, utility, or level of satisfaction that the individual derives. The basic economic paradigm assumes that individuals allocate their available income and time to achieve the greatest level of satisfaction. The value of a good—in terms of the utility or satisfaction it provides—is revealed by the tradeoffs that individuals make between that good and other goods, or between that good and income.[16] The term "willingness to pay" (WTP) is used by economists to represent the value of something, *i.e.*, the individual's willingness to trade money for that particular good, service, or state of the world.

Economists distinguish between market and non-market goods. Market goods are those that can be bought and sold in the market, and for which a price generally exists. Market behavior and, in particular, the prices that are paid for these goods, is a source of information on the economic value or benefit of these goods. The economic benefit—the amount that members of society would in aggregate be willing to pay

16 Although economists are careful to distinguish between the metrics of utility and money as distinct, valuation metric in dollar units (rather than units of utility) may be generally viewed as the outcomes of individual preference expressions among goods, income, and time.

for these goods—is related to, but frequently greater than, market prices.

Non-market goods are those that are not bought and sold in markets. Consequently, climate change impacts that involve non-market effects—such as health effects, loss of endangered species, and other effects—are difficult to value in monetary terms. Economists have developed techniques for measuring non-market values, by inferring economic value from behavior (including other market behavior), or by asking individuals directly.

A number of studies have attempted to value the range of effects of climate change. For the United States, some of the most comprehensive studies are the Report to Congress completed by U.S. EPA in 1989 (U.S. EPA, 1989), Cline (1992), Nordhaus (1994), Fankhauser (1995), Mendelsohn and Neumann (1999), Nordhaus and Boyer (2000), and a body of work by Richard Tol (*e.g.*, Tol, 2002 and Tol, 2005). In all of these studies, the focus is largely on market impacts, particularly the effects of climate change on agriculture, forestry, water resource availability, energy demand (mostly for air conditioning), coastal property, and in some cases, health.

Non-market effects, however, are less well characterized in these studies (Smith *et al.*, 2003); where comprehensive attempts are made, they usually involve either expert judgment or very rudimentary calculations, such as multiplying the numbers of coastal wetland acres at risk of inundation from sea level rise by an estimate of the average non-market value of a wetland. One such comprehensive attempt generated a value for 17 ecosystem services from 16 ecosystem types (Costanza *et al.*, 1997), but also generated controversy and criticism from many economists (Bockstael *et al.*, 2000; Toman, 1998; see National Research Council 2004 for a summary). Other analysts have attempted to define measures to reflect non-market ecosystem services in terms similar to those used for Gross Domestic Product (Boyd, 2006), or indicators of ecosystem health that reflect ecological contributions to human welfare (Boyd and Banzhaf, 2006).[17] While there are several well-done valuation analyses for non-market effects of climate change (as described later in this chapter), it is fair to characterize this literature as opportunistic in its focus; where data and methods exist, there are high quality studies, but the overall coverage of non-market effects remains inadequate.

4.3.2 Impacts Assessment and Monetary Valuation

The process of estimating the welfare effects of climate change involves four steps: (1) estimate climate changes; (2) estimate physical effects of climate change, (3) estimate the impacts on human and natural systems that are amenable to valuation and (4) value or monetize effects. The first step requires estimating the change in relevant measures of climate, including temperature, precipitation, sea level rise, and the frequency and severity of extreme events. The second step involves estimating the physical effects of those changes in climate. Such effects might include changes in ecosystem structure and function, human exposures to heat stress, changes in the geographic range of disease vectors, or flooding of coastal areas. In the third step, the physical effects of climate change are translated into measures that economists can value, for example the number and location of properties that are vulnerable to floods, or the number of individuals exposed to and sensitive to heat stress. Many analyses that reach this step in the process, but not all, also proceed on to the fourth step, valuing the changes in dollar terms.

The simplest approach to valuation would be to apply a unit valuation approach. For example, the cost of treating a nonfatal case of heat stress or malaria attributable to climate change is a first approximation of the value of avoiding that case altogether. In many contexts, however, unit values can misrepresent the true marginal economic impact of these changes. For example, if climate change reduces the length of the ski season, individuals could engage in another

recreational activity, such as golf. Whether they might prefer skiing to golf at that time and location is something economists might try to measure.

This step-by-step linear approach to effects estimation is sometimes called the "damage function" approach. A damage function approach might imply that we look at effects of climate on human health as separate and independent from effects on ecology and recreation, an assumption that ignores the complex economic interrelationships among goods and services and individual decisions regarding these. Recent research suggests that the damage function approach, under some conditions, may be both overly simplistic (Freeman, 2003) and sometimes subject to serious errors (Strzepek and Smith, 1995; Strezpek *et al.*, 1999).

Economists have a number of techniques available for moving from quantified effects to dollar values. In some cases, the values estimated in one situation—*e.g.*, one ecosystem or species—can be transferred and used to value another. For example, value or benefits transfer is commonly used by federal agencies such as the U.S. EPA and U.S. Forest Service to value recreation when there is insufficient time or budget to conduct original valuation studies (Rosenberger and Loomis, 2003). Techniques commonly used by economists to value non-market goods and services include:

---

17 Some political economists also emphasize the role of explicit recognition of non-market environmental values as an important step in improving the well-being of poor populations (Boyce and Shelley, 2003).

- *Revealed preference.* Revealed preference, sometimes referred to as the indirect valuation approach, involves inferring the value of a non-market good using data from market transactions (U.S. EPA, 2000; Freeman, 2003). For example, the value of a lake for its ability to provide a good fishing experience can be estimated by the time and money expended by the angler to fish at that particular site, relative to all other possible fishing sites. Likewise, the amenity value of a coastal property that is protected from storm damage (by a dune, perhaps) can be estimated by comparing the price of that property to other properties similar in every way but the enhanced storm protection.

- *Stated preference.* Stated preference methods, sometimes referred to as the direct valuation, are survey methods that estimate the value individuals place on particular non-market goods based on choices they make in hypothetical markets. The earliest stated preference studies involved simply asking individuals what they would be willing to pay for a particular non-market good. The best studies involve great care in constructing a credible, though still hypothetical, trade-off between money and the non-market good of interest (or bundle of goods) to discern individual preferences for that good and hence, willingness to pay WTP.

- *Replacement or avoided costs.* Replacement cost studies approach non-market values by estimating the cost to replace the services provided to individuals by the non-market good. For example, healthy coastal wetlands

may provide a wide range of services to individuals who live near them (such as filtering pollutants present in water). A replacement cost approach would estimate the value of these services by estimating market costs for replacing the services provided by the wetlands. Analogously, the cost of health effects can be estimated using the cost of treating illness and of the lost workdays, etc. associated with illness.

- *Value of inputs.* This approach calculates value based on the contribution of an input into some productive process. This approach can be used to determine the value of both market and non-market inputs, for example, fertilizer, water, or soil, in farm output and profits.

In the remainder of this section, we briefly discuss the relationship between climate change and four non-market effects (human health, ecosystems, recreation and tourism, and amenities), and discuss economic estimates of these effects using these techniques.

4.3.3 Human Health

In the United States, climate change is likely to measurably affect health outcomes known to be associated with weather and climate, including heat-related illnesses and deaths, health effects due to storms, floods, and other extreme weather events, health effects related to poor air quality, water- and food-borne diseases, and insect-, tick-, and rodent-borne diseases. In addition to changes in mortality and morbidity, climate change may affect health in more subtle ways. Good health is more than the absence of illness; it includes mental health, the ability to function physically (to climb stairs or walk a mile), socially (to move freely in the world), and in a work environment. See Chapter 2 of this report, for an overview of health effects that have been associated with climate change.

Despite our understanding of the pathways linking climate and health effects, there is uncertainty as to the magnitude and geographic and temporal variation of possible impacts on morbidity and mortality in the United States. This is primarily due to a poor understanding of many key risk factors and confounding issues, such as behavioral adaptation and variability in population vulnerability (Patz *et al.*, 2001). Even where our understanding of underlying climate and health relationships is better, few studies have attempted to explicitly link these findings to climate change scenarios to quantitatively estimate health impacts. Economists have relatively well established (although sometimes controversial) techniques for valuing mortality and some forms of morbidity, which could, in theory, be applied to quantified impacts assessments.

4.3.3.1 Overview of Health Effects of Climate Change

The United States is a developed country with a temperate climate. It has a well-developed health infrastructure and government and non-governmental agencies involved in disaster planning and response, both of which can help to mitigate potential health effects from climate change. Nevertheless, certain regions of the United States will face difficult challenges arising from some of the following health effects.

- *Illnesses and deaths due to heat waves.* A likely impact in the United States is an increase in the severity, duration, and frequency of heat waves (Kalkstein and Greene, 1997; IPCC, 2007c). This, coupled with an aging (and therefore more vulnerable) population, will increase the likelihood of higher mortality from exposure to excessive heat (see, for example, Semenza *et al.*, 1996, and Knowlton *et al.*, 2007).

- *Injuries and death from extreme weather events.* Climate change is projected to alter the frequency, timing, intensity, and duration of extreme weather events, such as hurricanes and floods (Fowler and Hennessey, 1995). The health effects of these extreme weather events range from the direct effects, such as loss of life and acute trauma, to indirect effects, such as loss of shelter, large-scale population displacement, damage to sanitation infrastructure (drinking water and sewage systems), interruption of food production, damage to the health care infrastructure, and psychological problems such as post traumatic stress disorder (Curriero *et al.*, 2001).

- *Illnesses and deaths due to poor air quality.* Climate change can affect air quality by modifying local weather patterns and pollutant concentrations (such as ground level ozone), by affecting natural sources of air pollution, and by changing the distribution of air-borne allergens (Morris *et al.*, 1989; Sillman and Samson, 1995). Many of these effects are localized and, for ozone, compounded by assumptions of trends in precursor emissions. Despite these uncertainties, all else being equal, climate change is projected to contribute to or exacerbate ozone-related illnesses.

- *Water- and food-borne diseases.* Altered weather patterns, including changes in precipitation, temperature, humidity, and water salinity, are likely to affect the distribution and prevalence of food- and water-borne diseases resulting from bacteria, overloaded drinking water systems, and increases in the frequency and range of harmful algal blooms (Weniger *et al.*, 1983; MacKenzie *et al.*, 1994; Lipp and Rose, 1997; Curriero *et al.*, 2001).

- *Insect-, tick-, and rodent-borne diseases.* Vector-borne diseases, such as plague, Lyme's disease, malaria, hanta virus, and dengue fever have distinct seasonal patterns, suggesting that they may be sensitive to climate-driven changes in rainfall and temperature (Githeko and Woodward, 2003). Moderating factors, such as housing quality, land-use patterns, vector control programs, and a robust public health infrastructure, are likely to prevent the large-scale spread of these diseases in the United States.

4.3.3.2 Quantifying the Health Impacts of Climate Change

A large epidemiological literature exists on the health effects associated with climate change, particularly the mortality effects associated with increases in average monthly or seasonal temperature, and with changes in the intensity, frequency, and duration of heat waves. As described in Chapter 2, there is considerable speculation concerning the balance of climate change-related decreases in winter mortality compared with increases in summer mortality, although researchers suspect that declines in winter mortality associated with climate change are unlikely to outweigh increases in summer mortality (McMichael *et al.*, 2001; Kalkstein and Greene, 1997; Davis, 2004).

Net changes in mortality are difficult to estimate because, in part, much depends on complexities in the relationship between mortality and the changes associated with global change. Using average temperatures to estimate cold-related mortality, for example, is complicated by the fact that many factors contribute to winter mortality (such as spread

of the influenza virus). Similarly, increased summer mortality may be affected not only by average temperature, but also by other temperature factors, such as variability in temperature, or the duration of heat waves. Moreover, quantifying projected temperature-related mortality requires going beyond epidemiology and projecting adaptive behaviors, such as the use of air conditioning, expanded public programs (such as heat warning systems), or migratory patterns.

Few studies have attempted to link the epidemiological findings to climate scenarios for the United States, and studies that have done so have focused on the effects of changes in average temperature, with results dependent on climate scenarios and assumptions of future adaptation.[18] Moreover, many factors contribute to winter mortality, making highly uncertain how climate change could affect mortality. No projections have been published for the United States that incorporate critical factors, such as the influence of influenza outbreaks. Below, we report the results of these studies in order to give a sense of the magnitude of mortality that might be associated with temperature changes due to climate change and, by intimation, the magnitude of potential changes in economic welfare. The conclusions should be considered preliminary, however, in part because of the complexities in estimating mortality under future climate scenarios. Moreover, none of the studies reported below traces through the quantitative implications of various climate scenarios for mortality in all regions of the United States using region-specific data, suggesting a clear need for future research.

---

18 McMichael *et al.* (2004) estimate the impact of climate change on DALYs (Disability-Adjusted Life Years) associated with waterborne and vector-borne illness for WHO regions. (DALYs represent the sum of life-years lost due to premature death and productive life years lost due to chronic illness or injury.) For the US, it is not anticipated that climate change will lead to loss of life or years of life due to chronic illness or injury from waterborne or foodborne illnesses. However, there will likely be an increase in the spread of several food- and water-borne pathogens among susceptible populations depending on the pathogens' survival, persistence, habitat range and transmission under changing climate and environmental conditions.

Quantifying the relationship between climate change and cases of injury, illness, or death requires an exposure-response function that quantifies the relationship between a health endpoint (*e.g.*, premature mortality due to cardiovascular disease (CVD), cases of diarrheal disease) and climate variables (*e.g.*, temperature and humidity). The exposure-response function can be used to compute the relative risk of illness or death due to a specified change in climate, *e.g.*, a temperature increase of 2.5°C. Applying this relative risk to the baseline incidence of the illness or death in a population yields an estimated number of cases associated with the climate scenario.

Two studies have attempted to link exposure-response functions to future climate scenarios and thereby develop temperature-related mortality estimates.[19] McMichael *et al.* (2004) estimate the effects of average temperature changes associated with projected climates resulting from alternative emissions scenarios, by WHO region. For the AMR-A region, which includes the United States, Canada, and Cuba, they estimate the impact on cardio-vascular mortality relative to baseline conditions in 1990. Effects are estimated for average temperature projections associated with three alternative emissions scenarios: (1) no control of GHG emissions,[20] (2) stabilization at 750 parts-per-million (ppm) of CO_2 equivalent by 2210, and (3) stabilization at 550 ppm CO_2 equivalent by 2170.[21]

McMichael *et al.* (2004) bases the estimates of the effects of average temperature changes on mortality from CVD for AMR-A on Kunst *et al.* (1993). Kunst *et al.* (1993) find CVD mortality rates to be lowest at 16°C, and to increase by 0.5 percent for every degree C below 16°C and increase by 1.1 percent for each degree above 16°C. In applying these results to future climate scenarios, McMichael *et al.*

(2004) assume that people will adjust to higher average temperatures; thus, the temperature at which mortality rates reach a minimum is adjusted by scenario. No adjustment is made for attempts to mitigate the effects of higher temperatures through (for example) increased use of air-conditioning. The effect of the climate scenarios for the AMR-A, reported for 2020 and 2030, is, on net, zero. Reductions in CVD mortality due to warmer winter temperatures cancel out increases in CVD mortality due to warmer summer temperatures.

Hayhoe *et al.* (2004) examine the impacts on climate and health in California of projected climate change associated with two emissions scenarios. The emissions scenarios are similar to those used in McMichael *et al.* (2004): (1) stabilization at 970 ppm of CO_2 and (2) stabilization at 550 ppm of CO_2.[22] In Los Angeles, by the end of the century, the number of heat wave days (3 or more days with temperatures above 32°C) increases fourfold under scenario B1 and six to eight times under scenario A1fi. From a baseline of 165 excess deaths in the 1990s, heat-related deaths in Los Angeles are projected to increase two to three times under scenario B1 and five to seven times under scenario A1fi by 2090.

These results can be compared with those of an earlier study that employed a composite climate variable to examine the impact of extreme temperatures on daily mortality under future climate scenarios. Kalkstein and Greene (1997)

---

19 These studies use climate scenarios that are associated with different emissions scenarios from IPCC (2000), the so-called SRES scenarios.

20 McMichael *et al.* (2004) represent unmitigated emissions using the IS92a emissions scenario presented in IPCC (2000).

21 Climate scenarios are projected for 2025 and 2050 using the HadCM2 model at a resolution of 3.75° longitude by 2.5° latitude and interpolated to other years.

---

22 Hayhoe uses two SRES (IPCC 2000) emissions scenarios: A1fi (corresponding to 970 ppm of CO_2) and B1 (corresponding to 550 ppm of CO_2).

analyzed the effect of temperature extremes (both hot and cold) on mortality for 44 U.S. cities in the summer and winter. They then applied these results to climate projections from two GCMs for 2020 and 2050. In 2020, under a no-control scenario, excess summer deaths in the 44 cities were estimated to increase from 1,840 to 1,981-4,100, depending on the GCM used. The corresponding figures for 2050 were 3,190-4,748 excess deaths.

4.3.3.3 Valuation of Health Effects

In cost-benefit analyses of health and safety programs, mortality risks are commonly valued using the "value of a statistical life" (VSL)—defined as the sum of what people would pay to reduce their risk of dying by small amounts that, together, add up to one statistical life. This approach allows valuation economists to focus on how people respond to and implicitly value mortality risk in their daily decisions, rather than attempting to value the lives lost, *per se* (U.S. EPA, 2000). This approach also responds to the type of data that is typically available; the excess deaths associated with a particular climate scenario are indeed the number of statistical lives that would be lost.

Willingness to pay for a current reduction in risk of death (*e.g.*, over the coming year) is usually estimated from compensating wage differentials in the labor market (a revealed preference method), or from contingent valuation surveys (a stated preference method) in which people are asked directly what they would pay for a reduction in their risk of dying. The basic idea behind compensating wage differentials is that jobs can be characterized by various attributes, including risk of accidental death. If workers are well-informed about risks of fatal and non-fatal injuries, and if labor markets are competitive, riskier jobs should pay more, holding worker and other job attributes constant (Viscusi, 1993). In theory, the impact of a small change in risk of death on the wage should equal the amount a worker would have to be compensated to accept this risk. For small risk changes, this is also what the worker should pay for a risk reduction.

For the compensating wage approach to yield reliable estimates of the VSL, it is necessary

that workers be informed about fatal job risks and that there be sufficient competition in labor markets for compensating wage differentials to emerge.[23] To measure these differentials empirically requires accurate estimates of the risk of death on the job—ideally, broken down by industry and occupation. The researcher must also be able to include enough other determinants of wages that fatal job risk does not pick up the effects of other worker or job characteristics. Empirical estimates of the value of a statistical life based on compensating wage studies conducted in the United States lie in the range of $0.6 million to $13.5 million (1990 dollars) (Viscusi, 1993; U.S. EPA, 1997), which is the rough equivalent of $0.7 million to $16.5 million in year 2000 dollars.[24]

The challenge in valuating health effects is compounded by the long-term nature of climate risks, which suggests that much of the premature mortality associated with higher temperatures will occur in the future. Indeed, McMichael *et al.* (2004) and Kalkstein and Greene (1997) estimate mortality based on climate effects around the years 2020 and 2050; Hayhoe *et al.* (2004) analyze impacts in 2070-2099.

It is also the case that the majority of the health effects of climate change will be felt by persons 65 and over. Recent attempts to examine how the VSL varies with worker age (Viscusi and Aldy, 2007) suggest that the VSL ranges from $9.0 million (2000 dollars) for workers aged 35-44 to $3.7 million for workers aged 55-62. Contingent valuation studies (Alberini *et al.*, 2004) also suggest that the VSL may decline with age. Further, economic theory suggests that, under some assumptions, persons are willing to pay less to reduce a risk they will face in the future (say, at age 65) than they are willing to pay to reduce a risk they face today (Cropper and Sussman, 1990). Both these factors may affect the economic value

---

23 Estimates of compensating wage differentials are often quite sensitive to the exact specification of the wage equation. Black *et al.* (2003), in a reanalysis of data from U.S. compensating wage studies requested by the USEPA, conclude that the results are too unstable to be used for policy.
24 Adjusted using the GDP implicit price deflator produced by the Bureau of Economic Analysis US Department of Commerce, available at http://www.bea.gov/national/nipaweb/TableView.asp#Mid.

that would be attached to excess mortality estimates, such as those derived by Kalkstein and Greene (1997).

The potential health effects associated with climate change are much broader than the changes in excess mortality discussed above. The effects of climate on illness have been examined in the literature, as indicated in the previous section; however, there have been few attempts to examine the implications of these studies for future climate scenarios. In addition to quantified estimates of mortality and morbidity, themselves indications of well-being and welfare, a range of economic techniques that have been developed for use in cost-benefit analyses of health and safety regulations could be applied to many of the endpoints that may be affected by climate change, as suggested by Table 4.3. Before these methods could be applied, however, the impacts of climate change must be translated into physical damages.

It is also the case that good health is more than the absence of illness. All of the dimensions of functioning measured in standard questionnaires (including various health outcomes surveys) (HCFR, 2004) may be affected by changes in climate. From a valuation perspective, we would expect changes in functional limitations (stiffness of joints, difficulty walking) not to be linked directly to climate or to weather, but rather to be instrumental in people's location decisions and, thus, reflected in wages and property values. The relationship between climate and wages and property values are discussed below in Section 4.3.6 on amenity values.

4.3.4 Ecosystems

Human welfare depends, in many ways, on the Earth's ecosystems and the services that they provide, where ecosystem services may be defined as "the conditions and processes through which natural ecosystems, and the species that make them up, sustain and fulfill human life" (Daily, 1997). These services contribute to human well-being and welfare by contributing to basic material needs, physical and psychological health, security, and economic activity, and in other ways (see Table 4.4). For example, a variety of ecosystem changes may be linked to changes in human health, from changes that encourage the expansion of the range of vector-borne diseases (discussed in Chapter 2) to the frequency and impact of floods and fires on human populations due to changes in protection afforded by ecosystems.

Table 4.3 Techniques to Value Health Effects Associated with Climate Change

Health Effect	Economic Valuation Tools
Premature mortality (associated with temperature changes, extreme weather events, and air pollution effects)	Use of revealed preference techniques to value changes in risk of death (e.g., compensating wage studies). Use of stated preference studies to value changes in risk of death. Use of foregone earnings as a lower bound estimate to the value of premature mortality.
Exacerbation of cardiovascular and respiratory morbidity; morbidity associated with water-borne or vector-borne disease	Use of stated preference methods to elicit WTP to avoid illness (e.g., asthma attacks) or risk of illness (heart attack risk) or injury. Estimation of medical costs and productivity losses (known as the cost-of-illness (COI)) as a lower bound estimate of the value of avoiding illness.
Injuries associated with extreme weather events	Use of stated preference methods to elicit WTP. Use of compensating wage studies that value risk of injury. Use of COI as a lower bound estimate.
Impacts of climate change on physical functioning; sub-clinical effects	Use of stated preference methods to estimate WTP to avoid functional limitation.

Table 4.4 Examples of Ecosystem Services Important to Human Welfare*

Service Category	Components of Service	Illustration of Service
Provisioning services	Food Fiber Fresh water Genetic resources Pharmaceuticals	Harvestable fish, wildlife, and plants Timber, hemp, cotton Water for drinking, hydroelectricity generation, and irrigation
Regulating services	Air quality regulation Erosion regulation Water purification Pest control Crop pollination Climate and water supply regulation Protection from natural hazards	Local and global amelioration of extremes Removal of contaminants by wetlands Removal of timber pests by birds Pollination of orchards by flying insects
Support services	Primary production Soil formation Photosynthesis Nutrient and water cycling	Conversion of solar energy to plant material Conversion of geological materials to soil by addition of organic material and bacterial activity
Cultural services	Recreation/tourism Aesthetic values Spiritual/religious values Cultural heritage	Natural sites for "green" tourism/recreation/nature viewing Existence value of rainforests and charismatic species, "holy" or "spiritual" natural sites

*Based on a classification system developed for the Millennium Ecosystem Assessment (MA, 2005).

The ability of the biosphere to continue providing these vital goods and services is being strained by human activities, such as habitat destruction, releases of pollutants, over-harvesting of plants, fish, and wildlife, and the introduction of invasive species into fragile systems. The recent Millennium Ecosystem Assessment reported that of 24 vital ecosystem services, 15 were being degraded by human activity (MA, 2005). Climate change is an additional human stressor that threatens to intensify and extend these adverse impacts to biodiversity, ecosystems, and the services they provide.

Changes in temperature, precipitation, and other effects of climate change will have *direct* effects on ecosystems. Climate change will also *indirectly* affect ecosystems, via, for example, effects of sea level rise on coastal ecosystems, decision-makers' responses to climate change (in terms of coastline protection or land use), or increased demands on water supplies in some locations for drinking water, electricity generation, and agricultural use. Understanding how these changes alter economic welfare requires identifying and potentially valuing changes in ecosystems resulting from climate change. Getting to the point of valuation, however, requires establishing a number of linkages—from projected changes in climate to ecosystem change, to changes in services, to changes in the value of those services—as illustrated in Figure 4.2. The scientific community has not, thus far, focused explicitly on establishing these linkages in the context of climate change. Consequently, the published literature is somewhat fragmented, consisting of discussions of climate effects on ecosystems and of valuation of ecosystems and their services (in only a few cases do the latter focus on climate change).

Already observed effects (see reviews in Parmesan and Yohe, 2003; Root *et al.*, 2003; Parmesan and Galbraith, 2004) and modeling results indicate that climate change is very likely to have major adverse impacts on ecosystems (Peters and Lovejoy, 1992; Bachelet *et al.*, 2001; Lenihan *et al.*, 2006; Galbraith *et al.*, 2006). It is also likely that these changes will

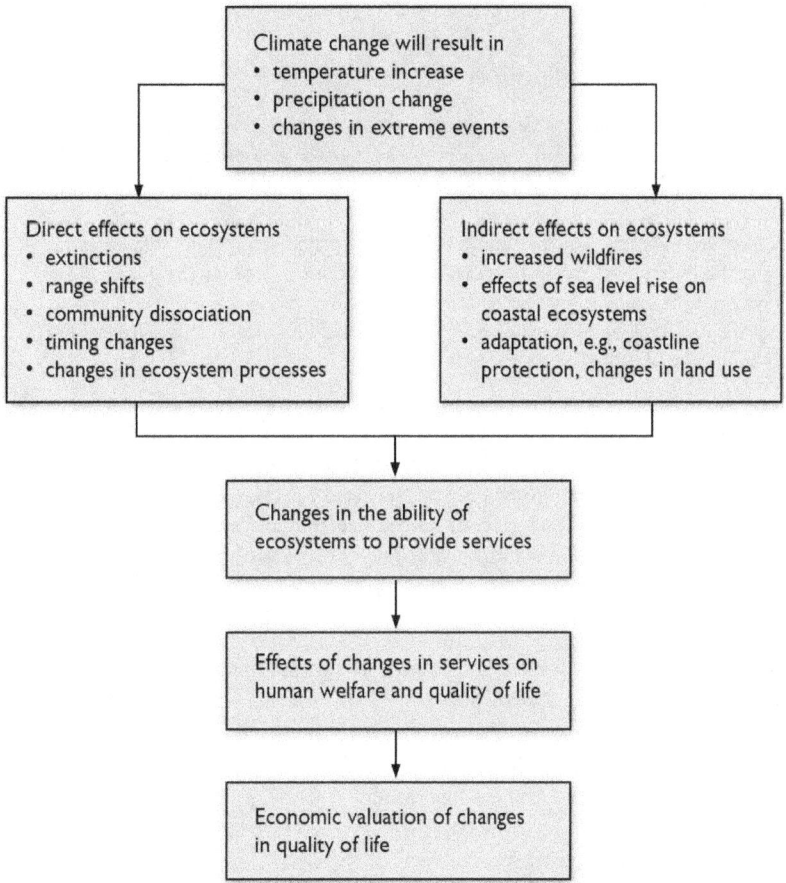

Figure 4.2 Steps from Climate Change to Economic Valuation of Ecosystem Services

adversely affect the services that humans and human systems derive from ecosystems (MA, 2005). Climate change may affect ecosystems in the United States within this century in the following ways.

Shifting, breakup, and loss of ecological communities. As climate changes, species that are components of communities will be forced to shift their ranges to follow cooler temperatures either poleward or upward in elevation. In at least some cases, this is likely to result in the breakup of communities as organisms respond to temperature change and migrate at different rates. In general, study projections include: northern extensions of the ranges of southern broadleaf forest types, with northward contractions of the ranges of northern and boreal conifer forests; elimination of alpine tundra from much of its current range in the United States; and the replacement of forests by grasslands, shrub-dominated communities,

and savannas, particularly in the south (*e.g.,* VEMAP, 1995; Melillo *et al.*, 2001; Lenihan *et al.*, 2006). Because of different intrinsic rates of migration, ecological communities may not move intact into new areas (Box 4.1).

Another potential ecological community effect of climate change is the facilitation of community penetration and degradation by invasive weeds that will replace more sensitive native species (Malcolm and Pitelka, 2000).

Extinctions of plants and animals and reduced biodiversity. While some species may be able to adapt to changing climate conditions, others will be adversely affected. It is very likely that one result of this will be to accelerate current extinction rates, resulting in loss in biodiversity. The most vulnerable species within the United States may be those that are currently confined to small, fragmented habitats that may be sensitive to climate change. This is the case with

BOX 4.1. Effects of Climate Change on Selected U.S. Ecosystems

At their most extreme, ecological community changes could result in the loss of entire habitats valued by the general public. For example, sea level rise puts much of the freshwater wetland that comprises Florida Everglades National Park at risk (Glick and Clough, 2006). Even relatively modest sea level rise projections could result in the conversion of much of this low-lying area to brackish or intertidal marine and mangrove habitats. Another such extreme example is alpine tundra habitat in mountain ranges in the contiguous states. Since tundra lies at the highest elevations, there is little or no opportunity for the plants and animals that comprise this ecosystem to respond to increasing temperatures by moving upward. Thus, one of the probable effects of climate change will be the further fragmentation and loss of this unique habitat (VEMAP, 1995; Root et al., 2003; Lenihan et al., 2006).

California already reports an example of how climate change might modify major marine ecological communities. Over the final four decades of the 20th century the average annual ocean surface temperature off the California coast warmed by approximately 1.5°C (Holbrook et al., 1997). Sagarin et al. (1999) found that the intertidal invertebrate community at Monterey has changed since first it was characterized in the 1930s. Many of the coolwater species have retracted their ranges northward, to be replaced by southern warm water species. The ecological community that exists there now is markedly different in its make-up from that which existed prior to warming of the coastal California Current.

Edith's checkerspot, a western butterfly species that is already undergoing local subpopulation extinctions due to climate change (Parmesan, 1996). Other potentially vulnerable organisms include those that are restricted to alpine tundra habitats (Wang et al., 2002), or to coastal habitats that may be inundated by sea level rise (Galbraith et al., 2002).

Range shifts. Faced with increasing temperatures, populations of plants and animals will attempt to track their preferred climatic conditions by shifting their ranges. Range shifts will be limited by factors such as geology (in the case of plants that are confined to certain soil types), or the presence of cities, agricultural land, or other human activities that block northward migration. Some individual species in North America and the United States are already undergoing range shifts (Root et al., 2003; Parmesan and Galbraith, 2004). The red fox in the Canadian arctic shifted its range northward by up to 600 miles during the 20th century, with the greatest expansion occurring where temperature increases have been the largest (Hersteinsson and Macdonald, 1992). More generally, a number of bird species have shifted their ranges northward in the United States over the past few decades. While some of these changes may be attributable to non-climatic factors, it is very likely that some

are due to climate change (Root et al., 2003; Parmesan and Galbraith, 2004).

Timing changes. The timing of major ecological events is often triggered or modulated by seasonal temperature change. Changes in timing may already be occurring in the breeding seasons of birds, hibernation seasons of amphibians, and emergences of butterflies in North America and Europe (Bebee, 1995; Crick et al., 1997; Brown et al., 1999; Dunn and Winkler, 1999; Root et al., 2003; Roy and Sparks, 2000). Disconnects in timing of interdependent ecological events may be accompanied by adverse effects on sensitive organisms in the United States. Such effects have already been observed in Europe where forest-breeding birds have been unable to advance their breeding seasons sufficiently to keep up with the earlier emergence of the arboreal caterpillars with which they feed their young. This has resulted in declining productivity and population reductions in at least one species (Both et al., 2006).

Changes in ecosystem processes. Ecosystem processes, such as nutrient cycling, decomposition, carbon flow, etc., are fundamentally influenced by climate. Climate change is likely to disrupt at least some of these processes. While these effects are difficult to quantify, some types of changes can—and have been observed. Increasing temperatures over

the past few decades on the North Slope of Alaska have resulted in a summer breakdown of the permanently frozen soil of the Alaskan Tundra and increased activity by soil bacteria that decompose plant material. This has accelerated the rate at which CO_2 (a breakdown product of the decomposition of the vegetation and also a greenhouse gas) is released to the atmosphere—changing the Tundra from a net sink (absorber) to a net emitter of CO_2 (Oechel et al., 1993; Oechel et al., 2000).

Indirect effects of climate change. Climate change may also result in "indirect" ecological effects as it triggers events (the frequency and intensity of fires, for example) that, in turn, adversely affect ecosystems. In U.S. forest habitats, increased temperatures are very likely to result in increased frequency and intensity of wildfires, especially in the arid west, leading to the breakup of contiguous forests into smaller patches, separated by shrub and grass dominated ecological communities that are more resistant to the effects of fire (Lenihan et al., 2006). Other major indirect effects are likely to include the loss of coastal habitat through sea level rise (Warren and Niering, 1993; Ross et al., 1994; Galbraith et al., 2002), and the loss of coldwater fish communities (and the recreational fishing that they support) as water temperatures increase (Meyer et al., 1999).

The linkages between these types of changes and the provision of ecosystem services are difficult to define. While ecologists have developed a number of metrics of ecosystem condition and functioning (e.g., species diversity, presence/absence of indicator species, primary productivity, nutrient cycling rates), these do not generally bear an obvious relation to metrics of services. In some cases, such as species diversity and bird population sizes, direct links might be drawn to services (in this case, opportunities for bird watching). However, in many, if not most cases, the linkages between stressor effects, change in ecosystem metrics, and service flows, are more obscure. For example, it is known that freshwater wetlands can remove contaminants from surface water (Daily, 1997) and this is an important service. However, the specific ways in which wetlands do this—in terms of the ecological processes and linkages within

the system—are not well understood, probably vary between different types of wetland (e.g., beaver swamps vs. cattail stands), and may vary spatially and temporally.

4.3.4.1 Economic Valuation of Effects on Ecosystems

Ecosystems are generally considered non-market goods: although land itself can be bought and sold, there is no market for ecosystem services per se, and so land value is only a partial measure of the value of the full range of ecosystem services provided. From the perspective of human welfare and climate change, however, we are concerned less with the ecosystems or the land on which they are located, than with the diverse services they provide, as illustrated in Table 4.4.

Economic valuation of changes in ecosystem services will be easier in cases where there are relationships between market goods and the ecosystem services being valued. For example, ecosystem changes may result in changes in the availability of goods and services that are traded on markets, as in the case of provisioning services, such as food, fisheries, pharmaceuticals, etc. In other cases, market counterparts to the services may exist, as in the case of regulating services; for example, insights into the value of water purification services can come from looking at the (avoided) cost of a water purification plant to substitute for the ecosystem service. Services, such as water purification, may also have relationships with market goods and services (e.g., as an input into the production process) that make it possible to estimate economic values at least in part or approximately.

Many ecosystem services however, are truly non-market, in that there are no market counterparts by which to estimate their value. Recreational uses of ecosystems fall into this category, and so economists have developed means of inferring values from behavior (e.g., travel cost), as discussed in the next section, and in other ways. Most of the support services and cultural values of ecosystems are also in the "true" non-market category. Value can arise even if a good or service is not explicitly

consumed, or an ecosystem even experienced.[25] Thus, it can be difficult to define, much less to measure the value of changes in these non-market services. To value these services, economists typically use stated preference (direct valuation) methods, a method that can be used not only for non-market services, but also to value services in other categories, such as the value that individuals place on clean drinking water or swimming facilities.

Below we report on the relevant literature in two categories. First, we report on studies that have looked at the non-market value of specific ecosystems or species. Since only a few of these studies attempt to value the impacts of climate change on ecosystems, we also highlight some non-market studies from the more general literature on ecosystem valuation, which can provide insights into the magnitude of potential values of services that might be vulnerable to climate change. Next we look at a different approach to valuation of ecosystems—a more "top-down" approach—that has been adopted both to look at the effects of climate change and more broadly at the total value of ecosystems. As the discussions indicate, the treatment of climate change, *per se* has been very sparse. Moreover, the lack of studies reflects, in part,

---

25 Economists have devoted much effort to defining the source of non-market values of ecosystems, coining such terms as "use" and "non-use" value, consumption value, existence value, and invoking, as reasons why people care about ecosystesm, the moral philosophies inherent in terms such as stewardship, spiritual values, etc. (see for example, Freeman (2003)).

a need to develop analytical linkages between the physical effects of climate on ecosystems, the services valued by humans, and appropriate techniques to value changes of the types, and with the breadth, indicated by studies of the effects of global change on ecosystems.

4.3.4.2 Valuation of the Effects of Climate Change on Selected Ecosystem and Species

Although climate change appears in a number of studies, it is often as a context for the scenario presented in the study for valuation, and so the study cannot be interpreted as valuation of climate change or climate effects per se. Only a few studies can be said to value the economic impacts of climate change on a particular ecosystem.

Two studies, Layton and Brown (2000) and Layton and Levine (2003) estimate total values for preventing Colorado (Rocky Mountain) forest loss due to climate change, based on data from the same stated choice or preference survey. The survey was conducted with Denver-area residents, who were expected to be familiar with forested regions in their nearby mountains. Respondents were given detailed information about climate change impacts on these forests, including changes in tree line elevation over both 60-year and 150-year time horizons, and asked to make choices between alternatives, allowing recovery of implied WTP. Layton and Brown (2000) found WTP in the range of $10 to $100 per month, per respondent, to prevent forest loss, with the range depending, in part, on the amount of forest lost. Layton and Levine (2003) reanalyzed the same data set, using a different approach that focuses on understanding respondents' least preferred, as well as most preferred, choices. They found that respondents' value of forest protection depends also on the time horizon—preventing effects that occur further into the future are valued less than nearer term effects.

Kinnell *et al.* (2002) designed and implemented several versions of stated preference studies that explored the impacts of wild bird (duck) loss due to either adverse agricultural practices, climate change, or both. The respondents consisted of Pennsylvania duck hunters, although the hypothetical ecosystem impacts occurred in the

Prairie Pothole region, which is in the northern midwestern states and parts of Canada. The authors considered a hypothetical loss in duck populations, with a scenario that presented some respondents with a 30 percent loss, and others with a 74 percent loss, some with a 40 year time horizon, and others with a 100 year time horizon. The study cannot be viewed as an estimate of WTP to avoid climate change; however, it is interesting because it suggests that recreational enthusiasts are willing to pay for ecosystem impacts that they do not necessarily expect to use. In addition, the study provides evidence that the context of climate change or other cause of ecosystem harm (in this case agricultural practices)—irrespective of the level of harm—may affect respondents' valuation of the harm.

Although very few studies have valued climate change impacts on ecosystems, economists have conducted numerous studies (primarily using direct valuation methods) of ecosystem values in particular geographic locations, often focusing on charismatic species, or specific types of ecosystems, such as wetlands, in a particular location. In some cases, the estimated values are linked to specific services that the species or ecosystem provide, but in many the services provided are somewhat ambiguous, and it is not always clear what aspect of the species, habitat, or ecosystem is driving the individual respondent's economic valuation.

A number of studies indicate that people value the protection of species or ecosystems. Some of these studies find potentially significant species values, ranging from a few dollars to hundreds of dollars per year, per person. For example, MacMillan *et al.* (2001) estimate the value of restoring woodlands habitat, and separately evaluate the reintroduction of the wolf and the beaver to Scottish highlands. In the United States, species such as salmon and spotted owls, as well as their habitat, have been examined in connection to their respective controversies.

Studies have also looked at the value of ecosystems or changes in ecosystems. In the former case, economists use either the value of productive output (harvest) as an indicator of value, or respondents value protecting the ecosystem. For example, numerous coastal

wetland and beach protection studies have used a variety of non-market valuation approaches. A survey of a number of these studies reports values ranging from $198 to approximately $1500 per acre (Woodward and Wui, 2001).

Some studies have looked explicitly at the services provided by ecosystems. For example, Loomis *et al.* (2000) consider restoration of several ecosystem services (dilution of wastewater, purification, erosion control, as fish and wildlife habitat, and recreation) for a 45-mile section of the Platte River, which runs east from the State of Colorado into western Nebraska. Average values are about $21 per month for these additional ecosystem services for the in-person interviewees. While these studies and their values are generally informative, transferring values from studies like the ones above to other ecosystems, and using the results to estimate values associated with climate change impacts, can be problematic.

4.3.4.3 Top-down Approaches to Valuing the Effects of Climate Change and Ecosystem Services

From the perspective of deriving values for ecosystem changes (or changes in ecosystem services) associated with climatic changes, one difficulty with the above studies is that the focus is on discrete changes to particular species or geographic areas. It is therefore difficult to know how these studies relate to, or shed light on, the types of widespread and far-reaching changes to ecosystems (and the services they

provide) that will result from climate change. Consequently, some studies have attempted to value ecosystems in a more aggregate or holistic manner. While these studies do not focus specifically on the United States, they are indicative of an alternative approach that recognizes the interdependence of ecosystems and their components, and therefore deserve some discussion.

Several models include values for non-market damages, worldwide, resulting from projected climate change. These impact studies have been conducted at a highly aggregated level; most of the models are calibrated using studies of the United States that are then scaled for application to other regions (Warren *et al.*, 2006).

A study of total ecosystem values, but not undertaken in the context of climate change, is the highly publicized study by Costanza *et al.* (1997), which offers a controversial look at valuing the "entire biosphere." Because their reported estimated average value of $33 trillion per year exceeds the global gross national product, economists have a difficult time reconciling this estimate with the concept of economic value. Ehrlich and Ehrlich (1996) and Pimental *et al.* (1997) are studies by natural scientists that have attempted to value ecosystems or in the case of the latter, biodiversity. These are important attempts to indicate the value of ecosystems, but the accuracy and reliability of the values are questionable. To paraphrase a study by several prominent environmental economists that is slightly critical of all of these studies, economists do not have any fundamental difference of opinion with these natural scientists about the importance of ecosystems and biodiversity, rather it is with the correct use of economic value concepts in these applications (see Bockstael *et al.*, 2000).

4.3.5 Recreational Activities and Opportunities

Ecosystems provide humans with a range of services, including outdoor recreational opportunities. In turn, outdoor recreation contributes to individual well-being by providing physical and psychological health benefits. In addition, tourism is one of the largest economic

sectors in the world, and it is also one of the fastest growing (Hamilton and Tol, 2004). The jobs created by recreational tourism provide economic benefits not only to individuals but also to communities.[26] A number of studies have looked at the qualitative effects of climate change on recreational opportunities (*i.e.*, resources available) and activities in the United States, but only a few have taken this literature the additional step of estimating the implications of climate change for visitation days or economic welfare. This section describes the results of this research into the impacts on several forms of recreation and reports the economic benefits and losses associated with these changes at the national level.

Slightly more than 90 percent of the U.S. population participates in some form of outdoor recreation, representing nearly 270 million participants (Cordell *et al.*, 1999), and several billion person-days spent each year in a wide variety of outdoor recreation activities. According to Cordell *et al.* (1999), the number of people participating in outdoor recreation is highest for walking (67 percent), visiting a beach or lakeshore or river (62 percent), sightseeing (56 percent), swimming (54 percent) and picnicking (49 percent). Most days are spent in activities such as walking, biking, sightseeing, bird-watching, and wildlife viewing (Cordell *et al.*, 1999), but the range of outdoor recreation activities in the United States is as diverse as its people and environment. While camping, hunting, backpacking and horseback riding attract a fraction of the people who go biking or bird-watching, these other specialized activities provide a very high value to their devotees. Many of these devotees of specialized outdoor recreation activities are people who "work to live," *i.e.*, specialized weekend recreation is one of their rewards for the 40+ hour workweek.

Climate change resulting from increasing average temperatures as well as changes in precipitation, weather variability (including more extreme weather events), and sea level rise, has the potential to affect recreation and tourism along two pathways. Figure 4.3 illustrates these direct and indirect effects of climate

---

26 Effects on jobs, income, and similar metrics are considered market impacts, and are not discussed here.

change on recreation. Since much recreation and tourism occurs out of doors, increased temperature and precipitation have a direct effect on the enjoyment of these activities, and on the desired number of visitor days and associated level of visitor spending (as well as tourism employment). Weather conditions are considered one of the four greatest factors influencing tourism visitation (Pileus Project, 2007). In addition, much outdoor recreation and tourism depends on the availability and quality of natural resources such as beaches, forests, wetlands, snow, and wildlife (Wall, 1998). Consequently, climate change can also indirectly affect the outdoor recreational experience by affecting the quality and availability of natural resources used for recreation.

Effects of climate change can be both positive and negative. The length of season for and desirability of several of the most popular activities—walking, visiting a beach, lakeshore, or river, sightseeing, swimming, and picnicking (Cordell *et al.*, 1999)—will likely be enhanced by small near- term increases in temperature. However, long-term higher increases in temperature may eventually have adverse effects on activities like walking, and result

in sufficient sea level rise to reduce publicly accessible beach areas, just at the time when demand for beach recreation to escape the heat is increasing. In contrast, some activities are likely to be unambiguously harmed by even small increase in global warming, such as snow and ice-dependent activities.

In some ways, one can interpret the direct effects of climate change as influencing the demand for recreation and the indirect effects as influencing the supply of recreation opportunities. For example, warmer temperatures make whitewater boating more desirable. However, the warmer temperatures may reduce river flows since there is less snowpack, higher evapotranspiration, and greater water diversions for irrigated agriculture. Some studies cited below look only at the direct effects, while others represent the combined effect of the direct and indirect pathways.

Direct effects. To date, most studies of the direct effects of climate change on recreation and tourism have been qualitative, although a few have been quantitative. Qualitatively, we would expect both positive and negative effects of climate change on different recreational

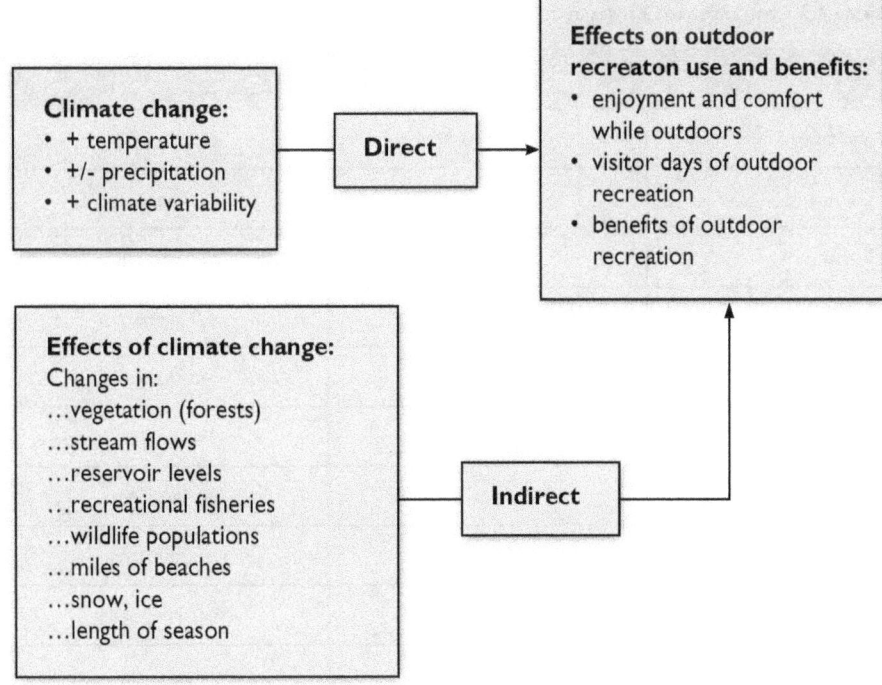

Figure 4.3 Direct and Indirect Effects of Climate Change on Recreation

The U.S. Climate Change Science Program

activities. Many of the qualitative studies rely simply on intuition to suggest that increases in air and water temperatures will have a positive effect on outdoor recreation visitation in two ways: (a) more enjoyment from the activity, and (b) a longer season in which to enjoy the activity (DeFreitas, 2005; Scott and Jones, 2005; Scott, Jones and Konopek (2007). Hall and Highman (2005) note that climate change may provide more days of "ideal" temperatures for water-based recreation activities and some land based recreation activities such as camping, picnicking and golf.

The recreational activities most obviously harmed by warmer climate are sports that require snow or cold temperatures, such as downhill and cross country skiing, snowmobiling, ice fishing, and snowshoeing. Reductions in visitor use (see, for example, the studies reported in Table 4.5) occur primarily from shorter season, particularly early in the year at such traditional times as Thanksgiving and spring break. But with warmer temperatures, there is also less precipitation as snow and more as rain on snow, which contributes to a much shallower snowpack and harder snow. Further, recreating

in freezing rain or slushy temperatures is not a pleasant experience, reducing benefits from skiing, snowshoeing, and snowmobiling, further reducing use.

Some recreation areas that are already hot during the summer recreation season will see decreases in use. For example, the Death Valley National Park, Joshua Tree National Park, and Mesa Verde National Park are all projected to be "intolerably hot," reducing visitation (Saunders and Easley, 2006).

Most quantitative studies of the effects of climate change on recreation evaluate specific projected changes in temperature and/or precipitation, such as a 2.5°C increase in temperature over the next fifty years. Two quantitative studies look at effects of temperature change in Canadian recreation.[27] Scott and Jones (2005) project that the golf season in Banff, Canada could be extended by at least one week and up to

---

27 Scott and Jones (2005) used +1°C to +5°C in their scenarios and Scott *et al.* (2006) used +1.5°C to +3°C in their low impact scenario and +2°C to +8°C in their high impact scenario.

Table 4.5 Comparison of Changes in U.S. Visitor Days

Activity	Loomis and Crespi (1999)	Mendelsohn and Markowski (1999)
Boating	9.2 percent	36.1 percent
Camping	-2.0 percent	-12.7 percent
Fishing	3.5 percent	39.0 percent
Golf	13.6 percent	4.0 percent
Hunting	-1.2 percent	no change
Snow Skiing	-52.0 percent	-39.0 percent
Wildlife Viewing	-0.1 percent	-38.4 percent
Beach Recreation	14.1 percent	not estimated
Stream Recreation	3.4 percent	included in boating
Gain in Visitor Benefits (in Billions)	$2.74	$2.80

eight weeks. The combined effect of warmer temperatures lengthening the golfing season, and the increasing desirability of golfing during the existing season, together result in an increase in the rounds of golf played by between 50 percent and 86 percent. Similar increases might be expected for golf in northern states of the United States such as Minnesota, Wisconsin, New York, etc. with longer golf seasons. Scott *et al.* (2006) and Scott and Jones (2005) suggest that some of the previously projected large (30 percent to 50 percent) reductions in length of ski seasons at northern ski areas (*e.g.*, in Canada, Michigan, and Vermont) can be reduced (to 5 percent to 25 percent) through the use of advanced snowmaking. While use of advanced snowmaking to minimize reductions in ski season seems plausible for the studied northern ski areas, it is doubtful that snowmaking would benefit ski areas in California, New Mexico, Oregon, and West Virginia where in some years the Thanksgiving and "Spring Break" periods are already too warm for successful snowmaking or retention of snow made.

Some studies have used natural variations in temperature to evaluate the effects of climate on recreation (including measures on monthly, seasonal, and inter-annual variation). Two of these have found that while visitation increases with initial increases in temperature, visitation actually decreases as temperature increases even further (Hamilton and Tol, 2004; Loomis and Richardson, 2006). Following the discussion of indirect effects two of the quantitative studies, which look not only at visitor days but also at monetary measures of economic welfare, are discussed in more detail below.

Indirect effects. While increased temperature may increase the demand for some outdoor recreation activities, in some cases climate change may reduce the supply of natural resources on which those recreational activities depend. As noted above, reduced snowpack for winter activities has been projected in the Great Lakes (Scott *et al.*, 2005), in northern Arizona (Bark-Hodgins and Colby, 2006) and

at a representative set of ski areas in the United States (Loomis and Crespi, 1999).[28]

For example, lower in-stream flows and lower reservoir levels have consistently been shown to reduce recreation use and benefits (Shaw, 2005). Thus, changes associated with climate can reduce opportunities for summer boating and other water sports. When less precipitation falls as snow in the winter, and more falls as rain in the spring, early spring season run-off will increase. Summer river flows will be correspondingly lower, at times when demand for whitewater boating is higher. Human responses to the physical changes associated with climate change may exacerbate natural effects reducing recreational opportunities. For example, many current reservoirs are not designed to handle huge spring inflows, and thus this water may be "spilled," which lowers reservoir levels during the summer season. These lower reservoir levels are then drawn down more rapidly as higher temperatures increase evapotranspiration and increase irrigation releases. In turn, the resulting lower reservoir may leave boat docks, marinas, and boat ramps inaccessible.

28 Higher temperatures (while they increase snowmelt reducing the snow skiing season) may have two subtle effects: (a) stimulating demand for snow skiing due to warmer temperatures, for those skiers who prefer "spring skiing" due to the warmer temperatures even if the snow conditions are less than ideal; and (b) reduced snowmelt opens up the high mountains for hiking, backpacking and mountain biking activities somewhat earlier than is the case now, which may lead to increases in those visitor use days.

Ecosystems that provide recreational benefits may also be at risk from climate change. Wetlands are another recreational environment that is at risk from climate change. Wetland based recreation include wildlife viewing and waterfowl hunting. With sea level rise, many existing coastal wetlands will be lost, and given existing development inland, these lost wetlands may not be naturally replaced (Wall, 1998). The higher temperatures and reduced water availability is also expected to adversely affect freshwater wetlands in the interior of the country. As such, waterfowl hunting and wildlife viewing may be adversely affected.

Higher water temperatures and lower stream flows are projected to reduce coldwater trout fisheries (U.S. EPA, 1995; Ahn *et al.*, 2000) as well as native and hatchery stocks of Chinook salmon in the Pacific Northwest (Anderson *et al.*, 1993). Given trout and Chinook salmon sensitivity to warm water temperatures, these affects are not surprising. However, Anderson *et al.*'s estimated magnitude of 50 percent to 100 percent reduction in Chinook spawning returns is quite large. Reductions of such magnitude will have a substantial adverse effect on recreational salmon catch rates, and possibly whether recreational fishing would even be allowed to continue in some areas of the Pacific Northwest. However, from a national viewpoint, fishing participation for trout, cool water species and warm water species dominates geographically specialized fishing like Chinook salmon. Warmer water temperatures are projected to eliminate stream trout fishing in

8-10 states and result in a 50 percent reduction in coldwater stream habitat in another 11-16 states depending on the GCM model used (U.S. EPA, 1995). This could adversely affect up to 25 percent of U.S. fishing days (Vaughan and Russell, 1982). This 25 percent loss may be an upper limit as some coldwater stream anglers may substitute to less affected coldwater lakes/reservoirs or switch to cool/warm-water species such as bass (U.S. EPA, 1995). Studies that better account for substitution effects, such as Ahn *et al.* (2000), indicate a 2-20 percent drop in benefits of trout fishing depending on the projected degrees of temperature increase which ranged from 1°C to 5°C.

Sea level rise reducing beach area and beach erosion are concerns with climate change that may make it difficult to accommodate the increased demand for beach recreation (Yohe *et al.*, 1999). In the near term, recreational forests may also be adversely affected by climate change. Although forests may slowly migrate northward and into higher elevations, in the short run there may be dieback of forests at the current forest edges (as these areas become too hot), resulting in a loss of forests for recreation. In the long term, however, several analyses suggest forest species composition and migration due as well as net increases in forest area due to carbon dioxide fertilization (Joyce *et al.*, 2001; Iverson and Matthews, 2007). Thus, eventually there may be resurgence in forest recreation.

Saunders and Easley (2006) find that natural resources of many western National Parks, National Recreation Areas, and National Monuments resources will be adversely affected by climate change. The most common adverse effects are reductions in some wildlife species, loss of coldwater fishing opportunities and increasing park closures due to wildfire associated with stressed and dying forest stands. Box 4.2 discusses in more detail potential effects of climate change on one park: Rocky Mountain National Park, which has been the subject of both ecological and economic analysis.

4.3.5.1 Economic Studies of Effects of Climate on Recreation

Changes in economic welfare due to the effects of climate change on non-market resources, such as recreation, can be evaluated in several ways. First, since decisions regarding recreational activities depend on both direct and indirect effects of climate, changes in human well-being (as a result of these changes) will be reflected in changes in visitor use. Social scientists believe changes in visitor use are motivated by people "voting with their feet" to maintain or improve their well-being. In the face of higher temperatures, people may seek relief, for example, by visiting the beach or water skiing at reservoirs more frequently to cool down. Similarly, reduced opportunities for recreation due to indirect effects of climate change will also be reflected in reduced visitation days. Thus, one metric of effects on human well-being is the change in visitation days.

Second, recreational trips—for example, to reservoirs and beaches—have economic implications to the visitor and the economy. Visitors allocate more of their scarce time and household budgets to the recreational activities that are now more preferred in a warmer climate. This reflects their WTP for these recreational activities, which is a monetary measure of the benefits they receive from the activity. Numerous economic studies provide estimates of the value of changes in diverse recreational activities, using various economic techniques (such as travel cost[29] analysis and stated preference methods) (see Section 3 of this chapter and the chapter Appendix for more information). While these studies typically do not focus directly on climate change, they can be used to extract values for the types of changes that are projected to be associated with climate change.

Third, some people who do not currently visit unique natural environments may value climate stabilization policies that preserve these natural

environments for future visitation. These people have what economists call a value for preserving their option—their ability— to visit the environments in the future (Bishop, 1982). This option value is much like purchasing trip insurance to guarantee that if one wanted to go in the future, that conditions would be as they are today.

As discussed below, economists have available a number of well-studied techniques to evaluate the impacts of climate change on at least some of the recreational service provided by ecosystems. However, only a few studies have looked explicitly at the effects of climate change on recreation in the United States. More research is needed to understand the linkages between weather and recreation, and to extrapolate results to the range of recreational activities throughout the United States.

Change in visitation days. Two studies (Loomis and Crespi, 1999; Mendelsohn and Markowski, 1999) have comprehensively examined the effects of climate on recreational opportunities for the entire United States. These studies both examined the effects of 2.5°C and 5°C increases in temperature, along with a 7 percent increase in precipitation. The studies used similar methodologies to estimate visitor days for a range of recreational opportunities. Each study looked at slightly different effects, but between them examined a mix of direct and indirect climate effects, including direct effects of higher temperatures on golf and beach recreation visitor days, and indirect effects of

---

29 The travel cost method traces out a demand curve for recreation using travel cost as proxies for the price of recreation, along with the corresponding number of trips individual visitors take at these travel costs. From the demand curve, the net willingness to pay or consumer surplus is calculated.

snow cover on skiing. Both studies estimate changes in visitation days due to climate change, and then use the results of a number of economic valuation studies to place monetary values on the visitation days. The studies find that, as expected, near-term climate change will increase participation in activities such as water-based recreation, and reduce participation in snow sports.

Table 4.5 presents the results of the two studies. The results suggest that relatively high participation recreation activities such as beach and stream recreation gain, and low participation activities like snow skiing lose. Although the percentage drop in visitor days of snow sports is much larger than the percentage increase in visitor days in water-based recreation, the larger number of water-based sports participants more than offsets the loss in the low participation snow sports. Thus, on net, there is an overall net gain in visitation associated with the assumed increases of 2.5°C in temperature and 7 percent in precipitation.[30]

The methods used to forecast visitation were slightly different between the two studies. To estimate visitor days for all recreation activities,

30 Geographic regions within the U.S. will experience different gains and losses. Currently hot areas with less access to water resources (*e.g.*, New Mexico) may suffer net overall reductions in recreation use to due higher heat that makes walking, sightseeing, and picnicking less desirable. States with substantial water resources (lakes, seashores) may gain visitor days and tourism. Currently cold areas such as the Dakotas and New York may see increases in some recreation due to longer summer seasons.

Mendelsohn and Markowski regressed state level data on visitation by recreation activity as a function of land area, water area, population, monthly temperature and monthly precipitation. The Loomis and Crespi study used a similar approach to Mendelsohhn and Markowski for some activities, such as golf. Other forecasting techniques were used for other activities. For example, for beach recreation, they used detailed data on two individual beaches in the northeastern, southern, and western United States to estimate three regional regression equations to project beach use, and the response of reservoir recreation to climate change was analyzed using visitation at U.S. Army Corps of Engineers reservoirs.

For some of the recreational activities, the Loomis and Crespi study included indirect, as well as direct, effects. For example, in addition to temperature and precipitation the reservoir models incorporated climate-induced reductions in reservoir surface area. Similarly, the estimate of visitor days for snow skiing used projected changes in the number of days of minimum snow cover to adjust skier days proportionally. In some cases, only indirect (supply) effects were included, as in the case of stream recreation, water fowl hunting, bird viewing, and forest recreation. Since these estimates do not include changes in visitation associated with direct effects of climate we have less confidence in the accuracy of these results than we do for reservoir recreation, which takes into account both demand and supply effects on recreation use.

Valuation of gains and losses in visitor days. Since different activities may have different levels of enjoyment provided to the visitor (and, therefore, different economic values), adding up changes in visitation days to produce a "net change" is not an accurate representation of the overall change in well-being. The two studies discussed above used net WTP as a measure of value of each day of recreation (Section 3 of this chapter provides a discussion of the concept of WTP as a common economic measure of changes in welfare).

To date there have been few original or primary valuation studies of climate change per se on recreation. The case study on Rocky Mountain

National Park presented below provides one of the few examples. Other studies include Scott and Jones (2005), which focused on Banff National Park, Scott *et al.* (2006), which looked at snow skiing, Scott *et al.* (2007), which focused on Waterton National Parks, and Pendleton and Mendelsohn (1998), which estimated values for fishing in the northeastern United States.[31] There have, however, been hundreds of recreation valuation studies. The values from these studies (generally travel cost or stated preference) can be applied to other applications using a "benefit transfer" approach, and applying average values of recreation from previous studies to value their respective visitor days.

Loomis and Crespi (1999) and Mendelsohn and Markowski (1999) estimate the overall net gain in visitor benefits, using the change in visitor days reported in Table 4.5 and estimated values of a visitation day reported in the literature. Loomis and Crespi (1999) adopt a disaggregated activity approach, and Mendelsohn and Markowski (1999) apply a state level approach.[32] Both of these studies find that temperature increases of 5°C and up result in increased benefits. However, as noted below, the case study of Rocky Mountain National Park suggests that extreme heat is likely (based on model results) to cause these visitor benefits to decrease at some point.

Visitors are somewhat adaptable to climate change in the recreation activities they choose and when they choose them. Thus, recreation represents one situation with opportunities to reduce the adverse impacts of climate change, or

increase its benefits, via adaptation. As noted by Hamilton and Tol (2004), warmer temperatures may shift visitors northward, and up into the mountains. Thus, currently cool areas (*e.g.,* Maine, Minnesota, Washington) may gain, and warm areas (*e.g.,* Florida, Arizona) may lose, tourism.

Some adaptive responses can be expensive, and may be of limited effectiveness; such as snowmaking at night, which is often mentioned as an adaptation for downhill skiing (Irland *et al.,* 2001). Other adaptive behavior may include moving some outdoor recreation activities indoors. For example, bouldering is now taking place in climbing gyms on artificial climbing walls. Running on a treadmill in an air-conditioned gym may be a substitute for running out of doors for some people, but casual observation suggests that many people prefer to run out doors when weather permits. Unless preferences adjust to increased temperatures, there may be a loss in human well-being from substituting the treadmill in the air conditioned gym for the out of doors. Box 4.2 summarizes a case study of the impacts of climate change on Rocky Mountain National Park.

4.3.6 Amenity Value of Climate

It is well established that preferences for climate affect where people choose to live and work. The desire to live in a mild, sunny climate may reflect health considerations. For example, people with chronic obstructive lung disease or angina may wish to avoid cold winters. Warmer

31 The three papers by Scott are discussed elsewhere in this paper. Pendleton and Mendelsohn use a random utility model of recreational fishing in the northeastern United States. They find that, while catch rates of rainbow trout would decrease, catch rates of other trout and pan fish would increase. On net, recreational fishing benefits (under a climate scenario associated with a doubling of atmospheric CO_2 concentrations) are reduced in the State of New York, but there are offsetting gains in more northern states like Maine.

32 As noted above, Mendelsohn and Markowski (1999) used state level regression modeling to estimate effects on all activities. In contrast, Loomis and Crespi (1999), used different regression models and different geographic scales for different recreation activities to take advantage of the more micro-level datasets available for beach and reservoir recreation.

BOX 4.2. Study of the Effects of Climate Change on Rocky Mountain National Park

One of the national parks most closely studied to determine the net effect of direct and indirect effect of climate change on visitation, visitor benefits, and tourism employment in Rocky Mountain National Park (RMNP) in Colorado. This alpine national park is located at elevations ranging from 7,000 to 14,000 feet above sea level. It is known for elk viewing, hiking, tundra flowers, snowcapped peaks, and one of Colorado's most visible and recognizable 14,000 foot peaks, Longs Peak.

Loomis and Richardson (2006) compared two approaches to estimating the effect of climate change on visitation and employment in RMNP. The first approach examined variations in monthly visitation in response to historic variations in temperature. The results of this first approach showed a statistically significant positive effect of temperature on visitation (see Loomis and Richardson (2006) for more details). However, increased visitation slowed as temperatures got hotter and hotter, and visitation even declined during one summer of very high temperatures (60 days over 80°F) by 7.5 percent.

The second approach used a survey that portrayed the direct effects (e.g., temperature) and indirect effects (e.g., changes in elk and ptarmigan (an alpine bird), or percent of the park in tundra). Visitors were then asked to indicate if they would change their visits to RMNP or length of stay in the park. The surveys used three climate change scenarios, one produced by the Canadian Climate Center (CCC) indicating a 4°F increase in temperature by 2020, a Hadley climate scenario that forecasted a 2°F temperature increase by 2020, and an extreme heat scenario designed to capture very hot future conditions (50 days with temperatures above 80°F, as compared to 3 days currently). All climate change scenarios were used with wildlife models to estimate the increase in elk populations and decrease in ptarmigan populations. The extreme heat survey found similar results to that of the monthly visitation model.

Table 4.6 shows the results of the CCC, Hadley, and Extreme Heat temperature scenarios on visitation, visitor benefits, and tourism employment as compared to current conditions. As indicated in the table, applying visitor survey estimates of visitation change yields a 13.6 percent increase with CCC and 9.9 percent increase with Hadley. Loomis and Richardson also report that applying the historic visitation patterns to the same scenarios yields an 11.6 percent increase in visitation with CCC and 6.8 percent with Hadley. Not only is there fairly good agreement between the two methods, but the warmer CCC climate change scenario produces larger increases in visitation. In the extreme heat scenario, however, visitations declines from current conditions.

climates may be more pleasant for persons with arthritis. Climate preferences may also reflect the desire to reduce heating and/or cooling costs. Certain climates may be complementary to leisure activities. For example, skiers may wish to live in colder climates, sunbathers in warmer ones. Alternatively, a particular climate may simply make life more enjoyable in the course of everyday life. Based on the evidence one would also expect that, in addition to preferring certain temperatures and more sunshine, people would prefer to reduce the risk of experiencing abrupt climate events such as hurricanes and floods.

While climate itself is not bought and sold in markets, the goods that are integral to location decisions—such as housing and jobs—are market goods. Consequently, economists look at behavior with regard to location choice (the prices that are paid for houses and the wages that are accepted for jobs) in order to determine how large a role climate plays in these decisions and, therefore, how valuable different climates are to the general public. The remainder of this section discusses methods that have been used to estimate the amenity values people attach to various climate attributes, as well as the value they attach to avoiding extreme weather events. Unfortunately, few studies have rigorously estimated climate amenity values (*e.g.*, the value of a 2°C change in mean January temperature) for the United States and then used these values to estimate the dollar value of various climate scenarios.

4.3.6.1 Valuing Climate Amenities

People's preferences for climate attributes should be reflected in their location decisions.

Table 4.6 Change in Visits, Jobs, and Visitor Benefits with Three Climate Change Scenarios

Climate Scenario	Annual Visits	Change	Tourism Jobs	Visitor Benefits (Millions)
Current	3,186,323		6,370	$1,004
CCC	3,618,856	13.6 percent	7,351	$1,216
Hadley	3,502,426	9.9 percent	7,095	$1,157
Extreme Heat	2,907,520	-8.7 percent	5,770	$959

Other things equal, homeowners should be willing to pay more for housing (and so bid up housing prices) in more desirable climates, and so property values should be higher in those climates. Similarly, workers should be willing to accept lower wages to live in more pleasant climates. If climate also affects firms' costs, however, actual wages may rise or fall due to the interaction between firms and workers (Roback, 1982).

Early attempts to estimate how much consumers will pay for more desirable climates start from the view that a good—such as housing or a job—is a bundle of attributes that are valued by the homeowner or worker. The price the consumer pays for the good (such as a house) is actually a composite of the prices that are implicitly paid for all the attributes of the good. Using a statistical technique (known as a hedonic value function), economists can estimate the price of a particular attribute, such as climate. The hedonic property value function, thus, describes how housing prices vary across cities as a function of housing characteristics and locational amenities, such as climate, crime, air quality, or proximity to the ocean. Similarly, the hedonic wage function relates observed wages to job characteristics (such as occupation and industry), worker characteristics (such as education and years of experience), and locational amenities.

The value of locational amenities—*i.e.*, how much individuals are willing to pay for amenities—can be inferred from these estimated hedonic wage and property value functions. Extracting this value, however, assumes that workers and homeowners are mobile, *i.e.*, that they can choose where to live

fairly freely within the United States. Similarly, it assumes that, in general, individuals have moved to where they would like to live (at the moment), so that housing and job markets are in what is said to be "equilibrium." It also assumes that workers and homeowners have good information about the location to which they are moving, and that sufficient options (in terms of jobs and houses and amenities) are available to them. The estimates of the value of a particular amenity—such as climate—will be more accurate the more nearly these assumptions are met.

A number of hedonic wage and property value studies have included climate, among other variables, in their analyses. See, for example, studies by Hoch and Drake (1974); Cropper and Arriaga-Salinas (1980); Cropper (1981); Roback (1982); Smith (1983); Blomquist *et al.* (1988); and Gyourko and Tracy (1991). The first four studies estimate only hedonic wage functions, while the last three estimate both wage and property value equations. As Moore (1998) and Gyourko and Tracy (1991) note, this literature suggests that climate amenities are reflected to a greater extent in wages than in property values.[33] Roback (1982), Smith (1983), and Blomquist *et al.* (1988) all find sunshine to be capitalized in wages as an amenity, while heating degree days are capitalized as a

---

33 The effect of weather variables on property values is mixed, with Blomquist *et al.* (1988) finding property values to be negatively correlated with precipitation, humidity and heating and cooling degree days, but Roback (1982) finding property values positively correlated with heating degree days. Gyourko and Tracy (1991) find heating and cooling degree days negatively correlated with housing expenditures, but humidity positively correlated.

disamenity (Roback, 1982, 1988; Gyourko and Tracy, 1991).

More recent studies using the hedonic approach include Moore (1998) and Mendelsohn (2001), who use their results to estimate the value of mean temperature changes in the United States associated with future climate scenarios. Moore uses aggregate wage data for Metropolitan Statistical Areas (MSAs) to estimate the responsiveness of wages with respect to climate variables for various occupations. Climate is captured by annual temperature, precipitation, and by the difference between average July and average January temperature. Moore estimates that a 4.5°C increase in mean annual temperature would be worth between $30 and $100 billion (in 1987 dollars) assuming that precipitation and seasonal variation in temperature remain unchanged.

Mendelsohn (2001) uses county-level data on wages and rents to estimate hedonic wage and property value models. Separate equations are estimated for wages in retail, wholesale, service, and manufacturing jobs. Climate variables, which include average January, April, June, and October temperature and precipitation, enter each equation in quadratic form. Warmer temperatures are generally associated with lower wages and lower rents, although the former effect is larger in magnitude. Mendelsohn uses the results of these models to estimate the impact of a uniform increase in temperature of 1°C, 2°C, and 3.5°C, paired, alternately with an 8 percent and a 15 percent

increase in precipitation. The results suggest that warming produces positive benefits in every scenario except the 3.5°C temperature change. Averaging across estimates produced by the 3 models for each of the 6 scenarios suggests annual net benefits (in 1987 dollars) of $25 billion.

Unfortunately, hedonic wage and property value studies have limitations that have caused them to be replaced by alternate approaches to analyzing data on location choices. One drawback of the hedonic approach is that, as mentioned above, it assumes that national labor and housing markets exist and are in equilibrium. As Graves and Mueser (1993) and Greenwood et al. (1991) point out, if national markets are not in equilibrium, inferring the value of climate amenities from hedonic wage and property value studies can lead to badly biased results. A second problem is that variables that are correlated with climate (e.g., the availability of recreational facilities) may be difficult to measure. Hence, climate variables may pick up their effects. In hedonic property value studies, for example, the use of heating and cooling degree days to measure climate amenities is problematic because their coefficients may capture differences in construction and energy costs as well as climate amenities *per se*. A related problem in hedonic wage equations is that more able workers may locate in areas with more desirable climates. If ability is not adequately captured in the hedonic wage equation, the coefficients of climate amenities will reflect worker ability as well as the value of climate.

Cragg and Kahn (1997) were the first to relax the national land and labor market equilibrium assumption by estimating a discrete location choice model. Using Census data, they model the location decisions of people in the United States who moved between 1975 and 1980. Movers compare the utility they would receive from living in different states—which depends on the wage they would earn and on the cost of housing, as well as on climate amenities—and are assumed to choose the state that yields the highest utility. This allows Cragg and Kahn to estimate the parameters of individuals' utility functions and thus infer the rate at which they will trade income for climate amenities.

The drawback of this study is that it estimates the preferences of movers, who may differ from the general population. An alternate approach (Bayer *et al.*, 2006; Bayer and Timmins, 2005) is to acknowledge that moving is costly and to explain the location decisions of all households, assuming that all households are in equilibrium, given moving costs. Unfortunately, the discrete choice literature has yet to provide reliable estimates of the value of climate amenities in the United States.

4.3.6.2 Valuing Hurricanes, Floods, and Extreme Weather Events

It is sometimes suggested that the value people place on avoiding extreme weather events can be measured by the damages that such events cause, or by the premiums that people pay for flood or disaster insurance. If people are risk averse, *ex post* losses associated with extreme weather events represent a lower bound to the value people place on avoiding these events. It is also the case that people can purchase insurance only against the monetary losses associated with floods and hurricanes. Thus, insurance premiums will not capture the entire value placed on avoiding these events.

Assuming that people are informed about risks, the value of avoiding extreme weather events should be reflected in property values, and, holding other amenities constant, houses in an area with high probability of hurricane damage should sell for less than comparable houses in an area with a lower chance of hurricane damage. To estimate the value of avoiding these events correctly is, however, tricky. It can be difficult, for example, to disentangle hurricane risk (a negative effect) from proximity to the coast (an amenity).

Recent studies use natural experiments to determine the value of avoiding hurricanes and floods. Hallstrom and Smith (2005) use property value data before and after hurricane Andrew in Lee County, Florida, a county that did not suffer damage from the hurricane, to determine the impact of people's perceptions of hurricane risk on property values. They find that property values in special flood hazard areas of Lee County declined by 19 percent after hurricane Andrew. The magnitude of

this decline is significant, and agrees with Bin and Polasky (2004). Bin and Polasky find that housing values in a flood plain in North Carolina declined significantly after hurricane Floyd, compared to houses not at risk. For the average house, the decline in price exceeded the present value of premiums for flood insurance, suggesting that the latter are, indeed, a lower bound to the value of avoiding floods.

4.4 CONCLUSIONS

The study of the impacts of climate change on human welfare, well-being, and quality of life, is still developing. Many studies of impacts on particular sectors—such as health or agriculture—discuss, and in some cases quantify, effects that have clear implications for welfare. Studies also hint at changes that are perhaps less obvious, but also have welfare implications (such as changes in outdoor activity levels and how much time is spent indoors) and point also to effects with far more dramatic consequences (such as the breakdown in public services and infrastructure associated with possible extreme events of the magnitude of Katrina). Adaptation, too, has welfare implications that studies do not always point out, such as the costs (financial and psychological) to the individual of changing behavior.

To our knowledge, no study has made a systematic survey of the myriad welfare implications of climate change, much less attempted to quantify, nor aggregate them. An almost bewildering

choice of typologies is available for categorizing effects on quality of life, well-being, or human welfare. The social science and planning literatures provide not only a range of typologies, but also an array of metrics that could be used to measure life quality.

This chapter explores one commonly used method: the social indicators approach. This approach generally divides life quality effects into broad categories, such as economic conditions or human health, and then identifies subcategories of important effects.

Most of the measures of well-being—including the social indicators approach—focus on individual measures of well-being, although measured at the society level. There is, however, another dimension to well-being—community welfare. Communities represent networks of households, businesses, physical structures, and institutions and so reflect the interdependencies and complex reality of human systems. Understanding how climate impacts communities, and how communities are vulnerable—or can be made more resilient—in the face of climate change, is an important component of understanding well-being and quality of life.

Economics offers one alternative to address the diversity of impacts: valuing welfare impacts in monetary terms, which can then be summed. Estimating value, however, requires completing a series of links—from projected climate change to quantitative measures of effects on commodities, services, or conditions that are linked to well-being, and then valuing those effects using economic techniques.

Regardless of the framework, estimating impacts on human well-being involves numerous and diverse effects. This poses several critical difficulties:

- The large number of effects makes the task of linking impacts to climate change—whether qualitatively or quantitatively—difficult.

- The interdependence of physical and human systems further complicates the process of quantification—both for community effects, and also for ecosystems, raising doubts about a piecemeal approach to estimation.

- The diversity of effects raises questions of how to aggregate effects in order to develop a composite measure of well-being or other metrics that can be used for policy purposes.

4.5 EXPANDING THE KNOWLEDGE BASE

Despite the potential for impacts on human well-being, little research focuses directly on understanding the relationship between well-being and climate change. Completely cataloging the effects of global change on human well-being or welfare would be an immense undertaking, and no well-accepted structure for doing so has been developed and applied. Moreover, identifying the potentially lengthy list of climate-related changes in lifestyle, as well as in other, more tangible, features of well-being (such as income), is itself a daunting task—and may include changes that are not easily captured by objective measures of well-being or quality of life.

This chapter has looked at the climate impacts and economics literature in four areas of welfare effects—human health, ecosystems, recreation, and climate amenities. For each of the non-market effects analyzed here, significant data gaps exist at each of the steps necessary to provide monetized values of climate impacts. Although the economics literature for only a few areas of effects is examined, it is probable that similar information gaps exist for the valuation of other impacts of climate change, particularly those that involve non-market effects (see Table 4.1). In addition, economic welfare—as with any other aggregative approach—does not adequately address the question of how to deal with effects that might not be amenable to valuation or with interdependencies among effects and systems.

Developing an understanding of the impacts of climate change on human welfare may require taking the following steps:

- Develop a framework for addressing individual and community welfare and well-being, including defining welfare/well-being for climate analysis and systematically categorizing and identifying impacts on welfare/well-being.

- Identify priority categories for data collection and research, in order to establish and quantify the linkage from climate to welfare effects.

- Decide which metrics should be used for these categories; more generally, which components of welfare/well-being should be measured in natural or physical units, and which should be monetized.

- Investigate methods by which diverse metrics can be aggregated into a synthetic indicator (*e.g.*, vulnerability to climate change impacts, including drought, sea level rise, etc.), or at least weighted and compared in policy decisions where aggregation is impossible.

- Develop an approach for addressing those welfare effects that are difficult to look at in a piecemeal way, such as welfare changes on communities or ecosystems.

- Identify appropriate top-down and bottom-up approaches for estimating impacts and value (whether economic or otherwise) of the most critical welfare categories.

- Identify situations in which evaluation following the above steps is likely to be prohibitively difficult, and determining alternative methods for approaching the topic of the impact of global change on well-being.

Together, these steps should enable researchers to make progress towards promoting the consistency and coordination in analyses of welfare/well-being that will facilitate developing the body of research necessary to analyze impacts on human welfare, well-being, and quality of life.

4.6 REFERENCES

Ahn, S., J. deSteiguer, R. Palmquist, and T. Holmes, 2000: Economic analysis of the potential impact of climate change on recreational trout fishing in the southern Appalachian Mountains. *Climatic Change*, **45**, 493-509.

Alberini, A., M. Cropper, A. Krupnick, and N. Simon, 2004: Does the value of a statistical life vary with age and health status? Evidence from the U.S. & Canada. *Journal of Environmental Economics and Management*, **48(1)**, 769-792.

Anderson, D., S. Shankle, M. Scott, D. Neitzel and J. Chatters. 1993: Valuing effects of climate change and fishery enhancement on Chinook salmon. *Contemporary Policy Issues*, **11(4)**, 82-94.

Arrow, K.J., M.L. Cropper, G.C. Eads, R.W. Hahn, L.B. Lave, R.G. Noll, P.R. Portney, M. Russell, R. Schmalensee, V.K. Smith, and R.N. Stavins, 1996: *Benefit-Cost Analysis in Environmental, Health, and Safety Regulation: A Statement of Principles*. American Enterprise Institute, The Annapolis Center, and Resources for the Future, Washington, DC.

Bachelet, D., R.P. Neilson, J.M. Lenihan, and R.J. Drapek, 2001: Climate change effects on vegetation distribution and carbon budget in the U.S. *Ecosystems*, **4**, 164-185.

Banuri, T. and J. Weyant, 2001: Setting the stage: climate change and sustainable development. *Climate Change 2001: Mitigation. Contributions of Working Group III to the Third Assessment Report of the Intergovernmental Panel on Climate Change.* [Metz, B., O. Davidson, R. Swart, and J. Pan (eds.)]. Cambridge University Press, UK, pp. 73-114.

Bark-Hodgins, R. and B. Colby, 2006: *Snow Days? Using Climate Forecasts to Better Manage Climate Variability in the Ski Industry.* Presentation at NOAA Climate Prediction Applications Science Workshop, held March 21-24 in Tucson, Arizona.

Bayer, P. and C. Timmins, 2005: On the equilibrium properties of locational sorting models. *Journal of Urban Economics*, **57(3)**, 462-477.

Bayer, P., N. Keohane and C. Timmins, 2006: *Migration and Hedonic Valuation: The Case of Air Quality.* NBER Working Paper #12106.

Beebee, T.J.C, 1995: Amphibian breeding and climate. *Nature*, **374**, 219-220.

Berke, P., D. Godschalk, and E. Kaiser, 2006: *Urban Land Use Planning.* University of Illinois Press, Urbana, Illinois.

Bin, O. and S. Polasky, 2004: Effects of flood hazards on property values: evidence before and after Hurricane Floyd. *Land Economics*, **80(4)**, 490-500.

Bishop, R., 1982: Option value: an exposition and extension. *Land Economics*, **58(1)**, 1-15.

Black, D.A., J. Galdo, and L. Liu, 2003: *How Robust Are Hedonic Wage Estimates of the Price of Risk?* Final Report to the U.S. Environmental Protection Agency [R 829-43-001].

Blomquist, G.C., M.C. Berger, and J.P. Hoehn, 1988: New estimates of quality of life in urban areas. *American Economic Review*, **78(1)**, 89-107.

Bockstael, N.E., A.M. Freeman, R. Kopp, P. Portney, and V.K. Smith, 2000: On measuring economic values for nature. *Environmental Science and Technology*, **34(8)**, 1384-1390.

Bolin, R., 1986: Disaster impact and recovery: a comparison of black and white victims. *International Journal of Mass Emergencies and Disasters (IJMED)*, **4**, 35-50.

Both, C., S. Bouwhuis, C.M. Lessells, and M.W. Visser, 2006: Climate change and population declines in a long-distance migratory bird. *Nature*, **441**, 81-83.

Bowling, A., 1997: *Measuring Health: A Review of Quality of Life Measurement Scales.* Open University Press, Philadelphia, Pennsylvania, 2nd edition.

Boyce, J. and B. Shelley, 2003: *Natural Assets: Democratizing Environmental Ownership.* Island Press, Washington, DC.

Boyd, J., 2006: *The Nonmarket Benefits of Nature: What Should Be Counted in Green GDP?* Resources for the Future Discussion Paper RFF DP 06-24, Washington, DC.

Boyd, J. and S. Banzhaf, 2006: *What Are Ecosystem Services? The Need for Standardized Environmental Accounting Units.* Resources for the Future Discussion Paper RFF DP 06-02, Washington, DC.

Brown, J.L., S.H. Li, and N. Bhagabati, 1999: *Long-Term Trend Toward Earlier Breeding in an American Bird: A Response to Global Warming?* Proceedings of the National Academy of Sciences, USA, **96**, 5565–5569.

Burby, R. (ed.), 1998: *Cooperating with Nature: Confronting Natural Hazards with Land Use Planning for Sustainable Communities.* Joseph Henry Press, Washington, DC.

Campbell, S., 1996: Green cities, growing cities, just cities? Urban planning contradictions of sustainable development. *Journal of the American Planning Association*, **62**, 296-312.

Cheshire, P. and S. Magrini, 2006: Population growth in European cities: weather matters – but only nationally. *Regional Studies*, **40(1)**, 23-37.

Cline, W.R., 1992: *The Economics of Global Warming*. Institute for International Economics, Washington DC.

Coleman, J.S., 1988: Social capital in the creation of human capital. *American Journal of Sociology Supplement*, **94**, S95-S120.

Coleman, J.S., 1990: *Foundations of Social Theory*. Harvard University Press.

Coleman, J.S., 1993: The rational reconstruction of society, *American Sociological Review*, **58**, 1-15.

Cordell, H.K., B. McDonald, R.J. Teasley, J.C. Bergstrom, J. Martin, J. Bason, and V.R. Leeworthy, 1999: Outdoor recreation participation trends. In: *Outdoor Recreation in American Life: A National Assessment of Demand and Supply Trends*. [Cordell, H.K. (ed.)]. Sagamore Publishing, Champaign, Illinois.

Costanza, R., R. d'Arge, R. deGroot, S. Farber, M. Grasso, B. Hannon, K. Limburg, S. Naeem, R. O'Neill, J. Paruelo, R.G. Raskin, P. Sutton, M. van den Belt, 1997: The value of the world's ecosystem services and natural capital. *Nature*, **387**, 253-255.

Costanza, R., B. Fisher, S. Ali, C. Beer, L. Bond, R. Boumans, N. Danigelis, J. Dickinson, C. Elliott, J. Farley, D.E. Gayer, L.M. Glenn, T. Hudspeth, D. Mahoney, L. McCahill, B. McIntosh, B. Reed, S. Abu Turab Rizvi, D.M. Rizzo, T. Simpatico, and R. Snapp, 2007: Quality of life: an approach integrating opportunities, human needs, and subjective well-being. *Ecological Economics*, **61**, 267-276.

Cragg, M. and M. Kahn, 1997: New estimates of climate demand: evidence from location choice. *Journal of Urban Economics*, **42(2)**, 261-284.

Crick, H.Q.P., C. Dudley, D.E. Glue, and D.L. Thompson, 1997: UK birds are laying eggs earlier. *Nature*, **388**, 526.

Cropper, M.L., 1981: The value of urban amenities. *Journal of Regional Science*, **21(3)**, 359-374.

Cropper, M.L. and A.S. Arriaga-Salinas, 1980: Inter-city wage differentials and the value of air quality. *Journal of Urban Economics*, **8(2)**, 236-254.

Cropper, M.L. and F. G. Sussman, 1990: Valuing future risks to life. *Journal of Environmental Economics and Management*, **19(2)**, 160-174.

Curriero, F.C., K.S. Heiner, J.M. Samet, S.L. Zeger, L. Strug, and J. Patz, 2002: Temperature and mortality in 11 cities of the eastern United States. *American Journal of Epidemiology*, **155(1)**, 80-87.

Curriero, F., J. Patz, J. Rose, and S. Lele, 2001: Analysis of the association between extreme precipitation and waterborne disease outbreaks in the United States: 1948-1994. *American Journal of Public Health*, **91(8)**, 1194-1199.

Curriero, F.C., J.M. Samet, and S.L. Zeger, 2003: Re:"on the use of generalized additive models in time-series studies of air pollution and health" and" temperature and mortality in 11 cities of the eastern United States". *American Journal of Epidemiology*, **158(1)**, 93-94.

Cutter, S., 2006: The geography of social vulnerability: race, class, and catastrophe, *Understanding Katrina: Perspectives from the Social Sciences*. Available at: http://understandingkatrina.ssrc.org/Cutter/.

Cutter, S., B. Boruff, and W. Shirley, 2003: Social vulnerability to environmental hazards. *Social Science Quarterly*, **84(1)**, 242-261.

Daily, G.C., 1997: *Nature's Services. Societal Dependence on Natural Ecosystems*. Island Press, Washington, DC.

Davis, R., P. Knappenberger, P. Michaels, and W. Novicoff, 2004: Seasonality of climate-human mortality relationships in US cities and impacts of climate change. *Climate Research*, **26**, 61-76.

DeFreitas, C.R., 2005: The climate-tourism relationship and its relevance to climate change impact assessment. In: *Tourism, Recreation and Climate Change*. [Hall, C.M. and J. Higham (eds.)]. Channel View Publications, Buffalo, New York, pp. 29-43.

Diamond, J., 2005: *Collapse: How Societies Choose to Fail or Succeed*. Viking, New York.

Diamond, P.A. and J. A. Hausman, 1993: On contingent valuation measurement of nonuse values. In: *Contingent Valuation: A Critical Assessment* [Hausman, J.A. (ed.)]. North-Holland, Amsterdam, Netherlands.

Diener, E. and C. Diener, 1995: The wealth of nations revisited: income and quality of life. *Social Indicators Research*, **36**, 275–286.

Diener, E., M. Diener, and C. Diener, 1995: Factors predicting the subjective well-being of nations. *Journal of Personality and Social Psychology*, **69(5)**, 851–864.

Dietz, T., E. Rosa, and R. York, In Press. Human driving forces of global change: examining current theories. In: *Human Dimensions of Global Change* (In press) [Rosa, E., A. Diekmann, T. Dietz, and C. Jaeger (eds.)]. MIT Press, Cambridge, Massachusetts.

Dunn, P.O. and D.W. Winkler, 1999: *Climate Change has Affected the Breeding Date of Tree Swallows Throughout North America*. Proceedings of the Royal Society of London, Series B, **266**, pp. 2487–2490.

Ehrlich, P. and A. Ehrlich, 1996: *Betrayal of Science and Reason*. Island Press, Washington DC.

Fagan, B., 2001: *The Little Ice Age*. Basic Books, New York.

Fankhauser, S., 1995: *Valuing Climate Change – The Economics of the Greenhouse*. EarthScan, London.

Florida, R., 2002a: The economic geography of talent. *Annals of the Association of American Geographers*, **92(4)**, 743–755.

Florida, R., 2002b: *The Rise of the Creative Class: And How It's Transforming Work, Leisure and Everyday Life*. Basic Books.

Fothergill, A., E. Maestas, and J.D. Darlington, 1999: Race, ethnicity, and disasters in the United States: a review of the literature. *Disasters*, **23(3)**, 156-173.

Fothergill, A. and L.A. Peek, 2004: Poverty and disasters in the United States: a review of recent sociological findings. *Natural Hazards*, **32**, 89-110.

Fowler, A.M., and K.J. Hennessey, 1995: Potential impacts of global warming on the frequency and magnitude of heavy precipitation. *Natural Hazards*, **11**, 283-303.

Frank, R.H., 1985: *Choosing the Right Pond*. Oxford University Press, Oxford.

Freeman, A.M., 2003: *The Measurements of Environmental and Resource Values: Theory and Methods*. Resources for the Future Press, Washington, DC, 2nd edition.

Frey, B.S. and A. Stutzer, 2002: *Happiness and Economics*. Princeton University Press, Princeton, New Jersey.

Galbraith, H., R. Jones, R. Park, J. Clough, S. Herrod-Julius, B. Harrington, and G. Page, 2002: Global climate change and sea level rise: potential losses of intertidal habitat for shorebirds. *Waterbirds*, **25**, 173–183.

Galbraith, H., J.B. Smith, and R. Jones, 2006: Biodiversity changes and adaptation. In: *The Impact of Climate Change on Regional Systems: a Comprehensive Analysis of California* [J.B. Smith and R. Mendelsohn (eds.)]. Edward Elgar, Massachusetts.

Githeko, A.K., and A. Woodward, 2003: International consensus on the science of climate and health: the IPCC Third Assessment Report. In: *Climate Change and Human Health: Risks and Responses* [McMichael, A.J., D.H. Campbell-Lendrum, C.F. Corvalan, K.L. Ebi, A. Githeko, J.D. Scheraga, et al., (eds.)]. World Health Organization, World Meteorology Organization, and UN Environment Programme, Geneva, Switzerland.

Glaeser, E.L., J. Kolko, and A. Saiz, 2001: Consumer city. *Journal of Economic Geography*, **1(1)**, 27-51.

Glick, P. and J. Clough, 2006: *An Unfavorable Tide: Global Warming, Coastal Habitats and Sportfishing in Florida*. National Wildlife Federation, Washington, DC.

Godschalk, D.R., 2003: Natural hazard mitigation: creating resilient cities. *Natural Hazards Review*, **4(3)**, 136-143.

Godschalk, D.R., 2007: Mitigation. In: *Emergency Management: Principles and Practice for Local Government*. Second edition. [Waugh, Jr., W.L., and K. Tierney (eds.)]. International City/County Management Association, Washington, DC, pp 89-112.

Graves, P.E. and P.R. Mueser, 1993: The role of equilibrium and disequilibrium in modeling regional growth and decline: a critical reassessment. *Journal of Regional Science*, **33(1)**, 69-84.

Greenwood, M., G. Hunt, D. Rickman, and G. Treyz, 1991: Migration, regional equilibrium, and the estimation of compensating differentials. *American Economic Review*, **81(5)**, 1382-1390.

Guyatt, G. H., D.H. Feeny, and D.L. Patrick, 1993: Measuring health-related quality of life. *Annals of Internal Medicine*. **118(8)**, 622-629.

Gyourko, J. and J. Tracy, 1991: The structure of local public finance and the quality of life. *Journal of Political Economy*, **99(4)**, 774-806.

Hall, C.M. and J. Highman, 2005: Introduction: tourism, recreation, climate change. In: *Tourism, Recreation and Climate Change* [Hall, C.M. and J. Higham (eds.)]. Channel View Publications, Buffalo, New York, pp. 3-28.

Hallstrom, D. and V.K. Smith, 2005: Market responses to hurricanes. *Journal of Environmental Economics and Management*, **50(3)**, 541-561.

Hamilton, J. and R. Tol, 2004: *The impact of climate change on tourism and recreation*. FNU-52 working paper, Research Unit Sustainability and Global Change, University of Hamburg, Germany.

Hamilton, J. and R. Tol, 2007: The impact of climate change on tourism in Germany, the UK and Ireland. *Regional Environmental Change*, **7**, 161-172.

Hayhoe, K., D. Cayan, C.B. Field, P.C. Frumhoff, E.P. Maurer, N.L. Miller, S.C. Moser, S.H. Schneider, K.N. Cahill, E.E. Cleland, L. Dale, R. Drapek, R.M. Hanemann, L.S. Kalkstein, J. Lenihan, C.K. Lunch, R.P. Neilson, S.C. Sheridan, and J.H. Verville, 2004: Emissions pathways, climate change, and impacts on California. Proceedings of the National Academy of Sciences of the United States of America, **101(34)**, 12,422-12,427.

Health Care Financing Review, 2004: Special Issue on the Health Outcomes Survey. **25(4)**, 1-119.

Hersteinsson, P. and D.W. Macdonald, 1992: Interspecific competition and the geographical distribution of red and arctic foxes, *Vulpes vulpes* and *Alopex lagopus*. *Oikos*, **64**, 505–515.

Hoch, I. and J. Drake, 1974: Wages, climate, and the quality of life. *Journal of Environmental Economics and Management*, **1**, 268-296.

Holbrook, S.J., R.J. Schmitt, and J.S. Stephens, 1997: Changes in an assemblage of temperate reef fishes associated with a climatic shift. *Ecological Applications*, **7**, 1299-1310.

IPCC, 2007a: *Climate Change 2007: Impacts, Adaptation and Vulnerability. Contribution of Working Group II to the Fourth Assessment Report of the Intergovernmental Panel on Climate Change*, M.L. Parry, O.F. Canziani, J.P. Palutikof, P.J. van der Linden and C.E. Hanson, Eds., Cambridge University Press, Cambridge, UK.

IPCC, 2007b: *Climate Change 2007: Summary for Policymakers of the Synthesis Report of the IPCC Fourth Assessment Report*. Draft Copy, 16 November 2007.

IPCC, 2007c: *Summary for Policymakers. In: Climate Change 2007: The Physical Science Basis. Contribution of Working Group I to the Fourth Assessment Report of the Intergovernmental Panel on Climate Change* [Solomon, S., D. Qin, M. Manning, Z. Chen, M. Marquis, K.B., Averyt, M. Tignor and h.L.Miller (eds.)]. Cambridge University Press, Cambridge, United Kingdom and New York, USA.

IPCC, 2000: *Special Report on Emissions Scenarios*, Coordinating Lead Author: Nebojsa Nakicenovic. Cambridge University Press, Cambridge, United Kingdom and New York, USA.

IPCC, 1996: *Climate Change 1995: Economic and Social Dimensions of Climate Change. Contribution of Working Group III to the Second Assessment Report of the Intergovernmental Panel on Climate Change* [Bruce, J., H. Lee, and E. Haites (eds.)]. Cambridge University Press, Cambridge, United Kingdom and New York, USA.

Irland, L., D. Adams, R. Alig, C. Betz, C.Chen, M. Hutchins, B. McCarl, K. Skog, and B. Sohngen, 2001: Assessing socioeconomic impacts of climate change on U.S. forests, wood-product markets, and forest recreation. BioScience, 51(9), 753-764.

Iverson, L. and S. Matthews, 2007: Potential changes in suitable habitat for 134 tree species in northeastern United States. *Mitigation and Adaptation Strategies for Global Change*. (in press).

Jacoby, H.D., 2004: Informing climate policy given incommensurable benefits estimates. *Global Environmental Change Part A*, **14(3)**, 287-297.

Joyce, L., J. Aber, S. McNulty, V. Dale, A. Hansen, L. Irland, R. Nelson and K. Skog, 2001: Potential consequences of climate variability and change for the forests of the United States. In: *Climate Change Impacts on the United States*. U.S. Global Change Research Program. Washington DC.

Kahneman, D., E. Diener, and N. Schwarz (eds.), 1999: *Well-Being: The Foundations of Hedonic Psychology*. Russel Sage Foundation, New York.

Kahneman, D. and A.B. Krueger, 2006: Developments in the measurement of subjective well-being. *Journal of Economic Perspectives*, **20(1)**, 3-24.

Kalkstein, L.S., 1989: The impact of CO2 and trace gas-induced climate changes upon human mortality. In: *The Potential Effects of Global Climate Change in the United States* [Smith, J. B. and D. A. Tirpak (eds.)]. Document no. 230-05-89-057, Appendix G, U.S. Environmental Protection Agency, Washington, DC.

Kalkstein, L.S. and R. E. Davis, 1989: Weather and human mortality: an evaluation of demographic and interregional responses in the United States. *Annals of the Association of American Geographers*, 79(1), 44-64.

Kalkstein, L.S. and Greene J.S., 1997: An evaluation of climate/mortality relationships in large U.S. cities and the possible impacts of a climate change. *Environmental Health Perspectives*, **105(1)**, 84-93.

Kinnell, J., J.K. Lazo, D.J. Epp, A. Fisher, J.S. Shortle, 2002: Perceptions and values for preventing ecosystem change: Pennsyvlania duck hunters and the prairie pothole region. *Land Economics*, **78(2)**, 228-244.

Knowlton, K., B. Lynn, R. Goldberg, C. Rosenzweig, C. Hogrefe, J. K. Rosenthal and P. L. Kinney, 2007: Projecting heat-related mortality impacts under a changing climate in the New York City region. *American Journal of Public Health*, **97(11)**, 2028-2034.

Kunst, A.E., C.W.N. Looman, and J.P. Mackenbach, 1993: Outdoor air temperature and mortality in the Netherlands: a time series analysis. *American Journal of Epidemiology*, **137(3)**, 331-341.

Lambiri, D., B. Biagi, and V. Royuela, 2007: Quality of life in the economic and urban economic literature. *Social Indicators Research*, **84(1)**, 1-25.

Layton, D. and G. Brown, 2000: Heterogeneous preferences regarding global climate change. *Review of Economics and Statistics*, **82(4)**, 616-624.

Layton, D. F. and R. Levine, 2003: How much does the future matter? A hierarchical Bayesian analysis of the public's willingness to mitigate ecological impacts of climate change. *Journal of the American Statistical Association*, **98(463)**, 533-544.

Lenihan, J.M., R. Drapek, and R. Neilson, 2006: Terrestrial ecosystem changes. In: *The Impact of Climate Change on Regional Systems. A Comprehensive Analysis of California.* [J.B. Smith and R. Mendelsohn (eds.)]. Edward Elgar, Northampton, Massachusetts.

Lindell, M.K., and R.W. Perry, 2004: *Communicating Environmental Risk in Multiethnic Communities.* Sage, Thousand Oaks, California.

Lipp, E.K. and J.B. Rose, 1997: The role of seafood in foodborne diseases in the United States of America. *Revue Scientifique et Technique*, **16**, 620-640.

Liu, X., A. Vedlitz and L. Alston, 2008: Regional news portrayals of global warming and climate change. *Environmental Science and Policy.* Advance online publication: doi:10.1016/j.envsci.2008.01.002 (published online 4 March 2008).

Loomis, J. and J. Crespi, 1999: Estimated effects of climate change on selected outdoor recreation activities in the United States. In: *The Impact of Climate Change on the United States Economy* [Mendelsohn, R. and J. Neumann (eds.)]. Cambridge University Press, Cambridge, UK, pp. 289-314.

Loomis, J., P. Kent, L. Stange, K. Fausch, and A. Covich, 2000: Measuring the total economic value of restoring ecosystem services in an impaired river basin: Results from a contingent valuation survey. *Ecological Economics*, **33**, 103-117.

Loomis, J. and R. Richardson, 2006: An external validity test of intended behavior: Comparing revealed preference and intended visitation in response to climate change. *Journal of Environmental Planning and Management*, **49(4)**, 621-630.

MA, 2005: *Millennium Ecosystem Assessment: Ecosystems and Human Well-being – Synthesis.* Island Press, Washington, DC.

MacKenzie, W.R., W.L. Schell, K.A. Blair, D.G. Addiss, D.E. Peterson, N.J. Hoxie, J.J. Kazmierczak, and J.P. Davis, 1994: A massive outbreak in Milwaukee of cryptosporidium infection transmitted through the public water supply. *New England Journal of Medicine*, **331(3)**, 161-167.

MacMillan, D., E.I. Duff, and D. Elston, 2001: Modelling the non-market environmental costs and benefits of biodiversity projects. *Environmental and Resource Economics*, **18**, 391-401.

Malcolm, J.R., and L.F. Pitelka. 2000. *Ecosystems and Global Climate Change.* Pew Center on Global Climate Change, Washington, DC.

Martens, W.J.M., 1998: Climate change, thermal stress and mortality changes. *Social Science and Medicine*, **46(3)**, 331-344.

McMichael, A.J., D. Campbell-Lendrum, R.S. Kovats, S. Edwards, P. Wilkinson, N. Edmonds, N. Nicholls, S. Hales, F.C. Tanser, D. Le Sueur, M. Schlesinger, and N. Andronova, 2004: Global Climate Change. In: *Comparative Quantification of Health Risks: Global and Regional Burden of Disease due to Selected Major Risk Factor: Volume 2* [Ezzati M., A. Rodgers, and C. J. Murray (eds.)]. World Health Organization, Geneva, pp. 1543–1649.

McMichael, A.J. and A. Githeko, 2001: Human health. In: Climate Change 2001: Impacts, Adaptation, and Vulnerability. Contribution of Working Group II to the Third Assessment Report of the Intergovernmental Panel on Climate Change [McCarthy, J.J., O.F. Canziani, N.A. Leary, D.J. Dokken, and K.S. White (eds.)]. Cambridge University Press, Cambridge, UK and New York, USA, pp. 453-485.

Melillo, J., A. Janetos, D. Schimel, and T. Kittel, 2001: Vegetation and biogeochemical scenarios. In: *Climate Change Impacts on the United States. The Potential Consequences of Climate Variability and Change.* Report for the US Global Change Research Program, Cambridge University Press, Cambridge, UK.

Mendelsohn, R., 2001: A hedonic study of the non-market impacts of global warming in the US. In: *The Amenity Value of the Global Climate* [Maddison, D. (ed.)]. Earthscan, London, pp. 93-105.

Mendelsohn, R. and J.E. Neumann, 1999: *The Impact of Climate Change on the United States Economy.* Cambridge University Press, Cambridge, UK.

Mendelsohn, R. and M. Markowski, 1999: The impact of climate change on outdoor recreation. In: *The Impact of Climate Change on the United States Economy* [Mendelsohn, R. and J. Neumann (eds.)]. Cambridge University Press, Cambridge, UK, pp. 267-288.

Meyer, J.L., M.J. Sale, P.J. Mulholland, and N.L. Poff, 1999: Impacts of climate change on aquatic ecosystem functioning and health. In: *Potential Consequences of Climate Variability and Change to Water Resources of the United States* [D.B. Adams (ed).]. Conference papers 10-12 May 1999, Atlanta, GA.

Mileti, D.S., 1999: *Disasters by Design: A Reassessment of Natural Hazards in the United States.* Joseph Henry Press, Washington, DC.

Moore, T., 1998: Health and amenity effects of global warming. *Economic Inquiry*, **36(3)**, 471–488.

Morris, M. (ed.), 2006: *Integrating Planning and Public Health: Tools and Strategies to Create Healthy Places.* American Planning Association, Washington, DC.

Morris, R.E., M.S. Gery, M.K. Liu, G.E. Moore, C. Daly, and S.M. Greenfield, 1989: Sensitivity of a regional oxidant model to variations in climate parameters. In: *The Potential Effects of Global Climate Change on the United States* [J.B.Smith and D.A.Tirpak (eds.)]. U.S. EPA, Office of Policy, Planning and Evaluation, Washington, DC.

Multihazard Mitigation Council, 2005: *Natural Hazard Mitigation Saves: An Independent Study to Assess the Future Savings from Mitigation Activities.* National Institute of Building Sciences, Washington, DC.

Murphy, C. and P. Gardoni, 2008: The acceptability and the tolerability of societal risks: a capabilities-based approach. *Science and Engineering Ethics*, **14(1)**, 77-92.

National Research Council, 2004: *Valuing Ecosystem Services: Toward Better Environmental Decision-Making.* Committee on Assessing and Valuing the Services of Aquatic and Related Terrestrial Ecosystems, The National Academies Press, Washington, DC.

Ng, Y.K., 2003: From preferences to happiness: toward a more complete welfare economics. *Social Choice and Welfare*, **20**, 307-350.

Nordhaus, W., 1994: *Managing the Global Commons: The Economics of Climate Change.* MIT Press, Cambridge, Massachusetts.

Nordhaus, W. and J. Boyer, 2000: *Warming the World: Economic Modeling of Global Warming.* MIT Press, Cambridge, Massachusetts.

Northbridge, M., E. Sclar, and P. Biswas, 2003: Sorting the connections between built environment and health: a conceptual framework for navigating pathways and planning healthy cities. *Journal of Urban Health*, **80(4)**, 556-568.

Oechel, W.C., S.J. Hastings, G. Vourlitis, and M. Jenkins, 1993: Recent change of arctic tundra ecosystems from net carbon dioxide sink to a source. *Nature*, **361**, 520-523.

Oechel, W.C., G.L. Vourlitis, S.J. Hastings, R.C. Zulueta, L. Hinzman, and D. Kane, 2000: Acclimation of ecosystem CO_2 exchange in the Alaska Arctic in response to decadal warming. *Nature*, **406**, 978-981.

Parmesan, C., 1996: Climate and species' range. *Nature*, **382**, 765-766.

Parmesan, C., and H. Galbraith, 2004: *Observed Impacts of Global Climate Change in the U.S.* Pew Center on Global Climate Change, Arlington, Virginia.

Parmesan, C., and G. Yohe, 2003: A globally coherent fingerprint of climate change impacts across natural systems. *Nature*, **421**, 37-42.

Patz, J.A., M.A. McGeehin, S.M. Bernard, K.L. Ebi, P.R. Epstein, A. Grambsch, D.J. Gubler, P. Reiter, I. Romieu, J.B. Rose, J.M. Samet, and J. Trtanj, 2001: The potential health impacts of climate variability and change for the United States: executive summary of the report of the health sector of the U.S. National Assessment. *Journal of Environmental Health*, **64**, 20–28.

Peacock, W.G., 2003: Hurricane mitigation status and factors influencing mitigation status among Florida's single-family homeowners. *Natural Hazards Review*, **4(3)**, 1-10.

Pendleton, L.H. and R. Mendelsohn, 1998: Estimating the economic impact of climate change on the freshwater sportsfisheries of the Northeastern U.S. *Land Economics*, **74(4)**, 483-96.

Peters, R.L., and T.E. Lovejoy, 1992: *Global Warming and Biological Diversity.* Yale University Press, New Haven, Connecticut.

Pileus Project, Which factors are typically believed to have the greatest influence on tourists' behaviors? *Pileus Project: Climate Science for Decision Makers.* Retrieved May 28, 2008, from http://www.pileus.msu.edu/tourism/tourism_influencefactor.htm

Pimentel, D., C. Wilson, C. McCullum, R. Huang, P. Dwen, J. Flack, Q. Tran, T. Saltman, and B. Cliff, 1997: Economic and environmental benefits of biodiversity. *Bioscience*, **47(11)**, 747-757.

Ponting, C., 1991: *A Green History of the World: The Environment and the Collapse of Great Civilizations.* Penguin, New York.

Porter, D. (ed.), 2000: *The Practice of Sustainable Development.* Urban Land Institute, Washington, DC.

Putnam, R., 1993: *Making Democracy Work: Civic Traditions in Modern Italy*. Princeton University Press, Princeton, New Jersey.

Putnam, R., 1995: Tuning in, tuning out: the strange disappearance of social capital in America, *Political Science and Politics*, **28(4)**, 664-683

Putnam, R., 2000: *Bowling Alone: The Collapse and Revival of American Community*. Simon and Schuster, New York.

Rahman, T., 2007: Measuring the well being across countries. *Applied Economic Letters*, **14(11)**, 779-783.

Raphael, D., R. Renwick, I. Brown, and I. Rootman, 1996: Quality of life and health: current status and emerging conceptions. *Social Indicators Research*, **39**, 65-88.

Raphael, D., R. Renwick, I. Brown, B. Steinmetz, H. Sehdevc, and S. Phillips, 2001: Making the links between community structure and individual well-being: community quality of life in Riverdale, Toronto, Canada. *Health & Place*, **7**, 179-196.

Raphael, D., B. Steinmetz, R. Renwick, I. Rootman, I. Brown, H. Sehdev, S. Phillips, and T. Smith, 1999: The community quality of life project: a health promotion approach to understanding communities. *Health Promotion International*, **14(3)**, 197-210.

Roback, J., 1988: Wages, rents, and amenities: differences among workers and regions. *Economic Inquiry*, **26(1)**, 23-41.

Roback, J., 1982: Wages, rents, and the quality of life. *Journal of Political Economy*, **90(6)**, 1257-1279.

Root, T.L, J.T. Price, K.R. Hall, S.H. Schneider, C. Rozensweig, and J.A. Pounds, 2003: Fingerprints of global warming on wild animals and plants. *Nature*, **421**, 57-60.

Rose, A., 2004: Defining and measuring economic resilience to disasters. *Disaster Prevention and Management*, **13**, 307-314.

Rose, A., K. Porter, N. Dash, J. Bouabid, C. Huyck, J.C. Whitehead, D. Shaw, R.T. Eguchi, C. Taylor, T.R. McLane, L.T. Tobin, P.T. Ganderton, D. Godschalk, A.S. Kiremidjian, and K. Tierney, and C.T. West, 2007: Benefit-cost analysis of FEMA hazard mitigation grants. *Natural Hazards Review*, **8(4)**, 97-111.

Rosenberger, R. and J. Loomis, 2003: Benefit transfer. In: *A Primer on Non Market Valuation* [Champ, P., K. Boyle, and T. Brown (eds.)]. Kluwer Academic Publishers, Boston, Massachusetts, pp 445-482.

Ross, M.S., J.J. O'Brien, L. Da Silveira, and L. Sternberg, 1994: Sea level rise and the reduction in pine forests in the Florida Keys. *Ecological Applications*, **4**, 144-156.

Rothman, D.S., B. Amelung, and P. Poleme, 2003: *Estimating non-market impacts of climate change and climate policy*. Prepared for OECD Workshop on the Benefits of Climate Policy, Improving Information for Policy Makers. Available at www.oecd.org/dataoecd/6/30/2483779.pdf.

Roy, D.B., and T.H. Sparks, 2000: Phenology of British butterflies and climate change. *Global Change Biology*, **6**, 407-416.

Sagarin, R.D., J.P. Barry, S.E. Gilman, and C.H. Baxter, 1999: Climate-related change in an intertidal community over short and long time scales. *Ecological Monographs*, **69**, 465-490.

Saunders, S. and T. Easley, 2006: *Losing ground: Western National Parks Endangered by Climate Disruption*. Rocky Mountain Climate Organization and Natural Resources Defense Council, New York.

Scott, D. and B. Jones, 2005: *Climate Change and Banff National Park: Implications for Tourism and Recreation*. Faculty of Environmental Studies, University of Waterloo, Waterloo, Canada.

Scott, D., B. Jones, and J. Konopek. 2007. Implications of Climate and Environmental Change for Nature-Based Tourism in the Canadian Rocky Mountains. *Tourism Management*, **28(2)**, 570-579.

Scott, D., G. McBoyle and A. Minogue, 2006: Climate Change and the Sustainability of Ski-based Tourism in Eastern North America: A Reassessment. *Journal of Sustainable Tourism*, **14(4)**, 376-398.

Scott, D., G. Wall, G. McBoyle, 2005: The evolution of the climate change issue tourism sector. In: *Tourism, Recreation and Climate Change* [Hall, C.M. and J. Higham (eds.)]. Channel View Publications, Buffalo, New York, pp. 44-62.

Semenza, J.C., C.H. Rubin, and K.H. Falter, 1996: Heat-related deaths during the July 1995 heat wave in Chicago. *New England Journal of Medicine*, **335(2)**, 84-90.

Shaw, D., 2005: *Water Resource Economics and Policy*. Edward Elgar Publishing, Northampton, Massachusetts.

Sillman, S. and P.J. Samson, 1995: Impact of temperature on oxidant photochemistry in urban, polluted, rural, and remote environments. *Journal of Geophysical Research*, **100**, 11,497-11,508.

Smit, B, and O. Pilifosova, 2001: Adaptation to climate change in the context of sustainable development and equity. In: *Climate Change 2001: Impacts, Adaptation, and Vulnerability. Contribution of Working Group II to the Third Assessment Report of the Intergovernmental Panel on Climate Change* [McCarthy, J.J., O.F. Canziani, N.A. Leary, D.J. Dokken, and K.S. White (eds.)]. Cambridge University Press, Cambridge, UK and New York, USA.

Smith, J.B., J.K. Lazo, and B. Hurd, 2003: The difficulties of estimating global non-market damages from climate change. In: *Global Climate Change: The Science, Economics, and Politics* [Griffin, J.M. (ed.)]. Edward Elgar, Northampton, Massachusetts.

Smith, V.K., 1983: The role of site and job characteristics in hedonic wage models. *Journal of Urban Economics*, **13(3)**, 296-321.

Strzepek, K., D. Major, C. Rosenzweig, A. Iglesias, D. Yates, A. Holt, and D. Hillel, 1999: New methods of modeling water availability for agriculture under climate change: the U.S. Corn Belt. *Journal of the American Water Resources Association*, **35(6)**, 1639-1656.

Strzepek, K.M., and J.B. Smith (eds.), 1995: *As Climate Changes: International Impacts and Implications*. Cambridge University Press, Cambridge, UK.

Sufian, A.J.M., 1993: A multivariate analysis of the determinants of urban quality of life in the world's largest metropolitan areas. *Urban Studies*, **30(8)**, 1319-1329.

Tainter, J., 1998: *The Collapse of Complex Societies*. Cambridge University Press, Cambridge, UK.

Tierney, K., 1997: Impacts of recent disasters on businesses: the 1991 Midwest floods and the 1994 Northridge earthquake. In: *Economic Consequences of Earthquakes: Preparing for the Unexpected* [B. Jones (ed.)]. National Center for Earthquake Engineering Research, Buffalo, New York.

Tol, R.S.J., 2002: Welfare specifications and optimal control of climate change: an application of fund. *Energy Economics*, **24**, 367-376.

Tol, R.S.J., 2005: The marginal damage costs of carbon dioxide emissions: an assessment of the uncertainties. *Energy Policy*, **33(16)**, 2064-2074.

Toman, M.A. 1998: Why not calculate the value of the world's ecosystem services and natural capital? *Ecological Economics*, **25**, 57-60.

U.S. EPA, 1989: *The Potential Effects of Global Climate Change on the United States*. J.B. Smith and D.A.Tirpak (eds.). EPA-230-05-89-050, Washington, DC.

U.S. EPA. 1995. *Ecological Impacts from Climate Change: An Economic Analysis of Freshwater Recreational Fishing.* EPA 220-R-95-004. Washington DC.

U.S. EPA, 1997: *The Benefits and Costs of the Clean Air Act, 1970-1990.* Report to Congress, Washington DC.

U.S. EPA, 2000: *Guidelines for Preparing Economic Analyses*. EPA 240-R-00-003, Washington, DC.

Vaughan, W.J. and C.S. Russell, 1982: *Freshwater Recreational Fishing: The National Benefits of Water Pollution Control*. Resources for the Future, Washington DC.

Vedlitz, A., L.T. Alston, S.B. Laska, R.B. Gramling, M.A. Harwell and H.D. Worthen, 2007: *Use of Science in Gulf of Mexico Decision Making Involving Climate Change*. Project Final Report, Prepared for the U.S. Environmental Protection Agency under Cooperative Agreement No. R-83023601-0.

Veenhoven, R., 1988: The utility of happiness. *Social Indicators Research*, **20**, 333-354.

Veenhoven, R., 1996: Happy life-expectancy, a comprehensive measure of quality of life in nations. *Social Indicators Research*, **39**, 1-58.

Veenhoven, R., 2000: The four qualities of life. *Journal of Happiness Studies*, **1**, 1-39.

VEMAP Members, 1995: Vegetation/ecosystem modeling and analysis project: comparing biogeography and biogeochemistry models in a continental-scale study of terrestrial ecosystem responses to climate change and CO_2 doubling. *Global Biogeochemical Cycles*, **9**, 407-437.

Viscusi, W.K., 1993: The value of risks to life and death. *Journal of Economic Literature*, **31(4)**, 1912-1946.

Viscusi, W.K. and J. Aldy, 2007: Labor market estimates of the senior discount for the calue of statistical life. *Journal of Environmental Economics and Management*, **53**, 377-392.

Wall, G., 1998: Implications of global climate change for tourism and recreation in wetland areas. *Climate Change*, **40**, 371-389.

Wang, G., N.T. Hobbes, H. Galbraith, D.S. Ojima, K.M. Giesen, 2002: Signatures of large-scale and local climates on the demography of white-tailed ptarmigan in Rocky Mountain National Park, Colorado, USA. *International Journal of Biometeorology*, **46**, 197-201.

Warren, R., C. Hope, M. Mastrandrea, R. Tol, N. Adger, and I. Lorenzoni, 2006: *Spotlighting impacts functions in integrated assessment.* Research report prepared for the Stern review on the economics of climate change by the Tyndall Centre for Climate Change Research, Working paper 91, Norwich, UK.

Warren, R.S., and W.A. Niering, 1993: Vegetation change on a northeast tidal marsh: interaction of sea level rise and marsh accretion. *Ecology,* **74**, 96–103.

Weniger, B.G., M.J. Blaser, J. Gedrose, E.C. Lippy, and D.D. Juranek, 1983: An outbreak of waterborne giardiasis associated with heavy water runoff due to warm weather and volcanic ashfall. *American Journal of Public Health,* **73(8)**, 868-872.

Woodward, R.T. and Y.S. Wui, 2001: The economic value of wetland services: a meta-analysis. *Ecological Economics,* **37**, 257-270.

World Health Organization, 1997: *City Planning for Health and Sustainable Development.* European Sustainable Development and Health Series: 2.

Yohe, G., J. Neumann, and P. Marshall, 1999: The economic damage induced by sea level rise in the United States. In: *The Impact of Climate Change on the United States Economy.* [Mendelsohn, R. and J. Neumann (eds).]. Cambridge University Press, Cambridge, UK, pp. 178-208.

Zahran, S., S. Brody, A. Vedlitz, H. Grover, and C. Miller, 2008. Vulnerability and capacity: explaining local commitment to climate change policy. *Environment and Planning C: Government and Policy,* **26(3)**, 544-562.

4.7 APPENDIX
Human Welfare Economic Valuation: An Introduction to Techniques and Challenges

Assessments of the benefits and costs, whether explicit or tacit, underlie all discussion and debates over alternative actions regarding climate change. These assessments are frequently used to inform such questions as: What actions are justified to ease adaptation to changing climate? Or how much are we willing to pay to reduce emissions? (Jacoby, 2004). Ideally, such analyses would be undertaken with complete and reliable information on benefits, converted into a common unit, commensurable with costs and with each other (Jacoby, 2004). In reality, however, while many impacts can be valued, some linkages from climate change to welfare effects are difficult to quantify, much less value. This appendix describes the steps in developing a benefits estimate, and the tools that economists have available for monetizing benefits. It also briefly discusses some of the challenges in monetizing benefits, and weaknesses in the approach.

Estimating the Effects of Climate Change

The process of estimating the effects of climate change, including effects on human welfare, involves up to four steps, illustrated in Figure 4A.1. Moving down from the top of Figure 4A.1, the gray area occupies a smaller portion of each box, indicating (in rough terms) that at each stage it is more and more difficult to develop quantified, rather than qualitative, results. The first step is to estimate the change in relevant measures of climate, including temperature, precipitation, sea level rise, and the frequency and severity of extreme events. This step is usually accomplished by atmospheric scientists; some form of global circulation model (GCM) is typically deployed. Some analyses stop after this step.

The second step involves estimating the physical effects of those changes in climate in terms of qualitative changes in human and natural systems. These might include changes in ecosystem structure and function, human exposures to heat stress, changes in the

geographic range of disease vectors, melting of snow on ski slopes, or flooding of coastal areas. A wide range of disciplines might be involved in carrying out those analyses, deploying an equally wide range of tools. Many analyses are complete once this step is completed; for example, we may be unable to say anything more than that increases in precipitation will change an ecosystem's function.

The third step involves translating the physical effects of changes in climate into metrics indicating quantitative impacts. If the ultimate goal is monetization, ideally these measures should be amenable to valuation. Examples include quantifying the number and location of properties that are vulnerable to floods, estimating the number of individuals exposed to and sensitive to heat stress, or estimating the effect of diminished migratory bird populations on bird-watching participation rates. Many analyses that reach this step in the process, but not all, also proceed on to the fourth step.

The fourth step involves valuing or monetizing the changes. The simplest approach would be to apply a unit valuation approach; for example, the cost of treating a nonfatal case of heat stress or malaria attributable to climate change is a first approximation of the value of avoiding that case altogether. In many contexts, however, unit values can misrepresent the true marginal economic impact of these changes. For example, if climate change reduces the length of the ski season, individuals could engage in another recreational activity, such as golf. Whether

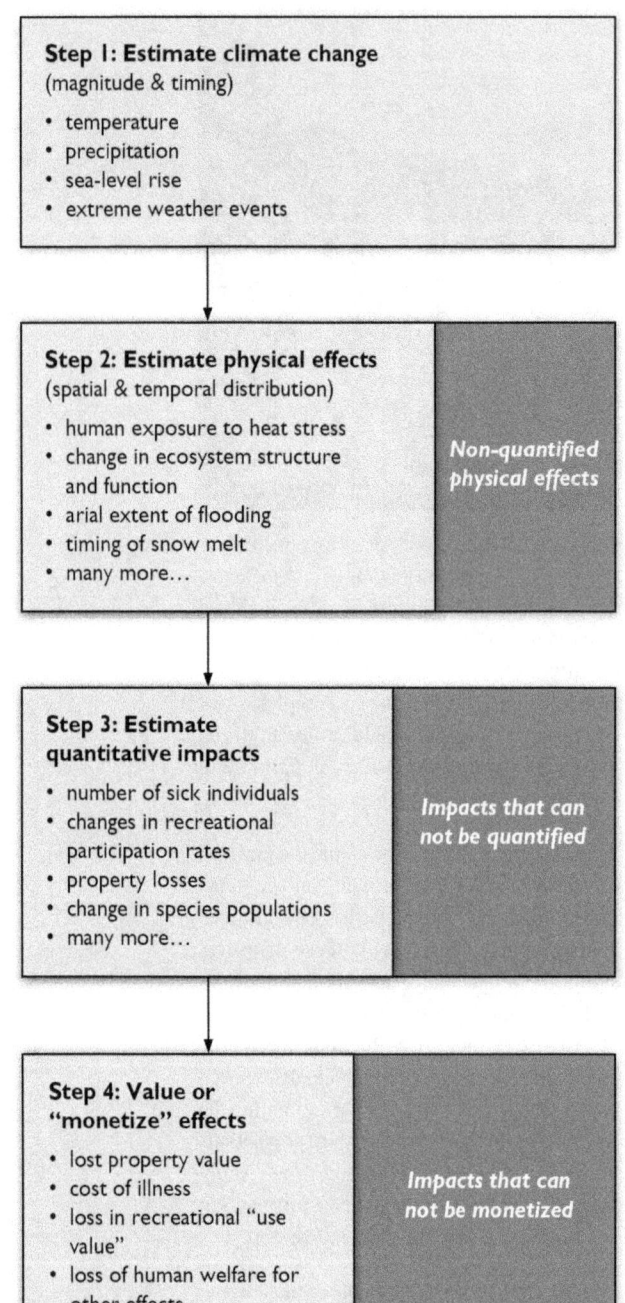

Step 1: Estimate climate change
(magnitude & timing)

- temperature
- precipitation
- sea-level rise
- extreme weather events

Step 2: Estimate physical effects
(spatial & temporal distribution)

- human exposure to heat stress
- change in ecosystem structure and function
- arial extent of flooding
- timing of snow melt
- many more...

Non-quantified physical effects

Step 3: Estimate quantitative impacts

- number of sick individuals
- changes in recreational participation rates
- property losses
- change in species populations
- many more...

Impacts that can not be quantified

Step 4: Value or "monetize" effects

- lost property value
- cost of illness
- loss in recreational "use value"
- loss of human welfare for other effects

Impacts that can not be monetized

Figure 4A.1 Estimating the Effects of Climate Change

they might prefer skiing to golf at that time and location is something economists might try to measure.

This step-by-step linear approach to effects estimation is sometimes called the "damage function" approach. One practical advantage of the damage function approach is the separation of disciplines—scientists can complete their work in steps 1 and 2, and sometimes in step 3, and then economists do their work in step 4. The linear process can work well in cases where individuals respond and change their behavior in response to changes in their environment, without any "feedback" loop.

The linear approach is not always appropriate, however. A damage function approach might imply that we look at effects of climate on human health as separate and independent from effects on ecology and recreation, but at some level they are inter-related, as health care and recreation both require resources in the form of income. In addition, responding to heat stress by installing air conditioning leads to higher energy demand, which in turn may increase greenhouse gas emissions and therefore contribute to further climate change. Recent research suggests that the damage function approach, under some conditions, may be both overly simplistic (Freeman, 2003) and subject to serious errors (Strzepek *et al.*, 1999; Strzepek and Smith, 1995).

Monetizing and Valuing Non-Market Goods

Economists have developed a suite of methods to estimate WTP for non-market goods (see text for a discussion of the market vs. non-market distinction). These methods can be grouped into two broad categories, based largely on the source of the data: revealed preference and stated preference approaches (Freeman, 2003; U.S. EPA, 2000). Revealed preference, sometimes referred to as the indirect valuation approach, involves inferring the value of a non-market good using data from market transactions. For example, a lake may be valued for its ability to provide a good fishing experience. This value can be estimated by the time and money expended by the angler to fish at that particular site, relative to all other possible fishing sites. Or, the amenity value of a coastal property that is protected from storm damage (by a dune, perhaps) can be estimated by comparing the price of that property to other properties similar in every way but the enhanced storm protection.

Stated and Revealed Preference Approaches

Accurate measurement of the non-market amenity of interest, in a manner that is not inconsistent with the way market participants perceive the amenity, is critical to a robust estimate of value.

Revealed preference approaches include recreational demand models, which estimate the value of recreational amenities through time and money expenditures to enjoy recreation; hedonic wage and hedonic property value models, which attempt to isolate the value of particular amenities of property and jobs not themselves directly traded in the marketplace based on their price or wage outcomes; and averting behavior models, which estimate the value of time or money expended to avert a particular bad outcome as a measure of its negative effect on welfare.

Stated preference approaches, sometimes referred to as *direct valuation* approaches, are survey methods that estimate the value individuals place on particular non-market goods based on choices they make in hypothetical markets.[34] The earliest stated preference studies involved simply asking individuals what they would be willing to pay for a particular non-market good. The best studies involve great care in constructing a credible, though still hypothetical, trade-off between money and the non-market good of interest to discern individual preferences for that good and hence, WTP. For example, economists might construct a hypothetical choice between multiple housing locations, each of which differs along the dimensions of price and health risk. Repeated choice experiments of this type ultimately map out the individual's tradeoff between money and the non-market good. The major challenges in

stated preference methods involve study design, particularly the construction of a reasonable and credible market for the good, and estimation of a valuation function from the response data.

In theory, if individuals understand the full implications of their market choices, in real or constructed markets, then both revealed and stated preference approaches are capable of providing robust estimates of the total value of non-market goods. When considering the complex and multidimensional implications of climate change in the application of revealed and stated preference approaches, it can be extraordinarily challenging to ensure that individuals are sufficiently informed that their observed or stated choices truly reflect their preferences for a particular outcome. As a result, these methods are most often applied to a narrowly defined non-market good, rather than to a complex bundle of non-market goods that might involve multiple tradeoffs and synergistic or antagonistic effects that would be difficult to disentangle.

In addition to market or non-market goods that reflect some use of the environment, value can arise even if a good or service is not explicitly consumed, or even experienced. For example, very few individuals would value a polar bear for its ability to provide sustenance; those who do might not express that value through a direct market for polar bear meat, but by hunting for the bear. Whether through a market or in a non-market activity, those individuals have value for a consumptive use—once enjoyed, that good is no longer available to others to enjoy. In addition to the consumptive users, a small but somewhat larger number of individuals might travel to the Arctic to see a polar bear in its natural environment. These individuals might express a value for polar bears, and their "use" of the bear is non-consumptive, but in some sense it does nonetheless affect others' ability to view the bear—if too many individuals attempt to view the bears, the congestion might cause the bears to become frightened or, worse, domesticated, diminishing the experience of viewing them.

A third, perhaps much larger group of individuals will never travel to see a polar bear in the flesh. But many individuals in this group would

---

34 The contingent valuation method (CVM), or a modern variants, a stated choice model (SCM), are forms of the stated preference methods.

experience some diminishment in their overall quality of life if they knew that polar bears had become extinct. This concept is called "non-use value." Although there are several categories of non-use value, some individuals may wish to preserve the future option to visit the Arctic and see a bear, others to bequeath a world with polar bears to future generations, and others might value the mere existence of the bears out of a sense of environmental stewardship. While not all economists agree that non-use values ought to be relevant to policy decisions (Diamond and Hausman, 1993), there is broad agreement that they are difficult to measure, because the expression of non-use values does not result in measurable economic behavior (that is, there is no "use" expressed). Those that recognize non-use values acknowledge that they are likely to be of greatest consequence where a resource has a uniqueness or "specialness" and loss or injury is irreversible, for example in the global or local extinction of a species, or the distribution of a unique ecological resource (Freeman, 2003).

Other Methods of Monetizing

Analysts can employ other non-market valuation methods: avoided cost or replacement cost, and input value estimates. These methods do not measure WTP as defined in welfare economic terms, but because the methods are relatively straightforward to apply and the results often have a known relationship to WTP, they provide insights into non-market values. This chapter focuses on WTP measures, but recognizes that

alternative methods may provide insights and sometimes be more manageable (or appropriate) to estimate a particular non-market value, given data constraints and the limitations imposed by available methods.

Cost of illness studies estimate the change in health expenditures resulting from the change in incidence of a given illness. Direct costs of illness include costs for hospitalization, doctors' fees, and medicine, among others. Indirect costs of illness include effects such as lost work and leisure time. Complete cost of illness estimates reflect both direct and indirect costs. Even the most complete cost of illness estimates, however, typically underestimate WTP to avoid incidence of illness, because they ignore the loss of welfare associated with pain and suffering and may not reflect costs of averting behaviors the individuals have taken to avoid the illness. Some studies suggest that the difference between cost of illness and WTP can be large, but the difference varies greatly across health effects and individuals (U.S. EPA, 2000).

Replacement cost studies approach non-market values by estimating the cost to replace the services provided to individuals by the non-market good. For example, healthy coastal wetlands may provide a wide range of services to individuals who live near them; they may filter pollutants present in water; absorb water in times of flood; act as a buffer to protect properties from storm surges; provide nursery habitat for recreational and commercial fish; and provide amenities in the form of opportunities to view wildlife. A replacement cost approach would estimate the value of these services by estimating market costs for treating contaminants, containing floods, providing fish from hatcheries, or perhaps restoring an impaired wetland to health.

The replacement cost approach is limited in three important ways: 1) the cost of replacing a resource does not necessarily bear any relation to the welfare enhancing effect of the resource; 2) as resources grow scarce, we would expect their value would be underestimated by an average replacement cost; 3) complete replacement of ecological systems and services may be highly problematic. Replacement cost studies are most informative in those conditions where loss of the resource would certainly and without exception trigger the incidence of replacement costs - in reality, those conditions are not as common as they might seem, because in most cases there are readily available substitutes for those services, even if accessing them involves incurring some transition costs.

Finally, value can also be calculated using the contribution of the resource as an input into a productive process. This approach can be used for both market and non-market inputs. For example, it can be used to estimate the value of fertilizer, as well as water or soil, in farm output and profits. An ecosystem's service input into a productive process could, in theory, be used in this same way.

Issues in Valuation and Aggregation

The topic of issues in valuation is far larger than can be covered here. We focus only on identifying in a superficial way a few of the most important issues, in the context of climate change.

By virtue of the simple process of aggregation, the economic approach creates some difficulties. These difficulties are not specific to the economic approach, however; any method of aggregation would face the same limitations.

- Aggregation, by balancing out effects to produce a "net" effect, masks the positive and negative effects that comprise net effects, hides inequities in the distribution of impacts, or large negative impacts that fall on particular regions or vulnerable populations.

- Any method of aggregation must make an explicit assumption about how to aggregate over time, *i.e.*, whether to weight future benefits the same as current benefits (economic analyses generally discount the future, *i.e.*, weight it less heavily in decision making than the present, for a number of reasons).

- The method of putting diverse impacts on the same yardstick ignores differences in how we may wish to treat these impacts from a policy perspective, and assumes that all impacts are equally certain or

uncertain, despite differences in estimation and valuation methods. These differences may be particularly apparent, for example, for non-market and market goods.

Several potential criticisms of the economic approach in the context of climate change relate more directly to how economists approach the task of valuation. One issue is the assumption of stability of preferences over time. Economic studies conducted today, whether revealed or stated preference, reflect the actions and preferences of individuals today, expressed in today's economic, social, and technological context. For an issue such as climate change, however, impacts may occur decades or centuries hence. The valuation of impacts that occur in the future should depend on preferences in the future. For the most part, however, while there are some rudimentary ways in which economists model changes in technology or income, there is no satisfactory means of modeling changes in preferences over time.

A second issue is the treatment of uncertainty. Economic analysis under conditions of imperfect information and uncertainty is possible, but is one of the most difficult undertakings in economics. While some climate change impacts may be relatively straight-forward, valuation of many climate change impacts requires analysis and use of welfare measures that incorporate uncertainty. When imperfect information prevails, the valuation measure must factor in errors that arise because of it, and when risk or

uncertainty prevail, the most commonly used valuation measure is the option price. Two related concepts are option value, and expected consumer's surplus. All three concepts are more complicated than the discussion here can do justice to, but briefly:

- Expected consumer's surplus, E[CS] is just consumer's surplus (CS), or value in welfare terms, weighted by the probabilities of outcomes that yield CS. For example, if a hiker gets $5 of CS per year in a "dry" forest and $10 in a wet forest (one that is greener) and the probability of the forest being dry is 0.40 and of it being wet is 0.60, then the E[CS] = 0.40 X $5 + 0.60 X $10. Expected consumer's surplus is really an ex-post concept, because we must know CS in each state after it occurs.

- Option price (OP) is the WTP that balances expected utility (utility weighted by the probabilities of outcomes) with and without some change. It is a measure of WTP the individual must express before outcomes can be known with certainty, *i.e.*, a true ex ante welfare measure. For example, the hiker might be willing to pay $8 per year to balance her expected utility with conditions being wet, versus conditions being dry. The $8 might be a payment to support a reduction in dryness otherwise due to climate change.

- Option value (OV) is the difference between OP and E[CS]. A related concept is called quasi-option value and pertains to the value of waiting to get more information.

A third issue concerns behavioral paradoxes. Most economic analyses, particularly if they involve uncertain or risky outcomes, require rationality in the expression of preferences. Such basic axioms as treating gains and losses equally, reacting to a series of small incremental gains with equal strength to a single large gain of the same aggregate magnitude, and viewing gains and losses from an absolute rather than relative or positional scale are particularly important to studies that rely on expected utility theory—that individuals gain and lose welfare in proportion to the product of the likelihood of the gain or loss and its magnitude. Several social and psychological science studies, however, suggest that under many conditions individuals do not behave in a manner consistent with this definition of rationality. For example, prospect theory, often credited as resulting from the work of Daniel Kahneman and Amos Tversky, suggests that behavior under risk or uncertainty is better explained both by reference to a status quo reference point and acknowledgement of unequal treatment of risk aversion when considering losses and gains, even when it can be shown that a different behavior would certainly make the individual better off.

Finally, the issue of perspective—"whose lens are we looking through"—is critical to welfare analysis, particularly economic welfare. In health policy, for example, thinking about whether it is worthwhile to invest in mosquito netting to control malaria depends on whether you are at CDC, are a provider of health insurance, or are an individual in a place where malaria risk is high. In general, the perspective of valuation focuses on the valuation of individuals who are directly affected, and who are living today. The perspectives of public decision makers may be somewhat different from those of individuals, since they will take into account social and community consequences, as well as individual consequences.

Common Themes and Research Recommendations

Convening Lead Author: Janet L. Gamble, U.S. Environmental Protection Agency

Lead Authors: Kristie L. Ebi, ESS, LLC; Frances G. Sussman, Environmental Economics Consulting; Thomas J. Wilbanks, Oak Ridge National Laboratory

Contributing Authors: Colleen E. Reid, ASPH Fellow; John V. Thomas, U.S. Environmental Protection Agency; Christopher P. Weaver, U.S. Environmental Protection Agency

5.1 SYNTHESIS AND ASSESSMENT PRODUCT 4.6: ADVANCES IN THE SCIENCE

The Synthesis and Assessment Product 4.6 assesses the impacts of climate variability and change on human systems in the United States. Each of the assessment chapters has drawn on different bodies of literature, with generally more available scientific knowledge on impacts and adaptation related to human health, somewhat less related to human settlements, and still less related to human welfare.

Several themes recur across these chapters and point to advances in the science of climate impacts assessment and the development and deployment of adaptation responses.

1. The connections between climate change and other environmental and social changes are complex and dynamic. In some cases, climate change compounds the effects of other global changes. Socioeconomic factors can, in some cases, determine or moderate the impacts of climate change (5.1.1).

2. Extreme weather events will play a defining role, particularly in the near-term, shaping climate-related impacts and adaptive capacity. While impacts associated with changes in climate averages may be less important now, these averages are expected to have more pronounced long-run effects on sea level rise, permafrost melt, glacial retreat, drought patterns and water supplies, etc. (5.1.2).

3. Climate change will have a disproportionate impact on disadvantaged groups in communities across the United States. Some regions and some resources are more vulnerable to climate impacts, such as coastal zones, drought-prone regions, and flood-prone river basins (5.1.3).

4. Adaptation of infrastructure and services to climate change may be costly, but many communities will have adequate resources. However, for places already struggling to provide or maintain basic public amenities and services, the additional costs of adaptation will impose a potentially insupportable burden (5.1.4).

5. With such a complex scientific and policy landscape, an integrated multi-disciplinary framework is needed for climate change impacts to be measured in meaningful ways and for optimal mitigation and adaptation strategies to be identified, developed, and deployed (5.1.5).

5.1.1 Complex Linkages and a Cascading Chain of Impacts Across Global Changes

Climate is only one of a number of global changes that impact human well-being. The major effects of climate will be shaped by interactions with non-climate stressors. As such, climate change will seldom be the sole or primary factor determining a population's or a location's well-being. Moreover, the impacts of changes in climate will be tied to the effects

of socioeconomic variables, such as population growth, and how these influence key sectors and decisions, such as infrastructure development, habitat preservation, and access to health care. Consequently, while this assessment focuses on the mechanisms by which climate change could affect future health, well-being, and settlements in the United States, the extent of any impacts will depend on an array of non-climate factors, including:

- Demographic changes related to the location, size, age, and characteristics of populations;

- Population and regional vulnerabilities;

- Future social, economic, and cultural contexts;

- Availability of natural resources;

- Human, cultural, and social capital;

- Advances in science and technology;

- Characteristics of the built environment;

- Land use change;

- Public health and public utility infrastructures; and

- The capacity and availability of health and social services.

The effects of climate change very often spread from directly affected areas and sectors to other areas and sectors through extensive and complex linkages. The importance of climate change depends on the directness of the climate impact coupled with demographic, social, economic, institutional, and political factors, including the degree of preparedness. Consider, for example, the damage caused by Hurricanes Katrina and Rita in 2005. Damage was measured not only in terms of lives and

property lost, but also in the devastating impacts on infrastructure, neighborhoods, businesses, schools, and hospitals. In addition, there have been consequences from continuing disruptions to established communities, livelihoods, psychological well-being, and the exacerbation of chronic illnesses. While the impacts of a single hurricane are not readily linked to climate change, such an event demonstrates the disruptive capacity of extreme weather events.

5.1.2 Changes in Climate Extremes and Climate Averages

Past and present climates have been, and are, variable. This variability in all likelihood will continue into the future. Changes in climate occur as changes in particular weather conditions, including extremes, in specific places (unfortunately, projections of climate changes at small geographic scales remain highly uncertain). The meteorological variables of interest from an impacts perspective include changes in both average and extreme conditions. Gradual changes in average temperature and precipitation have the potential to strongly affect, both positively and negatively, human systems. For example, changes in the average length of the growing season can affect agricultural practices, and changes in the timing and amount of spring runoff can affect water resource management. Effects such as these will not be confined to a few individual sectors, nor are the effects across all sectors independent (*e.g.,* changes in water supplies can impact agricultural practices such as irrigation).

Changes in climate extremes, both those that accompany changes in mean conditions (*e.g.,* a shift in the entire temperature distribution) as well as changes in variability are very often of more concern than changes in climate averages. Unfortunately these types of changes (which include prolonged and intense heat waves and drought or severe storms) are especially difficult to project using climate change models. Many human systems have evolved to accommodate the "average climate" and some variation around this average. This evolution takes place in the context of a variety of dynamic social, economic, technological, biophysical, and political settings, which together determine the ability of human systems to cope. Rapid onset extreme weather events in particular can do serious damage to a settlement's infrastructure, public health, and overall community reputation and quality of life, from which recovery might take years.

Finally, key vulnerabilities are often defined by certain "thresholds," below which effects are incidental but beyond which effects quickly become major. The severity of impacts is therefore not only related to the rate and magnitude of climate change, but also to the presence or absence of thresholds. In general, these climate-related thresholds for human systems in the United States are not well-understood. Focused research on thresholds would substantially improve our understanding of climate impacts and our ability to cope with extreme events.

5.1.3 Vulnerable Populations and Vulnerable Locations

Impacts of climate variability and change on human systems are location- and population-specific. For instance, along densely developed coastlines, populations are especially vulnerable to tropical storms, storm surge, and flooding. Likewise, the very old and the very young residing in urban areas are susceptible to increases in cardiovascular and pulmonary morbidity and mortality caused by extreme heat coupled with degraded air quality. Native American peoples in Alaska and elsewhere are vulnerable because of their limited capacity to prepare for and respond to the impacts of climate change. Just as there are differences across populations, there are important differences in vulnerability across geographic regions, such as the exposure to extreme events along the Gulf Coast and water supply issues in the Southeast, the Southwest and the Inter-Mountain West.

With respect to health impacts from climate variability and change, specific subpopulations may experience heightened vulnerability for climate-related health effects associated with any or all of the following:

1. *Biological sensitivity* relates to age (especially the very young and the very old), the presence of pre-existing chronic medical conditions (such as the sensitivity of people with chronic heart and pulmonary conditions

to heat-related illness), developmental characteristics, acquired factors (such as immunizations from vaccines), the use of certain medications (*e.g.*, some antihypertensive and psychotropic medications), and genetic factors (such as those that play a role in vulnerability to air pollution effects).

2. *Socioeconomic factors* also play a critical role in determining vulnerability to environmental condition factors. The distribution of climate-related effects will vary among those who live alone, among those with limited rights (for instance, some in the immigrant communities), by economic strata, by housing type, and according to other elements that either accentuate or limit vulnerability. Socioeconomic factors can increase the likelihood of exposure to harmful agents, interact with biological factors that mediate risk (such as nutritional status), and/or lead to differences in the ability to adapt or respond to exposures or to early phases of illness and injury.

3. Given their *location*, the underlying vulnerability of some communities is inherently high just as their adaptive capacity is similarly limited. Populations in gently sloping coastal areas are particularly vulnerable to sea level rise, and settlements along floodplains of large rivers are particularly vulnerable to increased variability in precipitation. The potential for increased frequencies of drought put the increasing populations of desert Southwest cities at risk.

It is essential that public health interventions and preventions recognize populations that may experience interactive or synergistic effects of multiple risk factors for health problems. Poor communities and households are already under stress from climate variability and climate-related extreme events such as heat waves, hurricanes, and tropical and riverine flooding. Since they tend to be concentrated in relatively high-risk areas and have limited access to services and other resources for coping, they can be especially vulnerable to climate change. These differential effects raise concerns about social inequity and environmental justice and increase pressure for adaptive responses from local, state, and federal governments.

5.1.4 The Cost of and Capacity for Adaptation

The United States is capable of considerable adaptation. The success of adaptation plans and/or measures will depend heavily on the competence and capacity of individuals; communities; federal, state, and local governments; and available financial and other social resources. While adaptation to climate change will come at a cost that will likely reduce resources available to cope with other societal burdens, the potential for adaptation through technological and institutional development and behavioral changes is considerable, especially where such options meet other sustainable development needs.

With scarce resources, communities should also choose adaptation options with co-benefits that help ameliorate other issues or where they can easily add climate concerns to existing response plans. The focus on all-hazards response within public health agencies can simply add climate impacts to its list of hazards for which to prepare. This will likely improve their response plans to events in the near term such as storms that happen in a variable climate, whether or not they increase in frequency or intensity with a changing climate. Planting trees and creating green roofs can help reduce the urban heat island effect. In addition to creating more aesthetically pleasing locations, green roofs can also help with energy conservation in the buildings upon which they are located. Thus, some adaptation measures can also be considered mitigation measures.

5.1.5 An Integrative Framework

Human well-being is an emerging concept, and in theory could encompass human health and settlements—the two key focuses of this Product—as well as other critical aspects of the effect of climate change on human systems and the services provided by natural systems. As an organizing principle, human well-being could provide a paradigm for identifying and categorizing climate impacts, and may ultimately provide a framework for integrating multiple impacts into an internally consistent, coherent framework for assessing costs, benefits, and tradeoffs. As an integrating concept, human well-being can develop insights into the linkages between climate change impacts and human happiness. Just as health can be considered a component of well-being (*i.e.*, physical health is closely tied to individual measures of happiness, contentment, and quality of life) aspects of human settlements also determine well-being and could be incorporated into a broader framework of well-being or welfare.

The impacts of climate variability and change on human health and human settlements are fairly well characterized in broad terms, although additional research is needed to refine impact assessments and provide better decision support (particularly with respect to deploying adaptation measures). However, the potential for utilizing concepts of human well-being to develop an integrating framework is not yet mature. Additional conceptual work and research will be needed, such as developing valuation methodologies (in the case of economic welfare), or developing metrics of well-being or quality of life (in the case of a place-based indicators, or similar, approach).

An alternative integrating framework could revolve around settlements or the more expansive concept of communities (see Section 4.2.3 for an elaborated discussion). There is a growing awareness that the built environment can have a profound impact on our health and quality of life.[1] A major goal of community design is to create more vibrant and livable communities, making sure that they address the needs of residents and improve their quality of life. More specifically, "green communities," "smart communities," "smart growth," and "sustainable development" are intended to offer alternatives to traditional settlement patterns, aiming to meet the goals of creating livable, desirable communities while minimizing the collective footprint of communities on natural resources, ecosystems, and pollution.

As an integrating framework, communities could be evaluated based on how well they protect human health and welfare. Put slightly differently, adaptation could be realized as increasing resilience within communities. Resilience is measured by a community's capacity for absorbing climate changes and the shocks of extreme events without breakdowns in its economy, natural resources, and social systems. Resiliency, as a central concept in measuring the vulnerability and adaptability of communities and individuals, depends not only on physical infrastructure, but also on social infrastructure and the natural environment. As with welfare, these concepts involving settlements or communities as an integrating framework are not yet mature.

5.2 EXPANDING THE KNOWLEDGE BASE

The present state of the science suggests that opportunities remain for addressing critical research areas. SAP 4.6 concludes that climate observations and modeling are becoming increasingly important for a wide segment of public and private sector entities, such as water resource managers, public health officials, agribusinesses, energy providers, forest managers, insurance companies, and urban and transportation planners. In order to more accurately portray the consequences of climate change and support better-informed adaptation strategies, research efforts should focus on:

- Deriving socioeconomic scenarios that describe how the world may evolve in the future, including assumptions about changes in societal characteristics, governments, and public policy, as well as economic and technological development;

- Connecting socioeconomic scenarios to downscaled climate models in order to evaluate future actions that might address changes in climate, including the intensity and severity of extreme weather events, at the regional and local scales;

- Characterizing the costs of climate change, both those that relate to impacts and those that relate to response strategies (including adaptation and mitigation);

- Estimating the damages avoided by stabilizing or reducing emissions;

---

1 See for example, the CDC website on healthy places: www.cdc.gov/healthyplaces/.

- Determining the factors that contribute to synergies between adaptive capacity and sustainable development as well as synergies between adaptation and mitigation;

- Pursuing cross-disciplinary efforts that focus on the human dimensions of climate change in an integrated fashion;

- Improving capacity to incorporate scientific knowledge about climate, including uncertainty, into existing adaptation strategies;

- Conducting research at regional and sectoral levels that promotes understanding how human and natural systems respond to multiple stressors;

- Evaluating the adaptation strategies that effectively address challenges presented by current non-climate stressors (e.g., land use and population dynamics) and develop comprehensive estimates of the co-benefits of actions to address anticipated climate change;

- Investigating adaptation measures to address the near- and long-term responses to climate change, using regional and local stakeholders as key contributors for recommending effective, responsive, and timely adaptation policies;

- Advancing the concept of human welfare as an integrating framework by developing methods to achieve comparable and comprehensive valuations across diverse impacts and sectors;

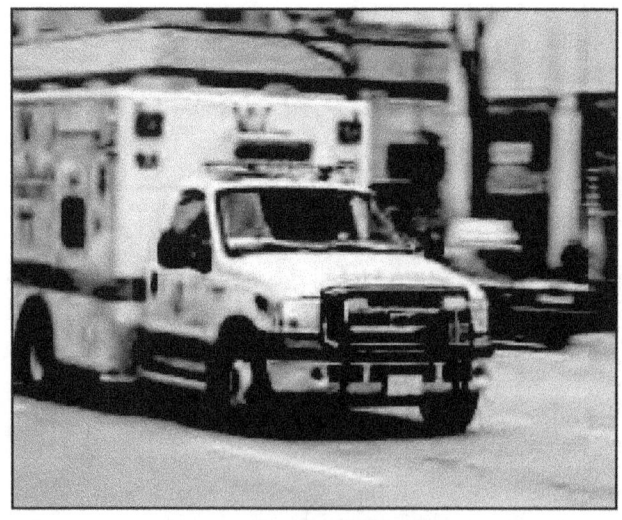

- Determining which climate impacts exhibit thresholds. Threshold-based damage functions can be fundamentally different in their nature and extent than continuous damage functions; and

- Supporting the development, implementation, and evaluation of adaptive responses, as well as expanding our understanding of impacts, by collecting high quality time-series measurements and other observations of both climate and human systems.

This report concludes that periodic assessments of the impacts of global change on human health, human settlements, and human welfare are necessary to support a rapidly developing knowledge base, especially related to impacts and adaptation. Gaps should be addressed that characterize exposure and sensitivity at the local or regional level. Research should evaluate the adaptive capacity of places and institutions to climate-induced risks. Key research and development areas should address short-term risk assessment and evaluation of the costs and effectiveness of near-term adaptive strategies as well as longer-term impacts and responses.

The following sections provide a more detailed discussion of research needs and recommendations by topic: human health, human settlements, and human welfare. There is significant overlap across topics with opportunities for investigating cross-disciplinary pursuits of research opportunities and adaptation responses.

5.2.1 Human Health Research Gaps

An important shift in perspective has occurred since the Health Sector Assessment of the First National Assessment in 2001. There is a greater appreciation of the complex pathways by which weather and climate affect individual and societal health and well-being. In the research community, there is a more finely honed understanding of the interaction of multiple non-climate, social, and behavioral factors and impacts on risks from injury and disease. While significant gaps remain, several gaps identified in the First National Assessment have been addressed, including:

- A better understanding of the differential effects of temperature extremes by community, demographic, and biological characteristics;

- Improved characterization of the exposure-response relationships for extreme heat; and

- Improved understanding of the public health burden posed by climate-related changes from heat waves and air pollution.

Despite these advances, the body of literature has only limited quantitative projections of future impacts. Research related to the human health impacts of climate change will lead to a better understanding in this area.

Specific suggestions for research on climate change and human health include the following.

- Increase the skill with which we characterize exposure-response relationships, including identifying thresholds and particularly vulnerable groups, considering relevant factors that affect the geographic range and incidence of climate-sensitive health outcomes, and including disease ecology and transmission dynamics;

- Develop quantitative models of possible health impacts of climate change that can be used to explore a range of socioeconomic and climate scenarios;

- Evaluate effectiveness of current adaptation projects, including the costs and benefits of interventions. For example, heat wave and health early warning systems have not been effective. Further research is needed to understand how public health messages can be made more helpful;

- Characterize with local stakeholders the local and regional scale vulnerability and adaptive capacity related to the potential risks and the time horizon over which climate risks might arise; and,

- Anticipate requirements for infrastructure such as may be needed to provide protection against extreme events, to alter urban design to decrease heat islands, and to maintain drinking and wastewater treatment standards and source water and watershed protection.

5.2.2 Human Settlements Research Gaps

Preceding chapters examine the vulnerabilities and impacts of climate change and variability on human settlements. The following list enumerates topics where a better understanding of the linkages between climate change and human settlements is appropriate.

- Advance the understanding of settlement vulnerabilities, impacts, and adaptive responses in a variety of different local contexts around the country.

- Develop plans for out-migration from vulnerable locations via realistic, socially acceptable strategies for shifting human populations away from vulnerable zones.

- Improve the understanding of vulnerable populations (such as the urban poor and native populations on rural, tribal lands) that have limited capacities for response to climate change in order to provide a basis for adaptation research that addresses social justice and environmental equity concerns.

- Improve the understanding of how urban decision-making is changing as populations become more heterogeneous and decisions become more decentralized, especially in so far as these changes affect adaptive responses.

- Improve researchers' abilities to associate projections of climate change in U.S. settlements with changes in other driving forces related to impacts, such as changes in metropolitan/urban patterns, changes in transportation infrastructure, and technological change. With continued growth in vulnerable regions, research is needed to consider alternative growth futures and to minimize the vulnerability of new development, to insure that communities adopt measures to manage significant changes in sea level, temperature, rainfall, and extreme weather events.

- Improve the understanding of relationships between settlement patterns (both regional and intra-urban) and resilience/adaptation.

- Improve the understanding of vulnerabilities of urban population inflows and outflows to climate change impacts.

- Improve the understanding of second- and third-order impacts of climate change in urban environments, including interactive effects among different aspects of the urban system.

- Review current policies and practices related to climate change responses to help inform community decision-makers and other stakeholders about potentials for relatively small changes that make a large difference.

Meeting these needs is likely to require well-developed partnerships across local, state, and federal governments; industry; non-governmental organizations; foundations; stakeholders; resource managers; urban planners; public utility and public health authorities; and the academic research community.

5.2.3 Human Welfare Research Gaps

Despite the potential for impacts on human well-being, little research focuses directly on understanding the relationship between well-being and climate change. Completely cataloging the effects of global change on human well-being or welfare would be an immense undertaking, and no well-accepted structure for doing so has been developed and applied. Moreover, identifying the potentially lengthy list of climate-related changes in lifestyle, as well as in other, more tangible, features of well-being (such as income), is itself a daunting task—and may include changes that are not easily captured by objective measures of well-being or quality of life.

Developing an understanding of the impacts of climate change on human welfare will require steps designed to develop a framework for addressing individual and community welfare and well-being, as well as to fill the data gaps associated with the estimation and quantification of effects.

Regarding climate change and human welfare, there is a range of topics associated with human welfare impacts and adaptations where improved understanding would be useful.

- Design an appropriate method for systematically categorizing and identifying impacts on welfare/well-being.

- Identify priority categories for data collection and research in order to establish and quantify the linkage from climate to effects on welfare/well-being.

- Decide which metrics should be used for these categories and, more generally, which components of welfare/well-being should be measured in natural or physical units, and which should be monetized.

- Investigate methods by which diverse metrics can be aggregated, or at least weighted and compared in policy decisions where aggregation is impossible.

- Develop an approach for addressing those human welfare effects that are difficult to look at in a piecemeal way, such as welfare changes on communities or ecosystem services.

- Identify appropriate top-down and bottom-up approaches for estimating impacts and value (whether economic or otherwise) of the most critical categories of welfare/well-being.

Together, these steps should enable researchers to make progress towards promoting the consistency and coordination in analyses of welfare/well-being that will facilitate developing the body of research necessary to analyze impacts on human welfare, well-being, and quality of life.

Glossary and Acronyms

Convening Lead Author: Janet L. Gamble,
U.S. Environmental Protection Agency

Lead Authors: Kristie L. Ebi, ESS, LLC; Anne Grambsch,
U.S. Environmental Protection Agency; Frances G. Sussman,
Environmental Economics Consulting; Thomas J. Wilbanks,
Oak Ridge National Laboratory

Contributing Authors: Colleen E. Reid, ASPH Fellow

6.1 GLOSSARY

Sources: Derived from the Intergovernmental Panel on Climate Change Third and Fourth Assessment Reports, Working Group II and other sources as indicated.

Words in italics indicate that the term is also contained in this glossary.

A

Acclimatization

The physiological *adaptation* to climatic variations.

Adaptability

See *adaptive capacity.*

Adaptation

Adjustment in natural or *human systems* to a new or changing environment. Adaptation to *climate change* refers to adjustment in natural or *human systems* in response to actual or expected climatic *stimuli* or their effects, which moderates harm or exploits beneficial opportunities. Various types of adaptation can be distinguished, including anticipatory and reactive adaptation, private and public adaptation, and autonomous and planned adaptation.

Adaptation assessment

The practice of identifying options to adapt to *climate change* and evaluating them in terms of criteria such as availability, benefits, costs, effectiveness, efficiency, and feasibility.

Adaptation benefits

The avoided damage costs or the accrued benefits following the adoption and implementation of *adaptation* measures.

Adaptation costs

Costs of planning, preparing for, facilitating, and implementing *adaptation* measures, including transition costs.

Adaptive capacity

The ability of a system to adjust to *climate change* (including *climate variability* and extremes) to moderate potential damages, to take advantage of opportunities, or to cope with the consequences.

Aeroallergens[1]

Any of various air-borne substances, such as pollen or spores, that can cause an allergic response.

---

1 The American Heritage® Dictionary of the English Language, Fourth Edition. Retrieved November 21, 2007 from http://dictionary.reference.com/browse/aeroallergen.

Aerosol

Particulate matter (solid or liquid) that is larger than a molecule but small enough to remain suspended in the *atmosphere*. Natural sources include dust and clay particles from weathered rocks and salt particles from sea spray, both of which are carried upward by the wind. Aerosols are often considered pollutants and can be created through human activities. They are important in the both the *atmosphere* and the Earth's *climate system* as nuclei for condensation of water droplets and ice crystals, participants in various chemical cycles, and as absorbers/scatterers of *solar radiation*.

Aggregate impacts

Total impacts summed up across sectors and/ or regions. The aggregation of *impacts* requires knowledge of (or assumptions about) the relative importance of impacts in different sectors and regions. Measures of aggregate impacts include, for example, the total number of people affected, change in net primary productivity, number of systems undergoing change, or total economic costs.

Albedo

The fraction of *solar radiation* reflected by a surface or object, often expressed as a percentage. Snow-covered surfaces have a high albedo; the albedo of soils ranges from high to low; vegetation-covered surfaces and oceans have a low albedo. The Earth's albedo varies mainly through varying cloudiness, snow, ice, leaf area, and land-cover changes.

Algal bloom

A reproductive explosion of algae in a lake, river, or ocean.

Ancillary benefits

The ancillary or side effects, of policies aimed exclusively at *climate change mitigation*. Such policies have an impact not only on *greenhouse gas emissions*, but also on resource use efficiency, like reduction in *emissions* of local and regional air pollutants

associated with *fossil fuel* use, and on issues such as transportation, agriculture, *land-use* practices, employment, and fuel security. Sometimes these benefits are referred to as "ancillary impacts" to reflect that in some cases the benefits may be negative. From the perspective of policies directed at abating local air pollution, *greenhouse gas mitigation* may also be considered an ancillary benefit, but these relationships are not considered in this assessment.

Anthropogenic

Resulting from or produced by human beings.

Anthropogenic emissions

Emissions of *greenhouse gases, greenhouse gas precursors*, and *aerosols* associated with human activities. These include burning of *fossil fuels* for energy, *deforestation*, and *land use* changes that result in net increase in *emissions*.

Aquifer

A stratum of permeable rock that bears water. An unconfined aquifer is recharged directly by local rainfall, rivers, and lakes, and the rate of recharge is influenced by the permeability of the overlying rocks and soils.

Arid regions

Ecosystems with less than 250 mm precipitation per year.

Atmosphere

The gaseous envelop surrounding the Earth. The dry atmosphere consists almost entirely of nitrogen (78.1 percent volume mixing ratio) and oxygen (20.9 percent volume mixing ratio), together with a number of trace gases, such as argon (0.93 percent volume mixing ratio), helium, and radiatively active *greenhouse gases* such as *carbon dioxide* (0.035 percent volume mixing ratio) and *ozone*. In addition, the atmosphere contains water vapor, whose amount is highly variable but typically 1 percent volume mixing ratio. The atmosphere also contains clouds and *aerosols*.

B

Baseline

The baseline (or reference) is any datum against which change is measured. It might be a "current baseline," in which case it represents observable, present-day conditions. It might also be a "future baseline," which is a projected future set of conditions excluding the driving factor of interest. Alternative interpretations of the reference conditions can give rise to multiple baselines.

Biodiversity

A term that refers to the variety and variability among living organisms and the ecological systems in which they exist. Diversity can be measured by the number of different items and their relative frequencies. The items are organized at many levels, encompassing everything from *ecosystems*, to species, to genes.

Biofuel

A fuel produced from organic matter or combustible oils produced by plants. Examples of biofuel include alcohol, black liquor from the paper-manufacturing process, wood, and soybean oil.

Biogenic[2]

Produced by living organisms or biological processes.

Biomass

The total dry weight of all living organisms that can be sustained at each tropic level in a food chain. Also, all biological materials, including organic material (dead and living) from above and below ground, such as crops, grasses, roots, animals, animal waste, etc.

Biosphere

The region of Earth and the *atmosphere* where organisms exist. Also, a part of the global *carbon cycle* that includes living organisms and *biogenic* organic matter.

Bottom-up models

An approach to modeling that includes both technological and engineering details in the analysis.

C

Carbon cycle

The term to describe the flow of carbon through a system by various chemical, physical, geological, and biological processes. This cycle is usually thought of as a series of four main *reservoirs* of carbon interconnected by pathways of exchange. The *reservoirs* include the *atmosphere*, terrestrial *biosphere* (including freshwater systems), oceans, and sediments (including *fossil fuels*).

Carbon dioxide (CO_2)

A naturally occurring gas, and also a by-product of burning *fossil fuels* and *biomass*, as well as *land-use changes* and other industrial processes. It is the principal *anthropogenic greenhouse gas* that affects the Earth's radiative balance. It is the reference gas against which other *greenhouse gases* are measured and has a *Global Warming Potential* of 1.

Cholera

An intestinal infection that results in frequent watery stools, cramping abdominal pain, and eventual collapse from dehydration.

Chronic obstructive pulmonary disease (COPD)[3]

Chronic obstructive pulmonary disease, or COPD, refers to a group of diseases that cause airflow blockage and breathing-related problems. It includes emphysema, chronic bronchitis, and in some cases asthma.

---

2 The American Heritage Dictionary of the English Language, Fourth Edition. Retrieved November 21, 2007 from http://dictionary.reference.com/browse/biogenic.

3 Centers for Disease Control and Prevention (CDC), "Chronic Obstructive Pulmonary Disease." Retrieved November 21, 2007 from http://www.cdc.gov/nceh/airpollution/copd/copdfaq.htm.

Climate

Climate in a narrow sense is usually defined as the "average weather" or, more rigorously, as the statistical description in terms of the mean and variability of relevant quantities over a period of time ranging from months to thousands or millions of years. The classical period is 30 years, as defined by the World Meteorological Organization (WMO, 2003). These relevant quantities are most often surface variables such as temperature, precipitation, and wind. Climate in a wider sense is the state, including a statistical description, of the *climate system.*

Climate change

Climate change refers to any change in *climate* over time, whether due to natural variability or as a result of human activity. This usage differs from that in the *United Nations Framework Convention on Climate Change (UNFCCC),* which defines "climate change" as: "a change of climate which is attributed directly or indirectly to human activity that alters the composition of the global *atmosphere* and which is in addition to natural *climate variability* observed over comparable time periods" (IPCC, 2007). See also *climate variability.*

Climate change commitment

Due to the thermal *inertia* of the ocean and slow processes in the *biosphere,* the cryospere and land surfaces, the *climate* would continue to change even if the atmospheric composition was held fixed at today's *values.* Past changes in atmospheric position lead to a "committed" *climatic change,* which continues for as long as a radiative imbalance persists and until all components of the *climate system* have adjusted to a new state. The further change in temperature after the composition of the *atmosphere* is held constant is referred to as the committed warming or warming commitment. Climate change commitment includes other future changes, for example in the hydrological cycle, in *extreme weather events,* and in *sea level rise.*

Climate model (hierarchy)

A numerical representation of the *climate system* based on the physical, chemical, and biological properties of its components, their interactions and *feedback* processes, and accounting for all or some of its known properties. The *climate system* can be represented by models of varying complexity—that is, for any one component or combination of components a "hierarchy" of models can be identified, differing in such aspects as the number of spatial dimensions, the extent to which physical, chemical, or biological processes are explicitly represented, or the level at which empirical *parametrizations* are involved. Coupled *atmosphere*/ocean/sea-ice *general circulation models* (AOGCMs) provide a comprehensive representation of the *climate system.* There is an evolution towards more complex models with active chemistry and biology. Climate models are applied, as a research tool, to study and simulate the *climate,* but also for operational purposes, including monthly, seasonal, and interannual *climate predictions.*

Climate prediction

A climate prediction or climate forecast is the result of an attempt to produce a most likely description or estimate of the actual evolution of the *climate* in the future (e.g., at seasonal, interannual, or long-term *time scales*). See also *climate projection* and *climate (change) scenario.*

Climate projection

A *projection* of the response of the *climate system* to *emission* or concentration *scenarios* of *greenhouse gases* and *aerosols,* or *radiative forcing scenarios,* often based upon simulations by *climate models.* Climate projections are distinguished from *climate predictions* in order to emphasize that climate projections depend upon the emission/concentration/*radiative forcing scenario* used, which are based on assumptions, concerning, for example, future socio-economic and technological developments that may or may not be realized, and are therefore subject to substantial *uncertainty.*

Climate scenario

A plausible and often simplified representation of the future *climate*, based on an internally consistent set of climatological relationships, that has been constructed for explicit use in investigating the potential consequences of *anthropogenic climate change*, often serving as input to impact models. *Climate projections* often serve as the raw material for constructing climate scenarios, but climate scenarios usually require additional information such as about the observed current climate. A "climate change scenario" is the difference between a climate scenario and the current climate.

Climate sensitivity

The equilibrium response of the *climate* to a change in *radiative forcing*, such as a doubling of *carbon dioxide* concentrations.

Climate system

The climate system is the highly complex system consisting of five major components: the *atmosphere*, the *hydrosphere*, the cryospere, the land surface and the *biosphere*, and the interactions between them. The climate system evolves in time under the influence of its own internal dynamics and because of external forcings such as volcanic eruptions, solar variations, and human-induced forcings such as the changing composition of the *atmosphere* and *land use change*.

Climate variability

Climate variability refers to variations in the mean state and other statistics (such as standard deviations, the occurrence of extremes, etc.) of the *climate* on all *spatial and temporal scales* beyond that of individual weather events. Variability may be due to natural internal processes within the *climate system* (internal variability), or to variations in natural or *anthropogenic* external forcing (external variability). See also *climate change*.

Co-benefits

The benefits of policies that are implemented for various reasons at the same time—including *climate change mitigation*— acknowledging that most policies designed to address *greenhouse gas mitigation* also have other, often at least equally important, rationales (e.g., related to objectives of development, sustainability, and equity). The term co-impact is also used in a more generic sense to cover both the positive and negative sides of the benefits. See also *ancillary benefits.*

Communicable disease

An *infectious disease* caused by transmission of an infective biological agent (virus, bacterium, protozoan, or multicellular macroparasite).

Confidence

In this Report, the level of confidence in a statement is expressed using a standard terminology defined in the Introduction. See also *uncertainty.*

Coping range

The variation in climatic *stimuli* that a system can absorb without producing significant impacts.

Cost-effective

A criterion that specifies that a *technology* or measure delivers a good or service at equal or lower cost than current practice, or the least-cost alternative for the achievement of a given target.

Cryosphere

The component of Earth's *climate system* that includes snow, ice, and *permafrost* on or beneath land and ocean surfaces.

D

DALY (Disability-adjusted life years)[4]

The sum of years of life lost due to premature death and illness, taking into account the age of death compared with natural life expectancy and the number of years of life lived with a disability. The measure of number of years lived with the disability considers the duration of the disease, weighted by a measure of the severity of the disease.

Deforestation

Those processes that result in the conversion of forested lands for non-forest *land uses*. This process is often considered to be a major cause of enhanced *greenhouse effect*, because the burning/decomposition of wood releases *carbon dioxide*, and also because trees that once removed *carbon dioxide* from the *atmosphere* through the process of *photosynthesis* are no longer present.

Dengue fever

An infectious viral disease spread by mosquitoes often called breakbone fever because it is characterized by severe pain in joints and back. Subsequent infections of the virus may lead to dengue hemorrhagic fever and dengue shock syndrome, which may be fatal.

Desert

An *ecosystem* with less than 100 mm precipitation per year.

Desertification

According to the United Nations Convention to Combat Desertification (UNCCD), desertification is "land degradation in *arid, semi-arid*, and dry sub-humid areas resulting from various factors, including climatic variations and human activities" (United Nations, 2004). Further, the UNCCD defines land degradation as a reduction or loss in *arid*, semi-arid, and dry sub-humid areas of the biological or economic productivity and complexity of rain-fed cropland, irrigated cropland, or range, pasture, forest, and woodlands resulting from

---

4 Millennium Ecosystem Assessment, 2005 glossary.

land uses or from a process or combination of processes, including processes arising from human activities and habitation patterns, such as: (i) soil *erosion* caused by wind and/or water; (ii) deterioration of the physical, chemical, and biological or economic properties of soil; and (iii) long-term loss of natural vegetation.

Detection and attribution

Climate varies continually on all *time scales*. Detection of *climate change* is the process of demonstrating that *climate* has changed in some defined statistical sense, without providing a reason for that change. Attribution of causes of *climate change* is the process of establishing the most likely causes for the detected change with some defined level of *confidence*.

Disturbance regime

Frequency, intensity, and types of disturbances, such as fires, insect or pest outbreaks, floods, and *droughts*.

Diurnal temperature range

The difference between the maximum and minimum temperature during a day.

Dose-response function[5]

A mathematical relationship is established which relates how much a certain amount of *exposure* impacts production, capital, *ecosystems*, or human health.

Downscaling

A method that derives local- to regional-scale (10 to 100 km) information from larger-scale models or data analyses.

Drought

The phenomenon that exists when precipitation has been significantly below normal recorded levels, causing serious hydrological imbalances that adversely affect land resource production systems.

---

5 Modified from OECD Glossary of Statistical Terms. "Dose-response Function." Retrieved November 21, 2007 from http://stats.oecd.org/glossary/detail. asp?ID=6404.

E

Economic potential

The portion of *technological potential* for reductions in *greenhouse gas emissions* or energy efficiency improvements that could be achieved cost-effectively through activities like creation of markets, reduction of market failures, or increases in financial and technological transfers. In order to achieve economic potential, additional policies and measures must be established to remove *market barriers*.

Ecosystem

A system of interacting living organisms together with their physical environment. The boundaries of what could be called an ecosystem are somewhat arbitrary, depending on the focus of interest or study. Thus, the extent of an ecosystem may range from very small *spatial scales* to, ultimately, the entire Earth.

Ecosystem processes

The processes that underpin the integrity and functioning of ecosystems, such as decomposition, carbon cycling, or soil renewal, etc.

Ecosystem services

Ecological processes or functions that have monetary or non-monetary *value* to individuals or society. There are (i) supporting services such as productivity or *biodiversity* maintenance, (ii) provisioning services such as food, fibre, or fish, (iii) regulating services such as climate regulation or carbon *sequestration*, and (iv) cultural services such as tourism or spiritual and aesthetic appreciation.

El Niño Southern Oscillation (ENSO)

El Niño, in its original sense, is a warm water current that periodically flows along the coast of Ecuador and Peru, disrupting the local fishery. This oceanic event is associated with a fluctuation of the intertropical surface pressure pattern and circulation in the Indian and Pacific Oceans, called the *Southern Oscillation*. This coupled *atmosphere*-ocean phenomenon is collectively known as El Niño Southern Oscillation, or ENSO. During an El Niño event, the prevailing trade winds weaken and the equatorial countercurrent strengthens, causing warm surface waters in the Indonesian area to flow eastward to overlie the cold waters of the Peru current. This event has great impact on the wind, sea surface temperature, and precipitation patterns in the tropical Pacific. It has climatic effects throughout the Pacific region and in many other parts of the world. The opposite of an El Niño event is called *La Niña*.

Emissions

In the *climate change* context, emissions refer to the release of *greenhouse gases* and/or their *precursors* and *aerosols* into the *atmosphere* over a specified area and period of time.

Endemic

Restricted or peculiar to a locality or region. With regard to human health, endemic can refer to a disease or agent present or usually prevalent in a population or geographical area at all times.

Epidemic

Occurring suddenly in numbers clearly in excess of normal expectancy, said especially of *infectious diseases* but applied also to any disease, injury, or other health-related event occurring in such outbreaks.

Erosion

The wearing away of land surfaces by wind or water, which is intensified by land-clearing activities related to farming, road building, logging, or residential/industrial development.

Eutrophication

The process by which a body of water (often shallow) becomes (either naturally or by pollution) rich in dissolved nutrients with a seasonal deficiency in dissolved oxygen.

Evaporation

The process by which a liquid becomes a gas.

Evapotranspiration

The combined process of *evaporation* from the Earth's surface and *transpiration* from vegetation.

Exotic species

See *introduced species*.

Exposure

The nature and degree to which a system is exposed to significant climatic variations.

External cost

Used to define the costs arising from any human activity, when the agent responsible for the activity does not take full account of the impacts of his or her actions on others. Equally, when the impacts are positive and not accounted for in the actions of the agent responsible they are referred to as external benefits. *Emissions* of particulate pollution from a power station affect the health of people in the vicinity, but this is not often considered, or is given inadequate weight, in private decision making and there is no market for such impacts. Such a phenomenon is referred to as an "externality," and the costs it imposes are referred to as the external costs.

Externality

See *external cost*.

Extinction

The complete disappearance of an entire species.

Extirpation

The disappearance of a species from part of its range; local *extinction*.

Extreme weather event

An extreme weather event is an event that is rare within its statistical reference distribution at a particular place. Definitions of "rare" vary, but an extreme weather event would normally be as rare as or rarer than the 10th or 90th percentile. By definition, the characteristics of what is called extreme weather may vary from place to place. An extreme *climate* event is an average of a number of weather events over a certain period of time, an average which is itself extreme (e.g., rainfall over a season).

F

Feedback

In relation to the climate, feedback is an interaction mechanism between processes in the *climate system* that occur when the result of an initial process triggers a change in a second process, which in turn influences the initial one again. Positive feedback occurs when the original process is intensified, negative feedback occurs when it is reduced.

Foodborne illness[6]

An illness caused by consuming foods or beverages contaminated with any of many different disease-causing microbes, *pathogens*, poisonous chemicals, or other harmful substances.

Food security

A situation that exists when people have secure access to sufficient amounts of safe and nutritious food for normal growth, development, and an active and healthy life. Food insecurity may be caused by the unavailability of food, insufficient purchasing power, inappropriate distribution, or inadequate use of food at the household level.

---

6 Modified from CDC. "Foodborne Illness." Retrieved November 21, 2007 from http://www.cdc.gov/ncidod/dbmd/diseaseinfo/foodborneinfections_g.htm.

Footprint (ecological)[7]

An index of the area of productive land and aquatic *ecosystems* required to produce the resources used and to assimilate the wastes produced by a defined population at a specified material standard of living, wherever on Earth that land may be located.

Forecast

See *climate prediction* and *climate projection*.

Fossil fuel

The general term for combustible geologic deposits of organic material buried underground. Fossil fuels are formed from decayed plant and animal matter that have been exposed to heat and pressure in the Earth's crust for hundreds of millions of years. Crude oil, coal, and natural gas are all fossil fuels.

G

General circulation

The large scale motions of the *atmosphere* and the ocean as a consequence of differential heating on a rotating Earth, aiming to restore the energy balance of the system through transport of heat and momentum.

General Circulation Model (GCM)

See *climate model*.

GIS (Geographic Information System)[8]

A computerized system organizing data sets through a geographical referencing of all data included in its collections.

Glacier

A mass of land ice that flows downhill and is constrained by its surrounding topography (i.e. sides of a valley or surrounding peaks). Glaciers are maintained by accumulation of snow at high altitudes and balanced by melting at low altitudes or discharge into the sea.

---

7 Millennium Ecosystem Assessment, 2005 glossary.

8 Millennium Ecosystem Assessment, 2005 glossary.

Globalization

The growing integration and interdependence of countries worldwide through the increasing volume and variety of cross border transactions in goods and services; free international capital flows; and the more rapid and widespread diffusion of *technology*, information, and culture.

Global surface temperature

The global surface temperature is the area-weighted global average of (i) the sea surface temperature over the oceans (i.e., the sub-surface bulk temperature in the first few meters of the ocean) and (ii) the surface air temperature over land at 1.5 m above the ground.

Global Warming Potential (GWP)

An index used to translate the *emission* levels of various gasses into a common measure so that their relative *radiative forcing* may be compared without directly calculated changes in atmospheric concentrations. GWPs are calculated as the ratio of *radiative forcing* that would result from the *emissions* of one kilogram of a *greenhouse gas* to that from the *emission* of one kilogram of *carbon dioxide* over a period of time (usually 100 years).

Greenhouse effect

Greenhouse gases effectively absorb *infrared radiation*, emitted by the Earth's surface, by the *atmosphere* itself due to the same gases, and by clouds. Atmospheric radiation is emitted to all sides, including downward to the Earth's surface. Thus *greenhouse gases* trap heat within the surface-*troposphere* system. This is called the "natural greenhouse effect." Atmospheric radiation is strongly coupled to the temperature of the level at which it is emitted. In the *troposphere*, the temperature generally decreases with height. Effectively, *infrared radiation* emitted to space originates from an altitude with a temperature of, on average, -19°C, in balance with the net incoming *solar radiation*, whereas the Earth's surface is kept at a much higher temperature of, on average, +14°C. An increase in the concentration of *greenhouse gases* leads to an increased infrared opacity of the *atmosphere*, and therefore to an

effective radiation into space from a higher altitude at a lower temperature. This causes a *radiative forcing*, an imbalance that can only be compensated for by an increase of the temperature of the surface-*troposphere* system. This is the "enhanced greenhouse effect."

Greenhouse gas

Greenhouse gases are those gaseous constituents of the *atmosphere*, both natural and *anthropogenic*, that absorb and emit radiation at specific wavelengths within the spectrum of *infrared radiation* emitted by the Earth's surface, the *atmosphere*, and clouds. This property causes the *greenhouse effect*. Water vapor (H_2O), *carbon dioxide* (CO_2), nitrous oxide (N_2O), *methane* (CH_4), and *ozone* (O_3) are the primary greenhouse gases in the Earth's *atmosphere*. Moreover there are a number of entirely human-made greenhouse gases in the *atmosphere*, such as the *halocarbons* and other chlorine- and bromine-containing substances, dealt with under the *Montreal Protocol*. Besides CO_2, N_2O, and CH_4, the *Kyoto Protocol* deals with the greenhouse gases *sulfur hexafluoride* (SF_6), *hydrofluorocarbons* (HFCs), and *perfluorocarbons* (PFCs).

Gross Domestic Product

Gross Domestic Product (GDP) is the monetary value of all goods and services produced within a nation.

Gross National Product

Gross National Product (GNP) is the monetary value of all goods and services produced in a nation's economy, including income generated abroad by domestic residents, but without income generated by foreigners.

Groundwater recharge

The process by which external water is added to the zone of saturation of an *aquifer*, either directly into a formation or indirectly by way of another formation.

H

Habitat

The particular environment or place where an organism or species tend to live; a more locally circumscribed portion of the total environment.

Halocarbon

A chemical that consists of carbon, sometimes hydrogen, and either chlorine, fluorine, bromine, or iodine.

Hantavirus

A virus in the family Bunyaviridae that causes a type of hemorrhagic fever. It is thought that humans catch the disease mainly from infected rodents, either through direct contact with the animals or by inhaling or ingesting dust that contains aerosolized viral particles from their dried urine and other secretions.

Healthy Cities program[9]

The WHO Healthy Cities program engages local governments in health development through a process of political commitment, institutional change, capacity building, partnership-based planning, and innovative projects. It promotes comprehensive and systematic policy and planning with a special emphasis on health inequalities and urban poverty; the needs of vulnerable groups; participatory governance; and the social, economic, and environmental determinants of health. It also strives to include health considerations in economic regeneration and urban development efforts.

Heat exhaustion[10]

Heat exhaustion is a phenomenon caused by fluid loss, which in turn causes decreased blood flow to vital organs. Reduced blood flow from heat exhaustion can result in a form of shock.

9 World Health Organization. "Healthy Cities and Urban Governance." Retrieved November 21, 2007 from http://www.euro.who.int/healthy-cities.

10 U.S. Environmental Protection Agency. "Heat Island Effect Glossary." Retrieved November 21, 2007 from http://www.epa.gov/hiri/resources/glossary.html#h.

Victims of heat exhaustion often complain of flu-like symptoms hours after *exposure*.

Heat index[11]

The heat index (HI), given in degrees F, is a measure of how hot it feels when relative humidity (RH) is combined with the actual air temperature.

Heat island

An area within an urban area characterized by ambient temperatures higher than those of the surrounding area because of the absorption of solar energy by materials like asphalt.

Heat stroke[12]

Heat stroke occurs when the body's heat regulating mechanisms–including convection, sweating, and respiration–fail. The *likelihood* of heat stroke increases when air temperatures are higher than skin temperature, and when individuals are low on fluids. Body temperatures can be raised to the point at which brain damage and death can result unless cooling measures are quickly taken.

Human settlement

A place or area occupied by settlers.

Human system

Any system in which human organizations play a major role. Often, but not always, the term is synonymous with "society" or "social system" (e.g., agricultural system, political system, technological system, economic system).

Hydrofluorocarbons (HFCs)

Compounds that contain hydrogen, fluorine, chlorine, and carbon atoms. They have been introduced as temporary replacements for chloroflourocarbons (CFCs) as they are less potent at destroying stratospheric *ozone*. However, they are considered both *ozone* depleting substances and *greenhouse gases*.

Hydrological systems

The systems involved in movement, distribution, and quality of water throughout the Earth, including both the hydrologic cycle and water resources.

Hydrosphere

All the water present on Earth, including liquid water (oceans, fresh water, underground *aquifers*), frozen water (polar ice caps, floating ice, frozen upper layer of soil known as *permafrost*), and water vapor in the *atmosphere*.

Hyperthermia[13]

Unusually high body temperature.

I

Ice sheet

A mass of land ice that is sufficiently deep to cover most of the underlying bedrock topography, so that its shape is mainly determined by its internal dynamics (the flow of the ice as it deforms internally and slides at its base). An ice sheet flows outward from a high central plateau with a small average surface slope. The margins slope steeply, and the ice is discharged through fast-flowing ice streams or outlet *glaciers*, in some cases into the sea or into *ice shelves* floating on the sea. There are only two large ice sheets in the modern world, on Greenland and Antarctica, the Antarctic ice sheet being divided into East and West by the Transantarctic Mountains; during glacial periods there were others.

Ice shelf

A floating *ice sheet* of considerable thickness attached to a coast (usually of great horizontal extent with a level or gently undulating surface); often a seaward extension of *ice sheets*.

11 Modified from NOAA. "Heat Index." Retrieved November 21, 2007 from http://www.crh.noaa.gov/jkl/?n=heat_index_calculator.

12 U.S. EPA. "Heat Island Effect Glossary."

13 *The American Heritage Dictionary of the English Language, Fourth Edition.* Retrieved November 21, 2007 from http://dictionary.reference.com/browse/hyperthermia.

(Climate) Impact assessment

The practice of identifying and evaluating the detrimental and beneficial consequences of *climate change* on natural and *human systems*.

(Climate) Impacts

Consequences of *climate change* on natural and *human systems*. Depending on the consideration of *adaptation*, one can distinguish between potential impacts and residual impacts. Potential impacts: All impacts that may occur given a projected change in *climate*, without considering *adaptation*. Residual impacts: The impacts of *climate change* that would occur after *adaptation*. See also *aggregate impacts*, *market impacts*, and *non-market impacts*.

Indicator[14]

Information based on measured data used to represent a particular attribute, characteristic, or property of a system.

Indigenous peoples

People whose ancestors inhabited a place or a country when persons from another culture or ethnic background arrived on the scene and dominated them through conquest, settlement, or other means and who today live more in conformity with their own social, economic, and cultural customs and traditions than those of the country of which they now form a part (also referred to as "native," "aboriginal," or "tribal" peoples).

Industrial revolution

A period of rapid industrial growth with far-reaching social and economic consequences, beginning in England during the second half of the 18th century and spreading to Europe and later to other countries including the United States. The invention of the steam engine was an important trigger of this development. The industrial revolution marks the beginning of a strong increase in the use of *fossil fuels* and emission of, in particular, fossil *carbon dioxide*. In this report, the terms "pre-industrial" and "industrial" refer, somewhat arbitrarily, to

the periods before and after the year 1750, respectively.

Inertia

Delay, slowness, or resistance in the response of the *climate*, biological, or *human systems* to factors that alter their rate of change, including continuation of change in the system after the cause of that change has been removed.

Infectious disease

Any disease that can be transmitted from one person to another. This may occur by direct physical contact, by common handling of an object that has picked up infective organisms, through a disease carrier, or by spread of infected droplets coughed or exhaled into the air.

Infrared radiation

Heat energy emitted by the Earth's surface and its *atmosphere*, some of which is strongly absorbed by *greenhouse gases* and re-radiated back towards the Earth's surface, creating the *greenhouse effect*. Also describes the heat energy emitted from all solids, liquids and gases.

Infrastructure

The basic equipment, utilities, productive enterprises, installations, institutions, and services essential for the development, operation, and growth of an organization, city, or nation. For example, roads; schools; electric, gas, and water utilities; transportation; communication; and legal systems would be all considered as infrastructure.

Integrated assessment

A method of analysis that combines results and models from the physical, biological, economic, and social sciences, and the interactions between these components, in a consistent framework, to evaluate the status and the consequences of environmental change and the policy responses to it.

14 Millennium Ecosystem Assessment, 2005, Current State and Trends Assessment glossary.

Introduced species

A species occurring in an area outside its historically known natural range as a result of accidental dispersal by humans (also referred to as *"exotic species"* or "alien species").

Invasive species

An *introduced species* that invades natural *habitats*.

IPCC[15]

A panel set up by the United Nations in 1988 to review scientific information on *climate change*. This panel involves over 2,000 of the world's climate experts. Many of the *climate change* facts and future predictions we read about come from information reviewed by the IPCC.

K

Kyoto Protocol

The Kyoto Protocol was adopted at the Third Session of the Conference of the Parties to the *UN Framework Convention on Climate Change (UNFCCC)* in 1997 in Kyoto, Japan. It contains legally binding commitments, in addition to those included in the *UNFCCC*. Countries included in Annex B of the Protocol (most member countries of the Organization for Economic Cooperation and Development (OECD) and those with economies in transition) agreed to reduce their *anthropogenic greenhouse gas emissions* (CO_2, CH_4, N_2O, HFCs, PFCs, and SF_6) by at least 5 percent below 1990 levels in the commitment period 2008 to 2012. The Kyoto Protocol entered into force on 16 February 2005.

L

Landslide

A mass of material that has slipped downhill by gravity, often assisted by water when the material is saturated; rapid movement of a mass of soil, rock, or debris down a slope.

Land use

The total of arrangements, activities, and inputs undertaken in a certain land cover type (a set of human actions). The social and economic purposes for which land is managed (e.g., grazing, timber extraction, and conservation).

Land-use change

A change in the use or management of land by humans, which may lead to a change in land cover. Land cover and land-use change may have an impact on the *albedo, evapotranspiration, sources*, and *sinks* of *greenhouse gases*, or other properties of the *climate system*, and may thus have an impact on *climate*, locally or globally.

La Niña

See *El Niño Southern Oscillation*.

Lifetime (atmospheric)

The lifetime of a *greenhouse gas* refers to the approximate amount of time it would take for the atmospheric pollutant concentration to return to its natural level (assuming *emissions* cease), as a result of either being converted to another chemical compound or being taken out of the *atmosphere* via a *sink*. This length of time depends on both the pollutant's *sources* and *sinks*, and its level of reactivity. Average lifetimes can vary from a week (e.g., sulfate *aerosols*) to more than a century (e.g., CFCs, *carbon dioxide*). Long lifetimes allow the pollutant to mix throughout the *atmosphere*.

Likelihood

The likelihood of an occurrence, an outcome or a result, where this can be estimated probabilistically, is expressed in this Report using a standard terminology, defined in the Introduction. See also *uncertainty* and *confidence*.

15 Climate Change North. "Glossary." Retrieved November 21, 2007 from http://www.climat-echangenorth.ca/H1_Glossary.html.

Lyme disease

A *vector-borne disease* caused by the spirochete Borrelia burgdorferi and transmitted by Ixodes ticks, commonly known as deer ticks. Symptoms include skin lesions, fatigue, fever, and chills, and if left untreated may later manifest itself in cardiac and neurological disorders, joint pain, and arthritis.

M

Maladaptation

Any changes in natural or *human systems* that inadvertently increase *vulnerability* to climatic *stimuli*; an *adaptation* that does not succeed in reducing *vulnerability* but increases it instead.

Malaria

Endemic or *epidemic* parasitic disease caused by species of the genus Plasmodium (protozoa) and transmitted by mosquitoes of the genus Anopheles; produces high fever attacks and systemic disorders, and kills approximately 2 million people every year.

Market barriers

In the context of *mitigation* of *climate change*, conditions that prevent or impede the diffusion of *cost-effective* technologies or practices that would mitigate *greenhouse gas emissions*.

Market-based incentives

Measures intended to use price mechanisms (e.g., taxes and tradable permits) to reduce *greenhouse gas emissions*.

Market impacts

Impacts that are linked to market transactions and directly affect *Gross Domestic Product* (a country's national accounts)—for example, changes in the supply and price of agricultural goods. See also *non-market impacts*.

Market potential

The portion of *economic potential* for reductions in *greenhouse gas emissions* or improvements in energy-efficiency that could be achieved under forecast market conditions; assuming there are no new policies and measures

Methane (CH₄)

A hydrocarbon *greenhouse gas* produced through anaerobic (without oxygen) decomposition of waste in landfills, animal digestion, decomposition of animal wastes, production and distribution of natural gas and petroleum, coal production, and incomplete *fossil fuel* combustion.

Mitigation

An *anthropogenic* intervention to reduce the *sources* or enhance the *sinks* of *greenhouse gases*.

Mitigative capacity

The social, political, and economic structures and conditions that are required for effective *mitigation*.

Montreal Protocol

Titled Montreal Protocol on Substances that Deplete the *ozone layer*, this international agreement addresses the phase-out of *ozone* depleting substances production and use. Under the Protocol, several international organizations report on the science of *ozone* depletion, implement projects to help move away from *ozone* depleting substances, and provide a forum for policy discussions. In the United States, the Protocol is implemented under the Clean Air Act Amendments of 1990.

Morbidity

Rate of occurrence of disease or other health disorder within a population, taking account of the age-specific morbidity rates. Health outcomes include chronic disease incidence/prevalence, rates of hospitalization, primary care consultations, disability-days (i.e., days when absent from work), and prevalence of symptoms.

Mortality

Rate of occurrence of death within a population within a specified time period; calculation of mortality takes account of age-specific death rates, and can thus yield measures of life expectancy and the extent of premature death.

N

Net primary production (NPP)

Refers to the increase in plant *biomass* or carbon of a unit of landscape. NPP is equal to the Gross Primary Production minus carbon lost through autotrophic respiration.

Nitrogen oxides[16]

Compounds of nitrogen and oxygen produced by the burning of *fossil fuels*.

Non-linearity

A process is called "non-linear" when there is no simple proportional relation between cause and effect. The *climate system* contains many such non-linear processes, resulting in a system with a potentially very complex behavior. Such complexity may lead to *rapid climate change*.

Non-market impacts

Impacts that affect *ecosystems* or human *welfare*, but that are not directly linked to market transactions—for example, an increased risk of premature death. See also *market impacts*.

Non-point source pollution

A large, non-specific area that discharges pollutants into surface and sub-surface water flows, for example crop fields and urban areas.

No-regrets opportunities

See *no-regrets policy*.

No-regrets options

See *no-regrets policy*.

No-regrets policy

One that would generate net social benefits whether or not there is *climate change*. No-regrets opportunities for *greenhouse gas emissions* reduction are defined as those options whose benefits such as reduced energy costs and reduced *emissions* of local/

---

16 Energy Administration Information, "Glossary." Retrieved November 21, 2007 from http://www.eia. doe.gov/glossary/glossary_n.htm.

regional pollutants equal or exceed their costs to society, excluding the benefits of avoided *climate change*. No-regrets potential is defined as the gap between the *market potential* and the *socio-economic potential*.

North Atlantic Oscillation (NAO)

The North Atlantic Oscillation consists of opposing variations of barometric pressure near Iceland and near the Azores. On average, a westerly current, between the Icelandic low pressure area and the Azores high pressure area, carries cyclones with their associated frontal systems towards Europe. However, the pressure difference between Iceland and the Azores fluctuates on *time scales* of days to decades, and can be reversed at times. It is the dominant mode of winter *climate variability* in the North Atlantic region, ranging from central North America to Europe.

O

Ocean conveyor belt

The theoretical route by which water circulates around the entire global ocean, driven by wind and the *thermohaline circulation*.

Opportunity

An opportunity is a situation or circumstance to decrease the gap between the *market potential* of any *technology* or practice and the *economic potential, socio-economic potential,* or *technological potential*.

Opportunity cost

The cost of an economic activity forgone by the choice of another activity.

Ozone (O_3)

Ozone, the triatomic form of oxygen (O_3), is a gaseous atmospheric constituent. In the *troposphere*, it is created both naturally and by photochemical reactions involving gases resulting from human activities (*photochemical "smog"*). In high concentrations, tropospheric ozone can be harmful to a wide-range of living organisms. Tropospheric ozone acts as a *greenhouse gas*. In the *stratosphere*, ozone is created by the interaction between solar

ultraviolet radiation and molecular oxygen (O_2). Stratospheric ozone plays a decisive role in the stratospheric radiative balance. Its concentration is highest in the *ozone layer*. Depletion of stratospheric ozone, due to chemical reactions that may be enhanced by *climate change*, results in an increased ground-level flux of ultraviolet-B radiation. See also *Montreal Protocol* and *ozone layer*.

Ozone layer

A group of human-made chemicals composed only of carbon and fluorine, introduced as alternatives (like hydroflourocarbons) to *ozone* depleting substances. PFCs are emitted as by-products of industrial processes, and they are used in manufacturing. While PFCs do not harm the stratospheric *ozone layer*, they are powerful *greenhouse gases* which high *global warming potential*. Examples of PFCs are CF4 and C2F6. See *Montreal Protocol*.

P

Parameterization

In *climate models*, this term refers to the technique of representing processes-that cannot be explicitly resolved at the *spatial or temporal resolution* of the model (sub-grid scale processes)-by relationships between the area- or time-averaged effect of such sub-grid-scale processes and the larger scale flow.

Pareto criterion/Pareto optimum

A requirement or status that an individual's *welfare* could not be further improved without making others in the society worse off.

Particulates

Very small solid exhaust particles emitted during the combustion of fossil and *biomass* fuels. Particulates may consist of a wide variety of substances. Of greatest concern for health are particulates of less than or equal to 10nm and 2.5 nm in diameter, usually designated as PM10 and PM2.5, respectively.

Pathogen[17]

An agent that causes disease, especially a living microorganism such as a bacterium or fungus.

Perfluorocarbons (PFCs)

A group of human-made chemicals composed only of carbon and fluorine, introduced as alternatives (like hydroflourocarbons) to *ozone* depleting substances. PFCs are emitted as by-products of industrial processes, and they are used in manufacturing. While PFCs do not harm the stratosphereic *ozone layer*, they are powerful *greenhouse gases* which high *global warming potential*. Examples of PFCs are CF_4 and C_2F_6.

Permafrost

Perennially frozen ground that occurs wherever the temperature remains below 0°C for several years.

Photochemical smog

A mix of photochemical oxidant air pollutants produced by the reaction of sunlight with primary air pollutants, especially hydrocarbons.

Photosynthesis

Complex process that takes place in living green plant cells that combines radiant energy from the sun with water (H_2O) and *carbon dioxide* (CO_2) to produce oxygen (O_2) and sugar, such as glucose ($C_6H_{12}O_6$).

Point-source pollution

Pollution resulting from any confined, discrete *source*, such as a pipe, ditch, tunnel, well, container, concentrated animal feeding operation, or floating craft. See also *non-point source pollution*.

Precursors

A term in photochemistry meaning a compound antecedent to a pollutant. For example, *volatile organic compounds* (VOCs) and nitric oxides of nitrogen react in sunlight to form *ozone* or

---

17 *The American Heritage Dictionary, Fourth Edition.* Retrieved November 21, 2007 from http://dictionary. reference.com/browse/pathogen.

other photochemical oxidants. As such, VOCs and oxides of nitrogen are precursors.

Present value cost

The sum of all costs over all time periods, with future costs discounted.

Projection (generic)

A projection is a potential future evolution of a quantity or set of quantities, often computed with the aid of a model. Projections are distinguished from "predictions" in order to emphasize that projections involve assumptions concerning, for example, future socio-economic and technological developments that may or may not be realized, and are therefore subject to substantial *uncertainty*. See also *climate projection* and *climate prediction*.

Proxy

A proxy *climate indicator* is a local record that is interpreted, using physical and biophysical principles, to represent some combination of climate-related variations back in time. Climate-related data derived in this way are referred to as proxy data. Examples of proxies are tree ring records, characteristics of corals, and various data derived from ice cores.

Q

QALY (Quality Adjusted Life Year)[18]

A measure of the outcome of actions (either individual or treatment interventions) in terms of their health impact. If an action gives a person an extra year of healthy life expectancy, that counts as one QALY. If an action gives a person an extra year of unhealthy life expectancy (partly disabled or in some distress), it has a value of less than one. Death is rated at zero.

Quality of life[19]

A scientific measure of personal *well-being*. Categories used to define place-specific quality of life include the inter-related categories of economic conditions; natural resources, environment, and amenities; human health; public and private *infrastructure*; government and public safety; and social and cultural resources.

R

Radiative forcing

Radiative forcing is the change in the net vertical irradiance (expressed in Wm^{-2}) at the *tropopause* due to an internal change or a change in the external forcing of the *climate system*, for example, a change in the concentration of *carbon dioxide* or the output of the Sun. Usually, radiative forcing is computed after allowing for stratospheric temperatures to readjust to radiative equilibrium, but with all tropospheric properties held fixed at their unperturbed *values*.

Rangelands

Lands (mostly grasslands) that support the growth of plants that provide food (i.e., forage) for grazing or browsing animals.

Range shifts

Climate change-induced changes in the geographical distributions of plants, animals and *ecosystems*

Rapid climate change

The *non-linearity* of the *climate system* may lead to rapid *climate change*, sometimes called abrupt events or even surprises. Some such abrupt events may be imaginable, such as a dramatic reorganization of the *thermohaline circulation*, rapid deglaciation, or massive melting of *permafrost* leading to fast changes in the *carbon cycle*. Others may be truly unexpected, as a consequence of a strong, rapidly changing forcing of a non-linear system.

---

18 Australian Institute of Health and Welfare. "Australia's Health 1996" glossary. Retrieved November 21, 2007 from http://www.aihw.gov.au/publications/health/ah96/ah96-x04.html.

---

19 Modified from text within Chapter 5 of this document.

Reference scenario

See *baseline/reference*.

Reinsurance

The transfer of a portion of primary insurance risks to a secondary tier of insurers (reinsurers); essentially "insurance for insurers."

Relative sea level

Sea level measured by a tide gauge with respect to the land upon which it is situated. See also mean sea level.

Reservoir

A component of the *climate system*, other than the *atmosphere*, that has the capacity to store, accumulate, or release a substance of concern (e.g., carbon or a *greenhouse gas*). Oceans, soils, and forests are examples of carbon reservoirs. The term also means an artificial or natural storage place for water, such as a lake, pond, or *aquifer*, from which the water may be withdrawn for such purposes as irrigation or water supply.

Resilience

Amount of change a system can undergo without changing state.

Response time

The response time or adjustment time is the time needed for the *climate system* or its components to re-equilibrate to a new state, following a forcing resulting from external and internal processes or *feedbacks*. It is very different for various components of the *climate system*. The response time of the *troposphere* is relatively short, from days to weeks, whereas the *stratosphere* comes into equilibrium on a *time scale* of typically a few months. Due to their large heat capacity, the oceans have a much longer response time, typically decades, but up to centuries or millennia. The response time of the strongly coupled surface-*troposphere* system is, therefore, slow compared to that of the *stratosphere*, and mainly determined by the oceans. The *biosphere* may respond fast (e.g., to *droughts*), but also very slowly to imposed changes.

Revealed preference[20]

The use of the value of expenditure to "reveal" the preference of a consumer or group of consumers for the bundle of goods they purchase compared to other bundles of equal or smaller value.

Rodent-borne disease[21]

Disease that is transmitted between hosts by a rodent (e.g., bubonic plague, *hantavirus*).

Runoff

That part of precipitation that does not *evaporate* and is not *transpired*.

S

Salinization

The accumulation of salts in soils.

Salmonella[22]

There are many different kinds of Salmonella bacteria. They pass from the feces of people or animals to other people or other animals and can cause diarrheal illness in humans. For over 100 years, Salmonella have been known to cause illness. They were discovered by an American scientist named Salmon, for whom they are named.

Saltwater intrusion/encroachment

Displacement of fresh surface water or ground water by the advance of saltwater due to its greater density, usually in coastal and estuarine areas.

Scenario (generic)

A plausible and often simplified description of how the future may develop, based on a coherent and internally consistent set of assumptions about key driving forces (e.g., rate of *technology*

---

20 Deardorff's Glossary of International Economics. Retrieved November 21, 2007 from http://www-personal.umich.edu/~alandear/glossary/r.html.

21 Modified from definition of vector-borne disease.

22 Modified from information on the CDC's website retrieved November 21, 2007 from http://www.cdc.gov/nczved/dfbmd/disease_listing/salmonel-losis_gi.html.

change, prices) and relationships. Scenarios are neither predictions nor *forecasts* and sometimes may be based on a "narrative storyline." Scenarios may be derived from *projections*, but are often based on additional information from other sources. See also *SRES scenarios*, *climate scenario*, and *emission scenarios*.

Sea level rise

An increase in the mean level of the ocean. Eustatic sea level rise is a change in global average sea level brought about by an alteration to the volume of the world ocean. *Relative sea level* rise occurs where there is a net increase in the level of the ocean relative to local land movements. Climate modelers largely concentrate on estimating eustatic sea level change. *Impact* researchers focus on *relative sea level* change.

Seawall

A human-made wall or embankment along a shore to prevent wave *erosion*.

Semi-arid regions

Ecosystems that have more than 250 mm precipitation per year but are not highly productive; usually classified as *rangelands*.

Sensitivity

Sensitivity is the degree to which a system is affected, either adversely or beneficially, by climate-related *stimuli*. The effect may be direct (e.g., a change in crop yield in response to a change in the mean, range, or variability of temperature) or indirect (e.g., damages caused by an increase in the frequency of coastal flooding due to *sea level rise*).

Sequential decision making

Stepwise decision making aiming to identify short-term strategies in the face of long-term uncertainties, by incorporating additional information over time and making mid-course corrections.

Sequestration

The process of increasing the carbon content of a carbon *reservoir* other than the *atmosphere*. Biological approaches to sequestration include direct removal of *carbon dioxide* from the *atmosphere* through *land-use change*, afforestation, reforestation, and practices that enhance soil carbon in agriculture. Physical approaches include separation and disposal of *carbon dioxide* from flue gases or from processing *fossil fuels* to produce hydrogen and *carbon dioxide*-rich fractions, as well as long-term storage in depleted oil and gas reservoirs, coal seams, and saline *aquifers*.

Sink

Any process, activity or mechanism that removes a *greenhouse gas*, an *aerosol*, or a *precursor* of a *greenhouse gas* or *aerosol* from the *atmosphere*.

Smog[23]

Air pollution typically associated with oxidants.

Snowpacks

A seasonal accumulation of slow-melting snow.

Social cost

The social cost of an activity includes the *value* of all the resources used in its provision. Some of these are priced and others are not. Non-priced resources are referred to as externalities. It is the sum of the costs of these externalities and the priced resources that makes up the social cost. See *total cost*.

23 U.S. EPA. "Terms of Environment Glossary." Retrieved November 21, 2007 from http://www.epa.gov/OCEPAterms/sterms.html.

Social indicators[24]

Broad, standardized measures of the *quality of life* or other socio-economic conditions of geographic areas such as nations, metropolitan areas, or other areas; used to assess health conditions, educational levels, food availability, violence, and other conditions.

Socio-economic potential

Represents the level of *greenhouse gas mitigation* that would be reached by overcoming social and cultural obstacles to using technologies that are cost effective.

Socio-economic scenarios

Scenarios concerning future conditions in terms of population, *Gross Domestic Product* and other socio-economic factors relevant to understanding the implications of *climate change*. See *SRES scenarios*.

Solar radiation

Energy from the sun, including ultra-violet radiation, visible radiation, and *infrared radiation*; also referred to as short-wave radiation.

Source

Any process, activity, or mechanism that releases a *greenhouse gas*, an *aerosol*, or a *precursor* of a *greenhouse ga*s or *aerosol* into the *atmosphere*.

Southern Oscillation

See *El Niño Southern Oscillation*.

Spatial and temporal scales

Climate may vary on a large range of spatial and temporal scales. Spatial scales may range from local (less than 100,000 km²), through regional (100,000 to 10 million km²), to continental (10 to 100 million km²). Temporal scales may range from seasonal to geological (up to hundreds of millions of years).

---

24 Methods for Social Researchers in Developing Countries. Glossary. Retrieved November 21, 2007 from http://srmdc.net/glossary.htm#s.

SRES scenarios

SRES scenarios are *emissions scenarios* developed by Nakicenovic et al. (2000) and used, among others, as a basis for the *climate projections* in the IPCC WGI contribution to the Third Assessment Report (IPCC, 2001). The following terms are relevant for a better understanding of the structure and use of the set of SRES scenarios:

(Scenario) Family: *Scenarios* that have a similar demographic, societal, economic, and technical-change *storyline*. Four *scenario* families comprise the *SRES scenario* set: A1, A2, B1, and B2.

(Scenario) Group: *Scenarios* within a family that reflect a consistent variation of the *storyline*. The A1 *scenario family* includes four groups designated as A1T, A1C, A1G, and A1B that explore alternative structures of future energy systems.

In the Summary for Policymakers of Nakicenovic et al. (2000), the A1C and A1G groups have been combined into one "Fossil-Intensive" A1FI *scenario group*. The other three *scenario families* consist of one group each. The *SRES scenario* set reflected in the Summary for Policymakers of Nakicenovic et al. (2000) thus consist of six distinct *scenario groups*, all of which are equally sound and together capture the range of uncertainties associated with driving forces and emissions.

(Scenario) Illustrative: A *scenario* that is illustrative for each of the six *scenario groups* reflected in the Summary for Policymakers of Nakicenovic et al. (2000). They include four revised *scenario markers* for the *scenario groups* A1B, A2, B1, B2, and two additional *scenarios* for the A1FI and A1T groups. All *scenario groups* are equally sound.

(Scenario) Marker: A *scenario* that was originally posted in draft form on the SRES website to represent a given *scenario family*. The choice of markers was based on which of the initial quantifications best reflected the *storyline*, and the features of specific models. Markers are no more likely than other

scenarios, but are considered by the SRES writing team as illustrative of a particular *storyline*. They are included in revised form in Nakicenovic et al. (2000). These *scenarios* have received the closest scrutiny of the entire writing team via the SRES open process. *Scenarios* have also been selected to illustrate the other two *scenario groups*.

(Scenario) Storyline: A narrative description of a *scenario* (or family of *scenarios*) highlighting the main *scenario* characteristics, relationships between key driving forces, and the dynamics of their evolution.

Stabilization

The achievement of stabilization of atmospheric concentrations of one or more *greenhouse gases* (e.g., *carbon dioxide* or a CO_2-equivalent basket of *greenhouse gases*).

Stakeholder

A person or an organization that has a legitimate interest in a project or entity, or would be affected by a particular action or policy.

Stated preference[25]

Stated preference approaches, sometimes referred to as direct valuation approaches, are survey methods that estimate the value individuals place on particular non-market goods based on choices they make in hypothetical markets.

Stimuli (climate-related)

All the elements of *climate change*, including mean *climate* characteristics, *climate variability*, and the frequency and magnitude of extremes.

Storm surge

The temporary increase, at a particular locality, in the height of the sea due to extreme meteorological conditions (low atmospheric pressure and/or strong winds). The storm surge is defined as being the excess above the level expected from the tidal variation alone at that time and place.

---

25 SAP 4.6.

Storyline

See *SRES scenarios*.

Stratosphere

The highly stratified region of the *atmosphere* above the *troposphere* extending from about 10 km (ranging from 9 km in high latitudes to 16 km in the tropics on average) to about 50 km.

Streamflow

Water within a river channel, usually expressed in m³ sec⁻¹.

Submergence

A rise in the water level in relation to the land, so that areas of formerly dry land become inundated; it results either from a sinking of the land or from a rise of the water level.

Subsidence

The sudden sinking or gradual downward settling of the Earth's surface with little or no horizontal motion.

Subsidy

Direct payment from the government to an entity, or a tax reduction to that entity, for implementing a practice the government wishes to encourage. *Greenhouse gas emissions* can be reduced by lowering existing subsidies that have the effect of raising *emissions*, such as subsidies for *fossil fuel* use, or by providing subsidies for practices that reduce *emissions* or enhance *sinks* (e.g., for insulation of buildings or planting trees).

Sulfur hexafluoride (SF_6)

A colorless gas that is soluble in alcohol and ether, and slightly soluble in water. It is a very powerful *greenhouse gas* used primarily in electrical transmission and distribution systems, and as a dielectric in electronics.

Sustainable development

Development that meets the needs of the present without compromising the ability of future generations to meet their own needs.

T

Technological potential

The amount by which it is possible to create a reduction in *greenhouse gas emissions* or an improvement in energy efficiency by implementing a *technology* or practice that has already been demonstrated.

Technology

A piece of equipment or a technique for performing a particular activity.

Thermal erosion

The *erosion* of ice-rich *permafrost* by the combined thermal and mechanical action of moving water.

Thermal expansion

In connection with sea level, this refers to the increase in volume (and decrease in density) that results from warming water. A warming of the ocean leads to an expansion of the ocean volume and hence an increase in sea level.

Thermohaline circulation

Large-scale, density-driven circulation in the ocean, caused by differences in temperature and salinity. In the North Atlantic, the thermohaline circulation consists of warm surface water flowing northward and cold deepwater flowing southward, resulting in a net poleward transport of heat. The surface water sinks in highly restricted sinking regions located in high latitudes.

Threshold

The level of magnitude of a system process at which sudden or rapid change occurs. A point or level at which new properties emerge in an ecological, economic or other system, invalidating predictions based on mathematical relationships that apply at lower levels.

Time scale

The characteristic time it takes for a process to be expressed.

Time-series studies[26]

Studies done using a set of data that expresses a particular variable measured over time.

Top-down models

The terms "top" and "bottom" are shorthand for aggregate and disaggregated models. The top-down label derives from how modelers applied macro-economic theory and econometric techniques to historical data on consumption, prices, incomes, and factor costs to model final demand for goods and services, and supply from main sectors, like the energy sector, transportation, agriculture, and industry. Therefore, top-down models evaluate the system from aggregate economic variables, as compared to *bottom-up models* that consider technological options or project specific *climate change mitigation* policies. Some *technology* data were, however, integrated into top-down analysis and so the distinction is not that clear-cut.

Total cost

All items of cost added together. The total cost to society is made up of both the *external cost* and the private cost, which together are defined as *social cost*.

Trade effects

Economic impacts of changes in the purchasing power of a bundle of exported goods of a country for bundles of goods imported from its trade partners. Climate policies change the relative production costs and may change terms of trade substantially enough to change the ultimate economic balance.

Transient climate response

The globally averaged surface air temperature increase, averaged over a 20-year period, centered at the time of CO_2 doubling (i.e., at year 70 in a 1 percent per year compound CO_2 increase experiment with a global coupled *climate model*).

---

26 Modified from Millennium Ecosystem Assessment. 2005. "Time-Series Data."

Transpiration

The process by which water vapor is lost to the *atmosphere* from living plants; the term can also be used to describe the quantity of water dissipated as such.

Troposphere

The lowest part of the *atmosphere* from the surface to about 10 km in altitude in mid-latitudes (ranging from 9 km in high latitudes to 16 km in the tropics on average) where clouds and "weather" phenomena occur. In the troposphere, temperatures generally decrease with height.

Tropopause

The boundary between the *troposphere* and the *stratosphere*.

Tundra

A treeless, level, or gently undulating plain characteristic of arctic and subarctic regions.

U

Uncertainty

An expression of the degree to which a *value* (e.g., the future state of the *climate system*) is unknown. Uncertainty can result from lack of information or from disagreement about what is known or even knowable. It may have many types of sources, from quantifiable errors in the data to ambiguously defined concepts or terminology, or uncertain *projections* of human behavior. Uncertainty can therefore be represented by quantitative measures (e.g., a range of values calculated by various models) or by qualitative statements (e.g., reflecting the judgment of a team of experts). See Moss and Schneider (2000). See also *confidence* and *likelihood*.

Unique and threatened systems

Entities that are confined to a relatively narrow geographical range but can affect other, often larger entities beyond their range; a narrow geographical range points to *sensitivity* to environmental variables, including *climate*, and therefore attests to potential *vulnerability* to *climate change*.

United Nations Framework Convention on Climate Change (UNFCCC)

The Convention was adopted on 9 May 1992, in New York, and signed at the 1992 Earth Summit in Rio de Janeiro by more than 150 countries and the European Community. Its ultimate objective is the "*stabilization* of *greenhouse gas* concentrations in the *atmosphere* at a level that would prevent dangerous *anthropogenic* interference with the *climate system*." It contains commitments for all Parties. Under the Convention, Parties included in Annex I aim to return *greenhouse gas emissions* not controlled by the *Montreal Protocol* to 1990 levels by the year 2000. The Convention entered into force in March 1994. See also *Kyoto Protocol*.

Uptake

The addition of a substance into a *reservoir*. For example, the uptake of carbon-containing substances is often called (carbon) *sequestration*.

Urban heat island effect[27]

The urban *heat island* effect is a measurable increase in ambient urban air temperatures resulting primarily from the replacement of vegetation with buildings, roads, and other heat-absorbing *infrastructure*. The *heat island* effect can result in significant temperature differences between rural and urban areas.

Urbanization

The conversion of land from a natural state or managed natural state (such as agriculture) to cities; a process driven by net rural-to-urban migration through which an increasing percentage of the population in any nation or region come to live in settlements that are defined as "urban centers."

---

27 U.S. EPA. "Heat Island Glossary." Retrieved November 21, 2007 from http://www.epa.gov/hiri/resources/glossary.html#h.

V

Valley fever (Coccidiomycosis)[28]

An infectious respiratory disease of humans and other animals caused by inhaling the fungus Coccidioides immitis. It is characterized by fever and various respiratory symptoms. Also called coccidiomycosis.

Valuation[29]

The process of expressing a value for a particular good or service in a certain context (e.g., of decision-making) usually in terms of something that can be counted, often money, but also through methods and measures from other disciplines (sociology, ecology). See also *values*.

Value added

The net output of a sector after adding up all outputs and subtracting intermediate inputs.

Value of a statistical life (VSL)[30]

The sum of what people would pay to reduce their risk of dying by small amounts that, together, add up to one statistical life.

Values

Worth, desirability, or utility based on individual preferences. The total value of any resource, is the sum of the values of the different individuals involved in the use of the resource. The values, which are the foundation of the estimation of costs, are measured in terms of the willingness to pay (WTP) by individuals to receive the resource, or by the willingness of individuals to accept payment (WTA) to part with the resource.

Vector

An organism, such as an insect, that transmits a *pathogen* from one host to another. See also *vector-borne diseases*.

Vector-borne diseases

Disease that is transmitted between hosts by a *vector* organism such as a mosquito or tick (e.g., *malaria, dengue fever*, and leishmaniasis).

Volatile organic compounds (VOCs)[31]

Organic compounds that evaporate readily into the air. VOCs include substances such as benzene, toluene, methylene chloride, and methyl chloroform.

Vulnerability

The degree to which a system is susceptible to, or unable to cope with, adverse effects of *climate change*, including *climate variability* and extremes. Vulnerability is a function of the character, magnitude, and rate of climate variation to which a system is exposed, its *sensitivity*, and its *adaptive capacity*.

W

Waterborne diseases[32]

Diseases contracted through contact with water that is infected with any of numerous *pathogens* including Vibrio cholerae, Campylobacter, *Salmonella*, Shigella, and the diarrheogenic Escherichia coli.

Water consumption

Amount of extracted water irretrievably lost during its use (by *evaporation* and goods production).Water consumption is equal to *water withdrawal* minus return flow.

28 *The American Heritage Dictionary of the English Language, Fourth Edition*. Retrieved November 21, 2007 from http://dictionary.reference.com/browse/valley fever.

29 Millennium Ecosystem Assessment, 2005 glossary.

30 SAP 4.6.

31 Agency for Toxic Substances & Disease Registry (ATSDR). "ATSDR Glossary of Terms." Retrieved November 21, 2007 from http://www.atsdr.cdc.gov/glossary.html#G-T-.

32 Modified from CDC. "Preventing Bacterial Waterborne Diseases." Retrieved November 21, 2007 from http://www.cdc.gov/ncidod/dbmd/diseaseinfo/waterbornediseases_t.htm.

Watershed[33]

The land area that drains into a particular watercourse or body of water. Sometimes used to describe the dividing line of high ground between two catchment basins.

Water stress

A country is water-stressed if the available freshwater supply relative to *water withdrawals* acts as an important constraint on development. Withdrawal exceeding 20 percent of renewable water supply has been used as an indicator of water stress.

Water-use efficiency

Carbon gain in *photosynthesis* per unit water lost in *evapotranspiration*. It can be expressed on a short-term basis as the ratio of photosynthetic carbon gain per unit *transpirational* water loss, or on a seasonal basis as the ratio of *net primary production* or agricultural yield to the amount of available water.

Water withdrawal

Amount of water extracted from water bodies.

Welfare

An economic term used to describe the state of *well-being* of humans on an individual or collective basis. The constituents of *well-being* are commonly considered to include materials to satisfy basic needs, freedom and choice, health, good social relations, and security.

Well-being [34]

A context- and situation-dependent state, comprising basic material for a good life, freedom and choice, health and bodily well-being, good social relations, security, peace of mind, and spiritual experience.

West Nile virus[35]

West Nile virus (WNV) is a single-stranded RNA virus of the family Flaviviridae, genus Flavivirus. The main lifecycle of WNV is between birds and insects. Humans are most often infected by a bite from an infected mosquito. Most people infected with WNV don't show any symptoms, whereas those that do are often diagnosed with West Nile fever which can last up to two weeks.

Z

Zoonoses

Diseases and infections which are naturally transmitted between vertebrate animals and people. See also *zoonotic disease*.

Zoonotic disease

A disease that normally exists in other vertebrates but also infects humans, such as *dengue fever*, avian flu, *West Nile virus* and bubonic plague.

---

33 Millennium Ecosystem Assessment, 2005 glossary.

34 Modified from the Millennium Ecosystem Assessment, Current State and Trends Assessment Glossary, 2005.

35 Modified from CDC. "West Nile Virus." Retrieved November 21, 2007 from http://www.cdc.gov/nci-dod/dvbid/westnile/index.htm.

References

IPCC, 2001: *Climate Change 2001: The Scientific Basis. Contribution of Working Group I to the Third Assessment Report of the Intergovernmental Panel on Climate Change.* J. T. Houghton, Y. Ding, D.J. Griggs, M. Noguer, P. J. van der Linden and D. Xiaosu Eds. Cambridge University Press, UK. pp 944

IPCC, 2007: *Climate Change 2007: Mitigation. Contribution of Working Group III to the Fourth Assessment Report of the IntergovernmentalPanel on Climate Change.* B. Metz, O.R. Davidson, P.R. Bosch, R. Dave, L.A. Meyer (eds). Cambridge University Press, Cambridge, UK.

Moss, R.H., and S.H. Schneider, 2000: *Towards Consistent Assessment and Reporting of Uncertainties in the IPCC TAR.* In: Pachauri, R., and Taniguchi, T., eds., *Cross-Cutting Issues in the IPCC Third Assessment Report.* Global Industrial and Social Progress Research Institute (Tokyo) for IPCC.

Nakićenović, N., J. Alcamo, G. Davis, B. de Vries, J. Fenhann, S. Gaffin, K. Gregory, A. Grübler, T.Y. Jung, T. Kram, E.L. LaRovere, L. Michaelis, S. Mori, T. Morita, W. Pepper, H. Pitcher, L. Price, K. Raihi, A. Roehrl, H.-H. Rogner, A. Sankovski, M. Schlesinger, P. Shukla, S. Smith, R. Swart, S. van Rooijen, N. Victor and Z. Dadi, 2000: *Emissions Scenarios: A Special Report of Working Group III of the Intergovernmental Panel on Climate Change.* Cambridge University Press, Cambridge, and New York, 599 pp.

United Nations, 1994: *United Nations Convention to Combat Desertification (UNCCD).* UN document number A/AC.241/27, United Nations, New York, NY, USA, 58 pp.

World Meteorological Organization (WMO), 2003: *Climate: Into the 21st Century.* Burroughs, W. ed. Cambridge University Press, Cambridge, UK and New York, NY, USA.

6.2 ACRONYMS

AAG	Association of American Geographers
AAP	American Academy of Pediatrics
AIACC	Assessment of Impacts and Adaptations to Climate Change
AMR-A	North American Region
AR4	IPCC's Fourth Assessment Report
CCC	Canada Climate Center
CCP	ICLEI's Cities for Climate Protection
CCSP	Climate Change Science Program
CDC	Centers for Disease Control and Prevention
CLIMB	Climate's Long-Term Impacts on Metro Boston
CO$_2$	Carbon Dioxide
CVD	Cardiovascular Disease
DHS	Department of Homeland Security
ECHAM4	A model named after the European Centre for Medium Range Weather Forecasts (ECMRWF), (giving it the first part of the name—EC), which was developed in Hamburg (HAM)
EHE	Extreme Heat Event
ENSO	El Niño Southern Oscillation
EPA	Environmental Protection Agency
FDA	Food and Drug Administration
FEMA	Federal Emergency Management Agency
GCM	General Circulation Model
GDP	Gross Domestic Product
GHG	Greenhouse Gas
GIS	Geographic Information System
GISS	NASA Goddard Institute for Space Studies
ICLEI	International Council for Local Environmental Initiatives
IPCC	Intergovernmental Panel on Climate Change
MA	Millennium Assessment
MM5	Mesoscale Model
MSA	Metropolitan Statistical Area
NAAQS	National Ambient Air Quality Standards
NACC	U.S. National Assessment of Climate Change

NAS	National Academy of Sciences
NAST	National Assessment Synthesis Team
NEG/ECP	New England Governors and Eastern Canadian Premiers
NGO	Non-gGovernmental Organization
NO	Nitric Oxide
NOAA	National Oceanic and Atmospheric Administration
NRC	National Research Council
NYCHP	New York Climate and Health Project
PM	Particulate Matter
PM2.5	Particulate Matter (smaller than 2.5 micrometers)
PTSD	Post-traumatic Stress Disorder
RADM2	Regional Acid Deposition Model, Version 2
RCM	Regional Climate Model
RGGI	Regional Greenhouse Gas Initiative
RMNP	Rocky Mountain National Park
RPS	Renewable Portfolio Standards
SAP	Synthesis and Assessment Product
SHELDUS	Spatial Hazard Events and Losses Database for the United States
SRES	Special Report on Emissions Scenarios
TAR	IPCC's Third Assessment Report
TBE	Tick-borne Encephalitis
UHI	Urban Heat Island Effect
UNDP	United Nations Development Programme
UNEP	United Nations Environmental Programme
USBEA	United States Bureau of Economic Analysis
USDA	U.S. Department of Agriculture
USGCRP	United States Global Change Research Program
VBZ	Vector-borne and Zoonotic
VEMAP	Virtual Earth Map
VOC	Volatile Organic Compounds
VSL	Value of Statistical Life
WHO	World Health Organization
WTP	Willingness to Pay

CONTACT INFORMATION

Global Change Research Information Office
c/o Climate Change Science Program Office
1717 Pennsylvania Avenue, NW
Suite 250
Washington, DC 20006
202-223-6262 (voice)
202-223-3065 (fax)

The Climate Change Science Program incorporates the U.S. Global Change Research Program and the Climate Change Research Initiative.

To obtain a copy of this document, place an order at the Global Change Research Information Office (GCRIO) web site: http://www.gcrio.org/orders

CLIMATE CHANGE SCIENCE PROGRAM AND THE SUBCOMMITTEE ON GLOBAL CHANGE RESEARCH